ELECTROMAGNETIC AND QUANTUM MEASUREMENTS
A Bitemporal Neoclassical Theory

ELECTROMAGNETIC AND QUANTUM MEASUREMENTS

A Bitemporal Neoclassical Theory

by

Tore Wessel-Berg
Professor em
Department of Physical Electronics
The Norwegian University of Science and Technology
Norway

KLUWER ACADEMIC PUBLISHERS
Boston / Dordrecht / London

Distributors for North, Central and South America:
Kluwer Academic Publishers
101 Philip Drive
Assinippi Park
Norwell, Massachusetts 02061 USA
Telephone (781) 871-6600
Fax (781) 871-6528
E-Mail <kluwer@wkap.com>

Distributors for all other countries:
Kluwer Academic Publishers Group
Distribution Centre
Post Office Box 322
3300 AH Dordrecht, THE NETHERLANDS
Telephone 31 78 6392 392
Fax 31 78 6546 474
E-Mail <orderdept@wkap.nl>
Electronic Services <http://www.wkap.nl>

Library of Congress Cataloging-in-Publication Data

Wessel-Berg, Tore, 1923-
Electromagnetic and quantum measurements : a bitemporal neoclassical theory / by Tore Wessel-Berg.
p. cm.
ISBN 0-7923-7257-3 (alk paper)
1. Electromagnetic measurements. 2. Quantum theory. 3 Physical measurements. I. Title

QC670.W47 2000
530.14'1--dc21

00-049320

Printed on acid-free paper.

Printed in the United States of America

To Inger-Anne, Dag and Erik

Contents

Foreword

It is a pleasure to write a foreword for Professor Tore Wessel-Berg's book, "Electromagnetic and Quantum Measurements: *A Bitemporal Neoclassical Theory*." This book appeals to me for several reasons. The most important is that, in this book, Wessel-Berg breaks from the pack. The distinguished astrophysicist Thomas Gold has written about the pressures on scientists to move in tight formation, to avoid having their legs nipped by the sheepdogs of science. This book demonstrates that Wessel-Berg is willing to take that risk.

I confess that I do not sufficiently understand this book to be able to either agree or disagree with its thesis. Nevertheless, Wessel-Berg makes very cogent arguments for setting out on his journey. The basic equations of physics are indeed time-reversible. Our experience, that leads us to the concept of an "arrow of time," is derived from macroscopic phenomena, not from fundamental microscopic phenomena. For this reason, it makes very good sense to explore the consequences of treating microscopic phenomena on the assumption that forward time and backward time are equal.

Even if it turns out that Wessel-Berg's theory is wrong, this book will still be important. As Alfred North Whitehead once remarked, "The doctrines which best repay critical examination are those which for the longest period have remained unquestioned." To this, I would add that a wrong idea is far better than no idea. In breaking new ground, it often happens that a scientist, or a group of scientists, must explore several hypotheses that prove to be incorrect, before they stumble across, or invent, or recognize, the hypothesis that proves to be correct. We only learn to recognize the correct theory from the experience of recognizing several incorrect theories.

Having read many books on the philosophy of science written by philosophers, I have often wished that more practicing scientists would think about, and share their thoughts about, the fundamentals of science. Wessel-Berg is doing just that. In addition to questioning our ideas about time, he also questions our interpretation of probability. The role of probability in quantum mechanics has puzzled me, and - no doubt - many other scientists, greatly. It plays a fundamental role in the equations of quantum mechanics, but what relationship is there between the concept of probability in quantum mechanics and that of probability in statistics? In the latter, there are two schools of thought - the "frequentist" school and the "Bayesian" school. I tend to agree with the late physicist Edwin Jaynes (who had his legs nipped by many sheepdogs) that the latter is the more fundamental. It therefore appeals to me that Wessel-Berg's interpretation of probability seems to lean more towards the Bayesian, information-theory, type of approach than the frequentist approach.

It was with great relief that I read that, in Wessel-Berg's theory, the energy of the vacuum is zero, not the astronomically large number that appears in some theories of the vacuum. It will be interesting to learn whether Wessel-Berg's theory is compatible with the experimental evidence for the Casimir effect. There may be some resistance by some physicists that such a radically new hypothesis is being advanced by someone whose background is in applied physics and engineering rather than "fundamental" physics. If such there be, I would draw their attention to Michael Faraday, who was a very practical man, who nevertheless in his time revolutionized our basic physical concepts.

In my role as President of the Society for Scientific Exploration, I have learned that many interesting topics have been ignored by Twentieth Century science for inadequate reasons. We all have excellent hindsight. We can look back on Nineteenth Century science, and see just where it went wrong. We are not so good at seeing what changes are required of Twentieth Century science. It is to his credit that Wessel-Berg proposes a fundamental change and analyzes its consequences in some detail.

Is there any experimental or observational evidence that our conventional interpretation of time is in error? I think there is. In recent years a number of researchers have developed sophisticated laboratory experiments that provide evidence for what is known as "precognition." I have in mind Richard Bierman at the University of Amsterdam, Robert Jahn at Princeton University, Robert Morris at Edinburgh University, and Dean Radin at the Boundary Institute in Los Altos, California, to mention only a few. These experimental results are almost universally ignored, because they do not conform to our Twentieth Century "Standard Model of Reality." Perhaps that model needs tweaking. Wessel-Berg suggests one way in which it might be tweaked. Whether he is right or wrong, we are likely to learn something new by studying what he proposes.

Peter A. Sturrock
Emeritus Professor of Applied Physics, Stanford University
President of the Society for Scientific Exploration

Preface

The book presents an alternative interpretation of quantum effects based on a generalization of electromagnetic theory into the bitemporal domain where, by hypothesis, regular causal relations do not apply. The treatment follows an unexpected and unconventional procedure, taking classical electromagnetism as the initial platform and proceeding from there into the quantum world. The central theme in this new approach is the quoted hypothesis of bidirectional time effects occurring at the photonic or quantum level in regular microcosm. In a bitemporal microworld time is allowed to flow in both directions so that the regular causal relationships dominating the macroworld are abandoned, being replaced with time symmetric noncausal relations. The motivation and rationale behind this interpretation is the indication that strictly causal theories such as classical electromagnetism and orthodox quantum theory are deficient in satisfying the full set of boundary conditions required by photons, a failure shown to be rectified by inclusion of bitemporal effects.

Based on this platform the first part of the book presents a theoretical basis for this endeavor by generalizing classical electromagnetic theory to bidirectional time, presenting the fundamental extensions and changes in a theoretical framework referred to as the 'Neoclassical Electromagnetic Theory', which is put on a sound formal footing by noting that Maxwells field equations are valid for positive as well as negative time. From the neoclassical viewpoint classical electromagnetics is a subset of the more general bitemporal theory. The theory provides a merger of electromagnetic theory and quantum theory of photons by being applicable to macroscopic as well as microscopic phenomena. This is amply demonstrated in the second part of the book presenting analyses and discussions of how single photons manifest themselves in large scale everyday circuits as well as in experiments from the quantum domain. The inclusion of bitemporal effects lead to an entirely new and dramatically different interpretation of photonic quantum phenomena, completely void of the interpretational dilemmas and paradoxical puzzles penetrating orthodox quantum theory.

The underlying theme and approach have a consistent base, the launching of a reasonable physical hypothesis—a bitemporal microcosm—with subsequent investigation of the consequences of the hypothesis applied to theoretical predictions for a large selection of physical processes, and comparisons with available empirical facts. The investigations turned out to be a fascinating odyssey into largely unknown territory, a journey full of unexpected surprises and rewards in the way of resolutions of bizarre puzzles and familiar paradoxes of quantum mechanics, together with consistent agreements with existing experimental facts. Following a systematic development of the theoretical basis of bitemporal electromagnetics the book proceeds to investigate and analyze a number of well known and famous quantum experiments riddled by interpretational dilemmas, such as the double slit experiment for photons and electrons, the *EPR* experiment, and many more. Of examples from the macroscopic domain the theory proposes a photonic origin of the enigmatic $1/f$ noise in electronic circuits, a proposition supported by detailed analysis which appears to be consistent with existing empirical data.

The generalization of electromagnetic theory to include bitemporal effects turned out to be nontrivial, leading to the launching of new principles quite foreign to classical electromagnetism as well as quantum mechanics. Incorporation of negative time elements into the field equations gives rise to the concept of photon doublets, a composite entity consisting of a bitemporal photon pair having the peculiar property of possessing electromagnetic fields but zero time average energy. In the neoclassical theory the doublets are the missing variables making the overall boundary conditions for photons and electrons complete. The vanishing of time average energy and other related phenomena following from this principle touch fundamental interpretational problems in quantum physics and cosmology, a typical example being the concept of environmental quantum mechanical zero point energy which is dramatically modified through replacement by a distribution of photon doublets of zero time average energy. According to the neoclassical theory the principle of vanishing time average energy governs the behavior of photons and electrons in the most general electromagnetic environments, providing a direct link between the microscopic behavior of single photons and electrons and the macroscopic ensemble averages of electromagnetism and quantum theory.

The neoclassical formulation presented in the book represents an observer independent quantum theory of photons and electrons. Application of the theory to a number of typical quantum experiments on photons and electrons, with demonstrations of consistent agreement with existing empirical observations, should catch wide interest amongst quantum physicists. Many of the topics addressed in the book are touching interpretational problems in quantum physics, areas that today are in the forefront of attention of physicists and philosophers alike. The book is perhaps of particular interest to this more narrow segment of scientists. But the material presented in the book has a wider scope which should appeal to a larger audience in the physics and electronics communities. Electronic engineers and scientists might welcome the description of how photons behave in regular electronic circuits, with a proposed photonic basis of the enigmatic $1/f$ noise as perhaps the most intriguing item.

The topics discussed in the book began as a spare-time activity on my part, but soon developed into a more serious enterprise. Very early in the game its controversial nature with regard to standard quantum interpretations became quite obvious. Therefore, during the manuscript preparation I made a point of avoiding extended communications with fellow scientists and physicists because it was felt that such interaction might easily develop into futile discussions and corresponding time delays. Therefore, full responsibility for the basic theme of the book as well as possible errors and omissions are entirely my own.

Acknowledgments

I have had the fortune of spending my active career at several excellent academic institutions, all contributing indirectly to the present enterprise by providing an atmosphere conducive to original ideas, as did a number of electronic companies in consulting arrangements. An opportunity to spend the summer 1949 at Massachusetts Institute of Technology— as a graduate student—provided an early exposure to the tremendous advances in electromagnetic theory and microwave technology gained during the world war, leaving fond memories and setting a pattern on my part for further engagement in the general field of electromagnetics. A period of many years as Research Scientist at the Microwave Laboratory and at Stanford Linear Accelerator (SLAC), both of Stanford University, led to deeper involvement in electrodynamics with applications in microwave tubes and electronics. From this period I am particularly indebted to Marvin Chodorow, a great inspirator and long time associate, to the late Edward Ginzton, instigator of SLAC, and to other top scientists in the field of microwave electronics, too many to be mentioned individually. The general background of electromagnetics and electrodynamics gained from these settings and associations has certainly been a prerequisite for launching the ideas presented in this book.

I pay homage to my own Alma Mater, the Norwegian University of Science and Technology. I have fond memories of a long time association with the Institute of Physical Electronics and its staff members, providing the fertile environment and support that allowed the preparation of the manuscript for this book. I owe much to my long time colleagues and good friends Andreas Tonning and Kjell Bløtekjær for enlightening me on many subjects, professional and social. Thanks also to Joar Sæther whose expert knowledge of the intricacies of Latex made this book appear in time.

Chapter 1

THE CAUSAL ENIGMA

'If you can look into the seeds of time,
And say which grain will grow, and which
will not,
Speak then to me.'

William Shakespeare
in
Macbeth

1.1 MICROCOSM—IS IT CAUSAL OR BITEMPORAL?

Macroscopic time asymmetry, one of the cornerstones of classical physics, is an extremely difficult and elusive phenomenon to explain on a logical basis. The so-called 'arrow of time' is most often explained on the basis of the second law of thermodynamics and of entropy, a concept closely related to cosmological boundary conditions, which in a closed macroscopic environment is known to increase with the passing of time. Other attempts to provide acceptable explanations abound, but it appears that none has gained general recognition [12],[38],[32].

The amazing fact remains that no known physical laws involving functional development in time are asymmetric. Newton's laws of mechanics, Einstein's special and general theories of relativity, Hamilton's equations, Maxwell's equations—all are bitemporal in the sense that they permit the operation of time reversal letting $t \to -t$. It is also interesting to observe that in quantum theory the equations of isolated systems are time-symmetric. The irreversible aspects of quantum theory emanate only through interaction with the macroscopic environment. In all these cases time asymmetry is thrown in by hand as an addendum from considerations of strictly macroscopic concepts such as thermodynamics, entropy, and causality. The line of reasoning is from the microscopic to the macroscopic, with suitable adjustments made in the process so that strictly empirical macroscopic observations relating to the world we live in can be satisfied. It is well to keep in mind that this process is purely empirical and not based on any 'grand theory' of physics or any compelling logical arguments.

Even if a fully satisfactory and generally accepted understanding of time asymmetry is lacking, it should nevertheless be envisioned and interpreted as a strictly macroscopic concept, representing in some ways averages of microscopic phenomena. But then it must be legitimate to question the justification of taking strictly macroscopic concepts—the asymmetry of time and the adjunct of a causal world—to apply in the microworld as well. If we accept the precondition that time asymmetry is a strictly macroscopic phenomenon, it must be equally accepted that time is not necessarily asymmetric in the microcosm, i.e., in the quantum domain. At least, it is difficult to justify the á priori assumption of time asymmetry. There are simply no compelling reasons why the detailed temporal relations in the microworld should correspond to averages inherited from the macroworld. At the single event level the second law of thermodynamics and entropy do not come into play. Nor does the macroscopic 'arrow of time'.

In support of a cautious attitude in approaching the problem of how time should be conceived in microcosm there is ample direct and indirect empirical evidence of lack of time asymmetry in the quantum world. The evidence may be elusive, indirect, and difficult to interpret in the framework of standard quantum theory, which has not seriously addressed the problem of time symmetry. Phenomena that might point to time symmetry, at least for the unbiased observer, are traditionally taken to confirm the 'weirdness' or the paradoxical nature of quantum mechanics, which we are asked to accept at face value.

On the other hand, the idea of negative time is by no means entirely foreign in quantum physics. A number of physicists of known stature have approached the idea. J. A. Wheeler and R. P. Feynman in their 'Absorber Theory' [47], and R. P. Feynman [17], describing emission and absorption of photons, are expressively including negative time in the process, and so are several others, but in rather perfunctory ways without incorporating the proposition into an overall theoretical framework of noncausal theories. Of the more serious approaches Yakir Aharanov and Lev Vaidman [2] introduce two state vectors, one of which propagates forward in time whereas the second propagates backward in time and is governed by what happens to it in the future. Except for this supplement, their work is firmly anchored in the postulates of quantum theory. In his recent book Hugh Price [38] undertakes a profound discussion of the issue from a physical and philosophical viewpoint. The ideas of both these endeavors, extending time phenomena into the negative domain, are similar to those of the present book except that the platforms are different. They operate within the confines of standard quantum theory whereas the present theory is based on the platform of electromagnetic theory.

The irreversible nature of time and the adjunct of causality is deeply ingrained in our perception of the world around us. The reluctance of the physics community to accept ideas of possible time symmetry and noncausality in microcosm appears to be less due to negative time as such, but more to the implicit rejection of causality. The rejection of causality as an all-encompassing concept of the universe has many more ramifications than the acceptance of negative time. It touches on deeply embedded religious and psychological notions of the human mind such as consciousness and free will.

It is apparent that the consequences of negative time phenomena for interpretation of physical processes in microcosm have barely been touched. In particular, how might the principles of quantum theory be affected by such a radical step, and how would the modified physical picture of microcosm be perceived? If we proceed along these lines, accepting for a moment the possibility of existence of physical processes in the microworld where time is allowed to flow backward as well as forward, then we must give up micro-causality as well. Macroscopic causality is an adjunct of the 'arrow of time'. Therefore, if time asymmetry is a strictly macroscopic phenomenon, as we are led to believe, then causality is also a strictly macroscopic concept. Together with the relaxation of time asymmetry at the quantum level, we are compelled to discard causality in the ordinary sense. A frame of reference that is symmetric not only in space but also in time, with backward causation included, leads to a deeper form of 'causal' relations. In the symmetric reference frame the 'cause' of a disturbance is not simply propagating forward in time but backward as well, governed by strict adherence to general boundary conditions in space and bidirectional time. It is clear that rejection of the strict concept of causality gives rise to a conceptual split between macrocosm and microcosm, but this schism is certainly less serious than the existing dichotomy between classical physics and quantum physics.

In the mental transition from macrocosm to microcosm it would be well to exert a cautious attitude reflecting the inherently unknown terrain to be entered. In particular, attempts to implement purely macroscopic concepts should be viewed with a great deal of apprehension. Such cautious attitudes have not been particularly prevalent in the inception and development of quantum physics in the last century. For the uninhibited mind there exists considerable empirical evidence that the attempted theoretical mapping complies only meagerly with the microworld terrain. The discrepancies show up as the need for a number of corrective theoretical measures to explain experimental facts. These interpretational problems are regularly labeled as 'weird', or paradoxical, behavior. A growing number of scientists concerned with the interpretation of quantum physics shed away from such magic-like concepts, feeling that something is wrong with the interpretations and that some clarifications are in place. But many go considerably further, claiming that quantum mechanics is basically incomplete. Opinions centering on these problems are certainly not of new date, they are as old as quantum mechanics itself, demonstrated for instance in [5],[8],[15],[36],[38], and [48]. But with few exceptions discussions on the possible incompleteness of quantum mechanics studiously avoid the question of possible time symmetry and the adjunct of noncausal behavior in the quantum domain.

These very fundamental questions can not be settled by any kind of logical arguments alone, but rather by empirical facts relating to observed behavior at the microscopic level. And do we have such experimental observations? The answer to the question is affirmative, as witnessed by a large and increasing number of direct and indirect experimental evidence from selected areas at the quantum level, pointing to bidirectional time effects and causality violations. These may not be recognized as such in the framework of standard quantum theory, which has the habit of explaining them by entirely different labels such as quantum paradoxes, quantum weirdness, and magic-like behavior.

The studies and analyses presented in this book are a modest attempt to introduce bitemporal concepts into the physics of the microworld, from a platform of electromagnetic theory. As a working hypothesis, time reversal and causality violation are fully accepted in the theory of *single event behavior*. From there on the book investigates the consequences of how such a hypothesis will affect the behavior of photons, and to a lesser extent, electrons, under various circumstances. Many will undoubtedly argue that the hypothesis is violating empirical facts of what is observed in the real world around us, where causality and time irreversibility certainly are cornerstones of physics. But not for a moment does this book argue for the rejection of these concepts in a macroscopic environment. The theory and demonstrations presented in this book are concerned with the microscopic, or single event, behavior which by the hypothesis is not bound by strict macroscopic causality laws. The tricky problems of how to negotiate the border or transition between the alleged bitemporal microcosm and the time asymmetric macrocosm are not addressed. These questions, touching the very fundamental and difficult concepts of 'the arrow of time', form extremely important and fundamental problem areas but are way beyond the scope of the investigations.

Quantum theory sometimes refer to spontaneous phenomena such as emission from atoms as noncausal processes, thereby referring to the fact that they happen in a random fashion with no apparent *cause*. But once they are spawned the processes progress in the positive time direction. Throughout this book, however, the concept of noncausality is reserved strictly to bitemporal phenomena involving negative time.

1.2 THE NEOCLASSICAL ELECTROMAGNETIC THEORY

The book presents what I believe are objective analyses and interpretations of a considerable number of experimental situations from the quantum domain, applying the stated hypothesis of bidirectional time. The analysis follows an unexpected and nontraditional procedure, taking classical electromagnetism as the initial platform, and proceeding from there into the quantum world. The essential step in this process is a general extension of classical electromagnetic theory to what might be referred to as electromagnetic photonic theory. By this is meant a theory of single photon behavior in general electromagnetic circuits, which include regular everyday circuits as well as special circuits known from quantum mechanical experiments for the exploration of *single* quantum events. The generality maintained throughout the analysis allows us to apply the new photonic theory to a large number of esoteric experimental situations involving photons in quantum mechanics, for instance the double slit experiment for photons and electrons and the *EPR* (Einstein, Podolsky, Rosen) experiment [15], and its modern forms, in particular the Aspect experiment [3]. Extensive analyses of these and many more experiments are presented in the forthcoming chapters.

The central theme in this new approach, which will be referred to as the *Neoclassical Electromagnetic Theory*, is the quoted hypothesis of bidirectional time at the photonic or quantum level, together with additional conservational constraints dictated by the energy integrity of photons and charge conservation of electrons. The first part of the book provides the theoretical basis of the neoclassical electromagnetic theory applied to *single photons* at the macroscopic level as well as the microscopic quantum level. The major result of this study is the launching of a new general principle governing the photonic behavior in all electromagnetic circuits. This principle relates classical electromagnetism $(em)_{classical}$ with microscopic *photonic* electromagnetism $(em)_{photonic}$ through the *formal* relationship:

$$(em)_{photonic} = (em)_{neoclassical} = (em)_{classical} + (em)_{doublet} \tag{1.1}$$

where the last term $(em)_{doublet}$ represents a set of fluctuating electromagnetic components consisting of a new kind of composite photon entities dubbed *photon doublets.* A photon doublet consists of two photons of opposite energies oscillating in bidirectional time contributing to environmental energy fluctuations. Fluctuating environmental phenomena, the so-called zero point energy fluctuations (ZPE), are also predicted from standard quantum theory, but their origin is linked with a different physical mechanism associated with quantum mechanical ground states. The fluctuations predicted by the neoclassical theory have an entirely different origin; they are resonances in bidirectional time and bear no similarities to the ZPE fluctuations. This difference is demonstrated by a further theoretical result, just quoted here in anticipation of further explanations in subsequent chapters. If we take the time average, or ensemble average, of the equation above, the result is simply

$$\begin{gathered} < em >_{classical} = < em >_{photonics} = < em >_{neoclassical} \\ < em >_{doublet} = 0 \end{gathered} \tag{1.2}$$

The first of these relations simply says that the ordinary classical electromagnetics is the average of the photonic electromagnetics, i.e., neoclassical electromagnetics, which is no great surprise. The second relation tells us that the time average of the fluctuations of photon doublet energy vanishes. We shall find that this entirely new and perhaps surprising result emerges as a general principle of the neoclassical electromagnetic theory, guiding the behavior of individual photons in any electromagnetic circuit including free space. The principle of vanishing time average energy of the environmental fluctuations is in sharp contrast with the quantum mechanical notion that the so-called zero point energy fluctuations (ZPE) has infinite energy density (see Chapter 3.3). This somewhat unrealistic concept is a consequence of the standard quantum mechanical postulates, for instance expressed in Heisenberg's uncertainty relation applied to the ground states.

We shall find, in the demonstrations to follow throughout the book, that the concept of photon doublet energy and the vanishing of its time average is a central principle in the theory, governing the behavior of single photons and electrons in the most general electromagnetic environments. The significance and importance of this concept and the demonstrations appearing in the remainder of the book warrant the special name 'Photon Doublet Energy' shortened PDE, simply because the notion of photon doublet energy is conceptually much different from the quantum mechanical zero point energy ZPE, which in the neoclassical theory is replaced by the concept of photon doublet energy.

The neoclassical electromagnetic theory is applicable to both average and single photon behavior in macroscopic everyday electronic circuits as well as to single photon behavior in quantum experiments. Thus, the theory provides a merger of classical electromagnetics with quantum electromagnetics. It leads to a new interpretation of photonic quantum phenomena, markedly different from the standard theory. In order to avoid confusion I am not referring to the theory as a variant of quantum theory, instead using the name coined above: *Neoclassical Electromagnetic Theory.* The theory should not be confused with 'Quantum Electrodynamics' which is a pure quantum theory, or with 'Stochastic Electrodynamics' [9] which is based on classical fields in a background of quantum mechanical zero point energy fluctuations.

1.3 CAUSAL THEORIES—INCOMPLETE IN MICROCOSM

In the neoclassical theory the only ingredient inherited from quantum theory is the empirical fact of photon energy quantization and the adjunct of momentum quantization. Photon energy quantization, its inability to split or subdivide its energy between separate, independent absorbers, is imposing severe restrictions on photon behavior in all physical environments. These restrictions can not be met by the regular causal classical electromagnetic theory, which is unable to satisfy all the boundary conditions required by the photon, including energy integrity. The forthcoming chapters present ample examples illustrating the failure of classical electromagnetic theory to explain single photon behavior.

Whereas classical electromagnetic theory makes no serious attempts to explain single photon behavior, quantum theory certainly makes attempts in this direction. But because quantum theory is based on an ensemble platform, such efforts are tainted with paradoxical or weird interpretations which many physicists find difficult to accept. The discussions in the forthcoming chapters of typical quantum experiments on single events in light of the neoclassical theory will hopefully shed some fresh light on these problems. Supported by these forthcoming demonstrations I venture to put forward a tentative theorem which at present must be considered a mere conjecture, namely:

- *Empirical facts involving single events of microscopic entities in the microcosm can not be explained by any objective causal theory.*

In this conjecture the statements of *objectivity* and *causality* are central. Objectivity is the presumption that the microscopic entities are physically real, with physical attributes independent of any observational or measurement scheme. It is generally conceded, at least by a majority of quantum physicists, that the orthodox interpretation of quantum theory—the Copenhagen Interpretation—is nonobjective, although it is subject to controversial views in the physics community. In simple language, nonobjectivity means that the physical attributes of the entities do not really exist until they are measured. Before measurement the entities exist only in a superposed state of possibilities. This nonobjective formalism of quantum physics contrasts sharply with the neoclassical theory as presented in this book, where nonobjectivity is an entirely foreign concept.

In an objective microworld causal theories applied to single events are simply not able to account for all boundary conditions. The extra boundary conditions required in single events stem from additional conservation requirements not present in the average case. The photon must retain its energy without fragmentation regardless of the physical environment and regardless of the processes it is exposed to (at least in the linear domain). These requirements call for more boundary conditions and specifications than ensemble theories such as classical electromagnetism or orthodox quantum mechanics can supply.

The tentative theorem stated above as a mere conjecture is bound to gain considerable support from the theoretical developments and demonstrations presented in the forthcoming chapters. The first few chapters present a broad theoretical basis of the neoclassical theory, followed by detailed analyses and discussions of typical experimental configurations in the quantum as well as the macroscopic domains. The latter involve microwave or low frequency circuits and their response with regard to single photon behavior, seen in contrast to their macroscopic response. The chapter on $1/f$ noise in active low frequency circuits is a typical example of application of the theory to phenomena that are conventionally treated as being strictly macroscopic. But the majority of examples deal with central problem areas in the interpretation of quantum mechanical experiments in light of published experimental results. Running the risk of sounding overly presumptuous it is found that an amazing number of these experiments—essentially all that have been subject to investigation—are explained satisfactorily by the neoclassical theory, with no traces of paradoxical or weird explanations.

The travel into the largely unexplored territory of the bitemporal microworld told in the following chapters proved a venturesome and exciting journey, encountered with numerous surprising results and unexpected findings. And where does this lead us? If the neoclassical hypothesis of a noncausal microworld and the corresponding theoretical interpretations of experimental results on single events are accepted, a reassessment of some of the basic postulates of standard quantum theory of photons is a natural proposition. Discussions on these topics take place throughout the book when the particular quantum phenomena appear as relevant issues. In addition to these scattered discussions, Chapter 14 is devoted to an assessment of results from the neoclassical theory in relation to orthodox quantum theory, addressing some of the most fundamental differences and incompatibilities of the two theories.

Chapter 2

PHOTONS IN BITEMPORAL MICROCOSM

The Queen bawled out, 'He's murdering the time! Off with his head!'

Lewis Caroll
Alice in wonderland

2.1 INTRODUCTION

As an introduction to the microworld of photons in symmetric time frame the simplest kind of solutions of Maxwell's equations serve well as introductions to the basic new concepts. These solutions are shown in this chapter, whereas the comprehensive general theory is presented in the more extensive treatments in Chapters 3-6.

The simplest solutions of Maxwell's equations are a set of plane wave pulses, often presented as standard examples in textbooks on electromagnetic propagation. The examples in these books deal exclusively with classical electromagnetism and are therefore limited by the usual causal restriction. Relaxing this condition and accepting bidirectional time we expect to find supplementary solutions relating to negative time. And of course, this is exactly what happens. But it comes more as a surprise that the combined solutions, with ingredients of positive as well as negative time, exhibit properties that are not simply a superposition of the constituents. These composite solutions are the *photon doublets,* a name coined to convey the essential property of these new entities. A photon doublet is a composition of two types of photons; one is the regular photon belonging to the positive time domain; the other photon—the antiphoton—is associated with negative time, representing a new kind of photon entity that is absent in the classical theory. At this point some readers might object vigorously to the soundness of the concept, basing their arguments on the implicit rejection of macroscopic causality in these propositions. But such objections are not warranted, because the theory, with reference to (1.1) in Chapter 1, is not dealing with classical electromagnetism expressed by the formal $(em)_{classical}$, but rather with photonic electromagnetism $(em)_{photonic}$, i.e., the more fundamental bitemporal neoclassical theory of individual photon behavior. In this new theory the photon doublets are the missing entities serving to complement and aid the regular photons in their responses in general physical environments, including all kinds of electromagnetic circuits, either regular everyday circuits or more special configurations for single entity experiments in quantum mechanics. It will be demonstrated that the theory is equally applicable to photons in the macroworld as in the microworld.

The special standing of the composite photon doublet, rather than the more elementary antiphoton, is associated with the fact that its time average energy vanishes due to compensation of positive and negative energy components. We shall find that the doublets can be interpreted as *resonance phenomena* in bidirectional time, exhibiting quite unconventional properties. The resonances are characterized by instantaneous creation and annihilation—all very strange and unusual properties from the standpoint of classical electromagnetism.

As a preamble to the topic of photonic behavior, the configurations shown in Fig. 2.1 serve to illustrate the quandaries experienced by photons under quite simple everyday operating conditions. In Fig. 2.1a the incident optical beam is illuminating a polished metallic surface reflecting the major part of the beam. A small fraction of the beam power is absorbed and dissipated in the metal due to finite metallic conductivity. If the amplitude of the forward propagating beam component at terminal 2 (the plate) is b_2, classical macroscopic theory tells us that the reflected amplitude a_2 is specified by

$$a_2 = \rho_2 b_2 \tag{2.1}$$

where the reflection coefficient ρ_2 at the terminal is given by

$$\rho_2 = \frac{R_L - Z_0}{R_L - Z_0} \tag{2.2}$$

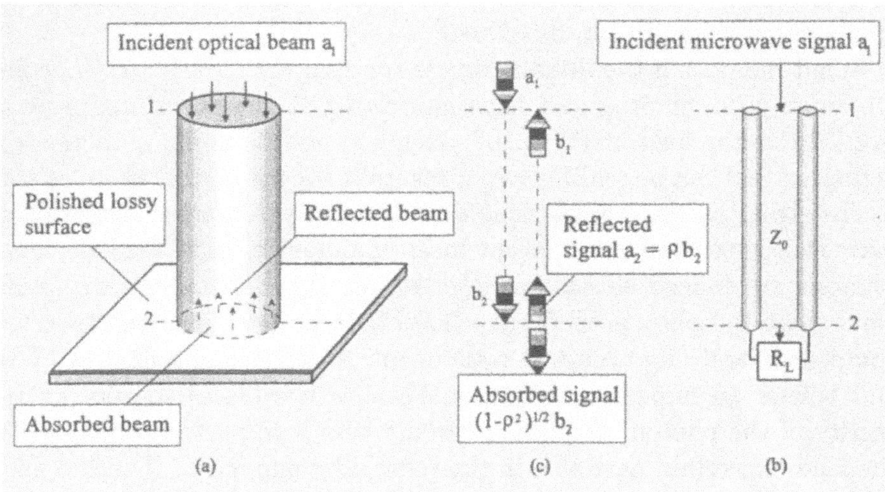

Figure 2.1. **Photon absorption and reflection in two simple configurations: (a) Optical beam reflected from a lossy polished metallic surface. (b) Low frequency two-wire transmission line with mismatched impedance $R_L \neq Z_0$. (c) Illustration of reflected and absorbed signals at the terminations.**

In this familiar relation R_L is the surface resistance of the plate and Z_0 is the characteristic impedance of the incident beam, i.e., the impedance of free space. The fractions of power transmitted and power reflected are given by

$$power_{tr} = 1 - \rho^2 \qquad power_{ref} = \rho^2 \tag{2.3}$$

The two-wire transmission line in Fig. 2.1b is a low frequency equivalent of the optical beam configuration. To avoid unessential details having nothing to do with the issues at hand the line is assumed lossless. With the characteristic impedance substituted for Z_0 in (2.2) all the formulae apply to the transmission line as well. This discussion, with the responses described by (2.1)-(2.3), is all classical electromagnetic theory dealing exclusively with averages of a large number of photons. Let the input consist of N photons per second at the given frequency ω so that each photon has the same energy $\hbar\omega$. Accordingly, the number ratio N_{tr}/N_{ref} of absorbed to reflected photons is given by

$$\lim_{N\to\infty}\left(\frac{N_{tr}}{N_{ref}}\right) = \left(\frac{power_{tr}}{power_{ref}}\right)^{1/2} = \left(\frac{1-\rho^2}{\rho^2}\right)^{1/2} \tag{2.4}$$

where $N = N_{tr} + N_{ref}$. The equation is one particular manifestation of the average nature of classical electromagnetism. Beyond the statement of relation (2.4) concerning large number averages no information on single event behavior can be offered.

What happens if the illumination is successively reduced to the point where only one photon at a time is arriving at the absorbing plate in Fig. 2.1a or the load in Fig. 2.1b? Such a condition is not a theoretical construct and can be realized experimentally for the optical beam as well as the signal on the transmission line. If we apply classical electromagnetic theory to these single event cases, maintaining that the individual photons are indeed electromagnetic waves, the classical theory would claim that the photon energy $a = \hbar\omega$ should be split into one reflected component and one absorbed component, exactly as specified in (2.3). But this is an impossible situation because it violates the energy integrity of the photon. A photon can not dissipate part of its energy in the load and retain part of it in the reflected component. If such a split were possible, the frequency of the reflected part would necessarily have to be reduced proportional to the energy reduction. And such frequency reduction is not possible in linear systems such as those discussed here. The only choices open to the photon is *either* full absorption *or* full reflection. But none of these choices can be offered by classical electromagnetics. Therefore, the classical relations (2.1)-(2.3) do not apply to single photons. Expressed in more general terms, Maxwell's equations in their classical casual form do not satisfy the full set of boundary conditions required by single photons. The forward and backward traveling wave components offered by the classical theory are simply not sufficient to satisfy the photon's needs. This really leaves the photon in a quandary, at least if we maintain that the photon is an honest and goodness wave satisfying the field equations. The photon simply needs supplementary solutions of Maxwell's equations in order to handle the described situation, and these can not be supplied by the regular classical theory. The required extra solutions come from the bitemporal form of Maxwell's equations, in this book referred to as the *neoclassical electromagnetic theory*, of which the classical theory is merely a subset. And we shall very soon demonstrate that these auxiliary solutions indeed supply the needed wave components for the single photons. Moreover, the overall resolution of the boundary value problem by the neoclassical theory predicts exactly the macroscopic number ratio (2.4) of transmitted and reflected photons.

Let us first discuss for a moment whether quantum theory can explain what happens to the single photons in the two configurations of Fig. 2.1. Quantum theory resolves the problem in a nonobjective way by invoking the principle of complementarity, or duality. According to this principle the photon sometimes is a wave satisfying Maxwell's equations and sometimes a particle. Quantum theory holds that the photon, in approaching terminal 2, is mysteriously transformed from a wave to a particle, upon which it dumps its full energy at the load, or is fully reflected. The theory has no explanation for the macroscopic number ratio (2.4) other than the so-called *correspondence* principle set forth by Niels Bohr, according to which the large number average simply *has* to correspond to the observed macroscopic average. The correspondence principle is just an heuristic statement of facts, offering no physical theory behind it. Details of explanations may vary in different quantum texts, but the essence of these are the same.

This book holds that the modeling of single events by quantum theory is incomplete because it is basically an ensemble theory. In handling single quantum events it is forced to introduce a number of principles beyond ensemble theory such as the duality and the correspondence principles. Beyond these principles the theory offers no explanation of the photon problems in the simple physical configurations of Fig. 2.1. With this difficulty of putting physical substance into the processes it should be abundantly clear that quantum theory is quite inept in explaining the quandary of the single photon in these simple configurations. It joins with classical electromagnetics in its inability to provide objective and realistic answers to the theoretical puzzles.

Maintaining the view expressed in this book that the photon is an objective entity quite independent of any observer, the conclusion drawn from this discussion is that neither classical electromagnetic theory nor quantum theory are able to resolve the single photon behavior to full satisfaction. The underlying reason for this inadequacy appear to stem from the causal restrictions of both theories; they are basically ensemble theories quite inept at handling single quantum events. With the assumption of a causal microcosm, characterized by physical events propagating exclusively in the forward time direction, the set of mode components available for satisfying boundary conditions is simply not sufficient. Quantum mechanics reacts to this deficiency by claiming that photons are nonobjective governed by the principle of complementarity, or duality. The neoclassical theory maintains that photons are indeed objective entities, in which case supplementary modes drawn from the bitemporal domain serve to make the overall set complete. In the following sections we shall demonstrate the neoclassical resolution of the stated problem of Fig. 2.1, first presenting the basic bitemporal solutions of Maxwell's equations giving rise to the supplementary wave components.

2.2 PLANE PHOTON PULSES IN THE BITEMPORAL TIME DOMAIN

This section's main objective is to introduce the new concepts of bitemporal solutions of Maxwell's equations and demonstrate their application in resolving single photon behavior in the simple configuration of Fig. 2.1a and in all essential respects that of Fig. 2.1b. For this reason the treatment is limited to plane wave solutions in the form of arbitrarily shaped pulses in the single spatial coordinate Z and bidirectional time T. With these specifications Maxwell's relations in free space reduce to the following simple partial differential equation:

$$\frac{\partial}{\partial T}\begin{bmatrix}\hat{e}(Z,T)\\ \hat{h}(Z,T)\end{bmatrix}+j\begin{bmatrix}0 & -1\\ 1 & 0\end{bmatrix}\frac{\partial}{\partial Z}\begin{bmatrix}\hat{e}(Z,T)\\ \hat{h}(Z,T)\end{bmatrix}=\mathbf{0} \tag{2.5}$$

where the normalized time is defined by $T{=}ct$, with c being the free space velocity of light. The normalized magnetic field $\vec{h}(Z,T)$ is specified by $\vec{h}(Z,T)\ = Z_0\vec{H}(Z,T)$, where $\vec{H}(Z,T)$ is the actual magnetic field, and $Z_0 = \sqrt{\mu_0/\varepsilon_0}$ is the impedance of free space. The amplitudes of the electric and magnetic transverse fields appearing in the equation are in complex form with real and imaginary parts specified by their X and Y components.

$$\begin{aligned}\hat{e}(Z,T) &= e_x(Z,T) + je_y(Z,T)\\ \hat{h}(Z,T) &= h_x(Z,T) + jh_y(Z,T)\end{aligned} \tag{2.6}$$

The polarization is specified by the direction of $\hat{e}(Z,T)$ in the complex X,Y plane. The general solution of (2.5) is given by

$$\begin{bmatrix}\hat{e}(Z,T)\\ \hat{h}(Z,T)\end{bmatrix}=\begin{bmatrix}\hat{F}(Z-T)+\hat{G}(Z+T)\\ j[\hat{F}(Z-T)-\hat{G}(Z+T)]\end{bmatrix} \tag{2.7}$$

where $\hat{F}(Z{-}T)$ and $\hat{G}(Z{+}T)$ are two arbitrary functions of $Z{-}T$ and $Z{+}T$, respectively. The two functions represent arbitrarily shaped pulses propagating with the velocity of light in the positive and negative Z-direction without changing their shape. Because (2.7) is entirely general, it contains *all* possible plane wave solutions of Maxwell's equations. As special cases the functions contain plane *harmonic* waves at some frequency ω. At no time during the derivations was T assumed to be positive, and the equations are therefore, at least formally, applicable in the entire time domain, i.e., for positive as well as negative time T. The power density $S_z(Z,T)$ is obtained from Poynting's vector:

$$S_z(Z,T) = \vec{E}(Z,T)\times\vec{H}(Z,T) = -\frac{1}{Z_0}\operatorname{Im}\left[\hat{e}(Z,T)\hat{h}(Z,T)^*\right] \tag{2.8}$$

which is evaluated to

$$S_z(Z,T) = \frac{1}{Z_0}\left[\hat{F}(Z-T)\hat{F}(Z-T)^* - \hat{G}(Z+T)\hat{G}(Z+T)^*\right] \quad (2.9)$$

The equation shows that the power density at the coordinates Z,T is simply the difference between the square of the pulse amplitudes. As noted above, the equations derived so far are quite general and applicable to positive as well as negative time. Let us first look at the classical case, maintaining positive time throughout, i.e., the flow of time is restricted to the positive direction.

2.2.1 Classical photon pulses in the regular time domain

The solutions in the classical case are represented by two arbitrarily shaped pulses propagating with the velocity of light in the positive as well as in the negative Z-direction, as illustrated in the space-time diagram in Fig. 2.2. The pulses shown in the diagrams have arbitrary shapes and polarizations. The forward type $\hat{F}(Z{-}T)$ and the backward type $\hat{G}(Z{+}T)$ are quite familiar from transmission line theory of classical electromagnetism. According to (2.9) the power density in the forward pulse is positive and equal to the product $\hat{F}(Z{-}T)\hat{F}(Z{-}T)^*$, whereas the power density in the backward pulse is seen to be negative and equal to $-\hat{G}(Z{+}T)\hat{G}(Z{+}T)^*$. As illustrated in the sketches of Fig. 2.3 both the forward and the backward pulses carry positive power in the *direction of propagation.*

Up to this point no negative time concepts have been considered, although it was noted that the general solution (2.7) is valid for positive as well as negative time, at least from a purely mathematical viewpoint. Let us proceed then to discuss the extra solutions of (2.7) in the negative time domain.

2.2.2 Photons in the negative time domain–parity reversal

If the time T in (2.7) is replaced by minus T, an operation that certainly is permissible from a mathematical viewpoint, the equation transforms to

$$\begin{bmatrix} \hat{e}(Z,-T) \\ \hat{h}(Z,-T) \end{bmatrix} = \begin{bmatrix} \hat{F}(Z+T)+\hat{G}(Z-T) \\ j[\hat{F}(Z+T)-\hat{G}(Z-T)] \end{bmatrix} \quad (2.10)$$

The operation converts the original forward traveling pulse $\hat{F}(Z{-}T)$ to a backward traveling pulse $\hat{F}(Z{+}T)$. Similarly, the original backward

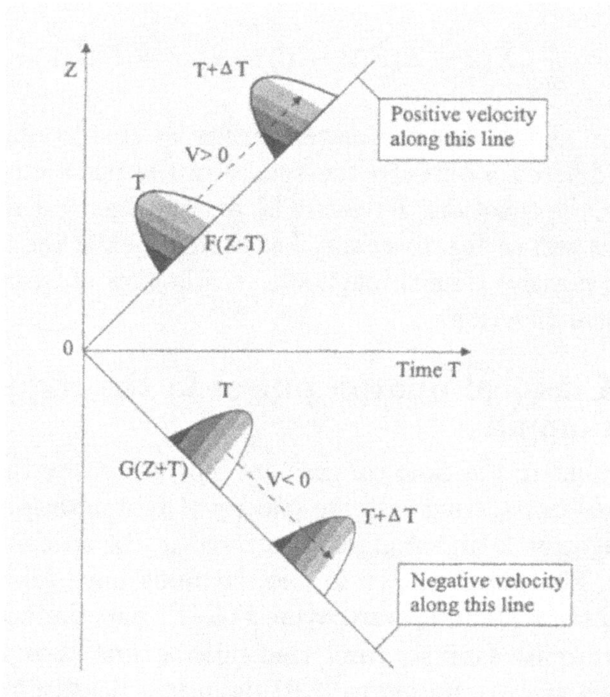

Figure 2.2. **Space-time diagram illustrating the two basic modes of propagation in regular time. The 45 degrees tilts of the displays indicate that the pulses propagate with the velocity of light.**

pulse $\hat{G}(Z+T)$ is converted to $\hat{G}(Z-T)$, which is a pulse traveling in the forward Z-direction. The power density is still specified by (2.9) which transforms to

$$S_z(Z,-T) = \frac{1}{Z_0}\left[\hat{F}(Z+T)\hat{F}(Z+T)^* - \hat{G}(Z-T)\hat{G}(Z-T)^*\right] \quad (2.11)$$

The equation shows that the power flux density is invariant with respect to time reversals. It follows then that these supplementary pulse solutions of Maxwell's equations have the peculiar property that the power density $S_z(Z,-T)$ is in the *opposite* direction of propagation. In the neoclassical theory of single photons it is expected that the transformed pulses have physical reality, representing the supplementary components needed by the photon to satisfy the lacking boundary conditions in the configurations of Fig. 2.1.

The graphical display in Fig. 2.4 illustrates the full set of solutions of the wave equation (2.5) in bidirectional time. In the negative time do-

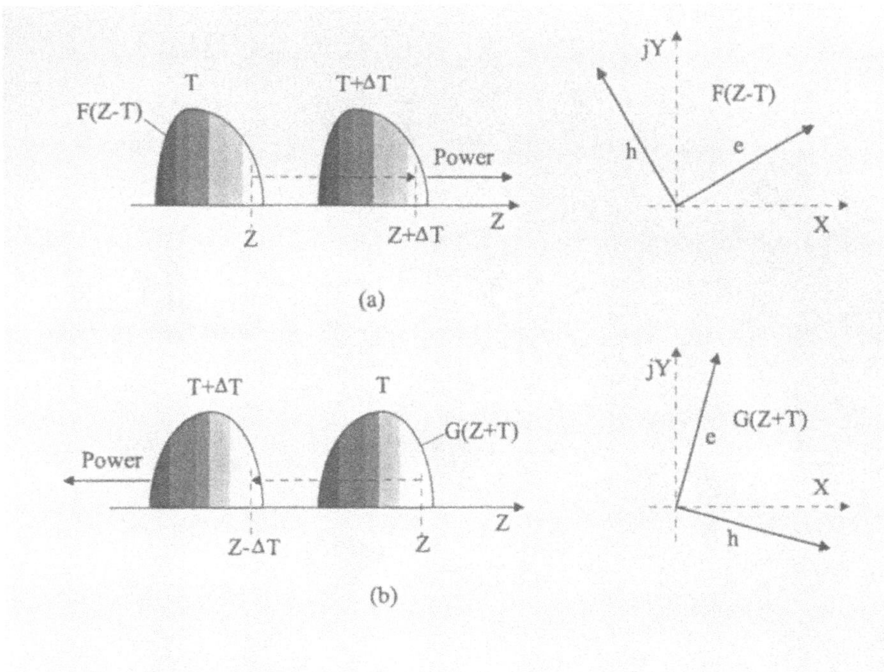

Figure 2.3. **Classical plane pulses of arbitrary shapes and polarizations propagating in regular time. (a) Forward propagating pulse $\hat{F}(Z-T)$. (b) Backward propagating pulse $\hat{G}(Z+T)$. In both pulses the power density points in the direction of propagation.**

main, localized in the left half plane, the two pulses—the antiphotons—supplement the two classical pulses of the right hand plane. These supplementary modes are certainly valid formal solutions of Maxwell's equations, but are they physically acceptable? Classical electromagnetism says that they are not, because they violate causality. However, in the electromagnetics of single photons—the $(em)_{photonic}$ of (1.1) in Chapter 1—macroscopic causality is not an issue and does not apply. The theory presented in this book argues that the supplementary negative power modes—the antiphotons—are physically real just as well as the regular, classical modes. In fact, they form the very foundations for the neoclassical theory of electromagnetism, to be introduced in the present chapter and, in considerable more generality, in the following chapters.

The space-time diagram in Fig. 2.4 can be visualized as being obtained through an operation similar to *parity reversal,* a familiar concept in particle physics. In this operation the diagram in Fig. 2.2 is turned into its mirror reflection with regard to the Z-axis and rotated by 180 degrees about the T-axis. It is well known in physics that the laws of

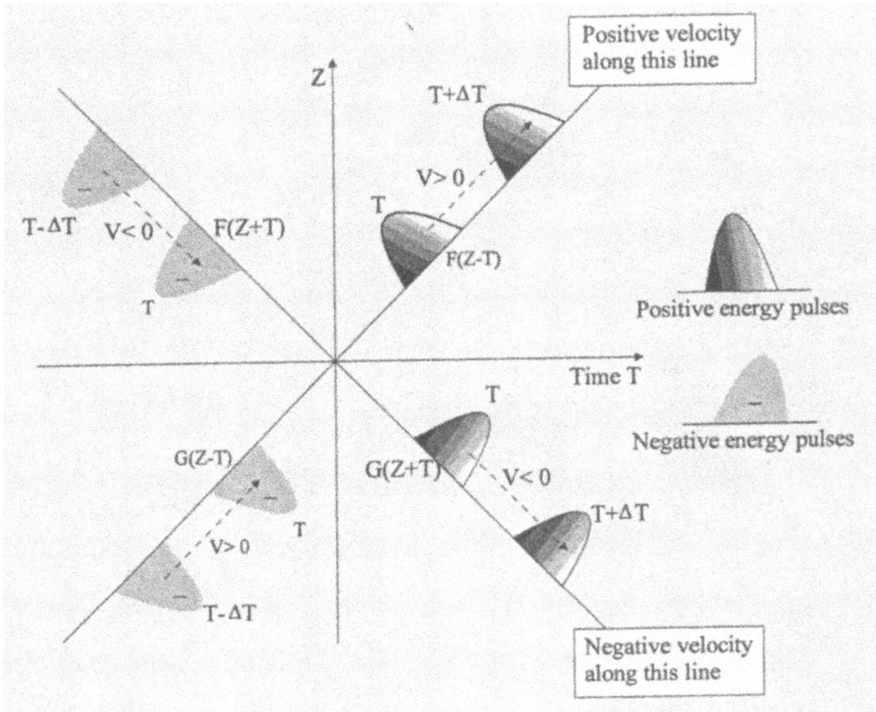

Figure 2.4. **Space-time diagram illustrating the four basic modes of propagation of arbitrarily shaped pulses in bidirectional space-time. The 45 degree tilts of the displays indicate that all pulses propagate with the velocity of light. The two pulses in the left half plane (graytone) are the new negative energy modes supplementing the regular positive energy modes (gradient tone) in the right half plane.**

electrodynamics are the same in a parity-reflected universe. The present exposition based directly on Maxwell's equations is an example of parity-reversal.

In the neoclassical theory the four photon types of Fig. 2.4 are the basic building blocs. Of these generalized photons the regular photons constitute half the set only. The other half contains the antiphotons carrying negative power. It turns out that these are not quite as interesting as a different photon entity which I shall refer to as a *photon doublet.* This new concept is a composite type of photons consisting of a union of one positive power pulse and one negative power pulse propagating in the same direction as a consolidated single pulse having the unusual property that it carries no net power.

The reader is reminded that the pulse modes specified in (2.7) to (2.10) energy-wise were described by their power density $S_z(Z,T)$ which is a vector, and not by the more desirable concept, the energy density $w(Z,T)$, although we would guess that they have the same sign. We shall show that this is indeed true, by establishing the general relation between $S_z(Z,T)$ and $w(Z,T)$ in the bitemporal time frame.

2.2.3 Generalization of photon pulse energy

Poynting's theorem, applied to the present one-dimensional configuration in bitemporal time frame, has the following form relating the power density $S_z(Z,T)$ and the energy density $w(Z,T)$:

$$\frac{\partial S_z(Z,T)}{\partial Z} + c\frac{\partial w(Z,T)}{\partial T} = 0 \tag{2.12}$$

The fact that the pulse shape is stationary implies that the energy density at some arbitrary stationary position within the pulse is constant. Therefore, the *total* time derivative must be zero.

$$\frac{dw(Z,T)}{dT} = \frac{\partial w(Z,T)}{\partial T} + c\frac{\partial w(Z,T)}{\partial Z} = 0 \tag{2.13}$$

The two equations taken together yields

$$\frac{\partial}{\partial Z}[S_z(Z,T) - cw(Z,T)] = 0 \tag{2.14}$$

Because he relation holds everywhere within the pulse the expression within the brackets must vanish identically, so that

$$S_z(Z,T) = cw(Z,T) \tag{2.15}$$

The equation shows that the power flux $S_z(Z,T)$ is always equal to the energy density $w(Z,T)$ multiplied by the pulse velocity c. As anticipated, the signs of the energy density and the power flux are the same for any of the four modes. For the two supplementary branches in the left half plane of Fig. 2.4 the power density was shown to be opposite the velocity of propagation. Therefore, the energy density of these pulses is negative. They carry *negative energy* in the direction of propagation. If the pulse were transformed to Fourier space their frequency would be negative. The association of negative pulse energy with negative frequencies is of course not surprising. It blends perfectly with photon energy being proportional to frequency. The significance and interpretation of negative energy modes are discussed more fully in Chapter 3 which presents the more general aspects of the neoclassical theory of electromagnetics.

2.2.4 The role of antiphotons in the matching process

Having established the four independent plane wave solutions in bidirectional time we shall return to solving the problem of single photon behavior in the configuration of Fig. 2.1. The photons are plane waves at a given frequency ω and as such belong to a subgroup of the general pulses $\hat{F}(Z\!-\!T)$ and $\hat{G}(Z\!+\!T)$. They are described by the conversions

$$\hat{F}(Z-T) = \hat{F}_0\, e^{jk(Z-T)} \qquad \hat{G}(Z+T) = \hat{G}_0\, e^{jk(Z+T)} \tag{2.16}$$

where $k\!=\!\omega/c$ is the free space wave number. The electromagnetic properties of the configurations in Fig. 2.1 are conveniently described by their scattering matrix $\mathbf{S}$, which by definition relates the *inward* variables contained in the column $\mathbf{a}$ and the *outward* variables in column $\mathbf{b}$ through the equation:

$$\mathbf{b} = \begin{bmatrix} b_1 \\ b_2 \end{bmatrix} = \mathbf{Sa} = \mathbf{S}\begin{bmatrix} a_1 \\ a_2 \end{bmatrix} \tag{2.17}$$

At this point the reader is asked to accept at face value some of the general results from Chapter 3, so that we can avoid the task of putting down all the details for the present special case. In Chapter 3 the scatter relation (2.17) is expressed in the slightly modified form

$$\begin{bmatrix} b_1 \\ b_2 \end{bmatrix} = \mathbf{S}\begin{bmatrix} \bar{a}_1 \\ \bar{a}_2 \end{bmatrix} + \mathbf{S}\begin{bmatrix} a_{1,in} \\ a_{2,in} \end{bmatrix} \tag{2.18}$$

where the inward column $\mathbf{a}$ in (2.17) is split into two parts, of which $\bar{\mathbf{a}}$ represents reflections from whatever loads are present and $\mathbf{a}_{in}$, to be defined below, is the circuit-independent input column. The 'reflected' part $\bar{\mathbf{a}}$ at terminals 1 and 2 can be expressed in terms of the corresponding reflection coefficients ρ_1 and ρ_2 specified in terms of the terminating loads by the general equation (2.2). In matrix form

$$\bar{\mathbf{a}} = \begin{bmatrix} \bar{a}_1 \\ \bar{a}_2 \end{bmatrix} = \boldsymbol{\rho}\begin{bmatrix} b_1 \\ b_2 \end{bmatrix} \tag{2.19}$$

where the diagonal *reflection matrix* $\boldsymbol{\rho}$ is defined by the components

$$\boldsymbol{\rho} = \begin{bmatrix} \rho_1 & 0 \\ 0 & \rho_2 \end{bmatrix} \tag{2.20}$$

The 'input' part $\mathbf{a}_{in}$ represents the circuit-independent part of the overall vector $\mathbf{a}$ in (2.17), and can be interpreted as constant amplitude generators at the two terminals. In the configurations of Fig. 2.1 the actual regular photon input $a_{2,in}$ at terminal 2 is of course zero, but for the developments to follow it is imperative to maintain generality. We shall indeed find that there are inputs at terminal 2, with contributions from left hand components amongst the four modes of Fig. 2.4.

Combinations of (2.18) and (2.19) results in a relation between the response $\mathbf{b}$ due to the input $\mathbf{a}_{in}$:

$$\begin{bmatrix} b_1 \\ b_2 \end{bmatrix} = (\tilde{\mathbf{S}}^* - \rho)^{-1} \begin{bmatrix} a_{1,in} \\ a_{2,in} \end{bmatrix} \tag{2.21}$$

In the present demonstration the inverse relation is more relevant.

$$\begin{bmatrix} a_{1,in} \\ a_{2,in} \end{bmatrix} = (\tilde{\mathbf{S}}^* - \rho) \begin{bmatrix} b_1 \\ b_2 \end{bmatrix} \tag{2.22}$$

This equation tells us what inputs $\mathbf{a}_{in}$ are needed for obtaining a specified output $\mathbf{b}$ in a general configuration characterized by the scattering matrix $\mathbf{S}$ and the loads at the terminals expressed by the reflection matrix ρ. The equation is in the proper form for resolving the photon routing problem.

Up to this point the scatter relations are straight forward applications of classical electromagnetism. Anticipating the general theory of Chapter 3, the scatter relations are valid also in the bitemporal domain with the scattering matrix $\mathbf{S}$ equal to the classical form. Let us proceed by first looking at the macroscopic case, where the responses are the averages of a large number of photons entering the configurations. In (2.21) there is no input at terminal 2 so that the vector $\mathbf{a}_{in}$ is given by

$$\begin{bmatrix} a_{1,in} \\ a_{2,in} \end{bmatrix} = \begin{bmatrix} a_{1,0} \\ 0 \end{bmatrix} \tag{2.23}$$

which specifies the macroscopic responses b_1 and b_2 in (2.21). With reference to Fig. 2.1c, the component b_1 is essentially the reflection from the load at terminal 2 (except for a phase shift over the transmission length), and b_2 is the forward traveling component. The point to bear in mind is that the *macroscopic* variables b_1 and b_2 are both nonzero in accordance with the requirements expressed by (2.3).

2.3 THE PHOTON ROUTING PROCESS

In the introduction to this chapter I discussed the special case of single photon inputs with the associated problems of satisfying boundary conditions. Returning now to these problems, it is clear that the macroscopic relations do not hold for the single photon because it refuses to split its energy into one transmitted part b_2 and one reflected part b_1.

We must pay lip service to the special needs of the photon so that it retains its energy integrity. The only choices open to the photon is either full absorption or full reflection. This condition puts limitations on the vector $\mathbf{b}$ in (2.22) which is restricted to only two values $\mathbf{b}'_{tr}$ and $\mathbf{b}'_{ref}$ representing full transmission or full reflection. A little deliberation shows that these can be expressed by

$$\mathbf{b}'_{tr} = \begin{bmatrix} 0 \\ b'_2 \end{bmatrix} = \begin{bmatrix} 0 \\ e^{-jk\ell} a'_{10} \end{bmatrix} \qquad \mathbf{b}'_{ref} = \begin{bmatrix} b'_1 \\ 0 \end{bmatrix} = \begin{bmatrix} e^{-j2k\ell} a'_{10} \\ 0 \end{bmatrix} \tag{2.24}$$

where $k\ell = \omega\ell/c$ is the normalized length between terminal 1 and terminal 2, and where a'_{10} is the input of the *single* photon. Equation (2.22) together with the specifications in (2.24) tells us exactly which input $\mathbf{a}_{in}$ is required in order to perform the routing task of sending the entire photon to the load, or alternatively, reflect it back, namely:

$$\begin{bmatrix} a_{1,in} \\ a_{2,in} \end{bmatrix}_{tr} = (\tilde{\mathbf{S}}^* - \rho) \begin{bmatrix} 0 \\ e^{-jk\ell} a'_{10} \end{bmatrix} \tag{2.25}$$

$$\begin{bmatrix} a_{1,in} \\ a_{2,in} \end{bmatrix}_{ref} = (\tilde{\mathbf{S}}^* - \rho) \begin{bmatrix} e^{-j2k\ell} a'_{10} \\ 0 \end{bmatrix} \tag{2.26}$$

According to (2.23) the *regular* photon input in this case is a single photon a'_{10} at terminal 1 and no photons a'_{20} at terminal 2, whereas (2.25) as well as (2.26) require inputs at both terminals. How then can the configuration be supplied with the inputs required by these equations? According to the neoclassical theory, worked out in detail in subsequent chapters, the regular input a'_{10} is supplemented as needed by extra plane wave components of the general types of Fig. 2.4. This supplementary input is in the form of photon doublet fluctuations (PDE) drawn from the environment. At the present time the reader is asked simply to accept this at face value, with postponement of the detailed physical background until the following chapters. Let these supplementary inputs, which add to the regular photon input, be the vectors $\boldsymbol{\alpha}_{tr}$ and $\boldsymbol{\alpha}_{ref}$ so that the overall input vectors are given by

$$\begin{bmatrix} a_{1,in} \\ a_{2,in} \end{bmatrix}_{tr} = \begin{bmatrix} a'_{10} \\ 0 \end{bmatrix} + \begin{bmatrix} \alpha_1 \\ \alpha_2 \end{bmatrix}_{tr} \tag{2.27}$$

$$\begin{bmatrix} a_{1,in} \\ a_{2,in} \end{bmatrix}_{ref} = \begin{bmatrix} a'_{10} \\ 0 \end{bmatrix} + \begin{bmatrix} \alpha_1 \\ \alpha_2 \end{bmatrix}_{ref} \tag{2.28}$$

The required environmental *PDE* supplements for routing the photon to terminal 2 or to terminal 1, respectively, are therefore specified by

$$\boldsymbol{\alpha}_{tr} = \begin{bmatrix} \alpha_1 \\ \alpha_2 \end{bmatrix}_{tr} = (\tilde{\mathbf{S}}^* - \boldsymbol{\rho}) \begin{bmatrix} 0 \\ e^{-jk\ell} a'_{10} \end{bmatrix} - \begin{bmatrix} a'_{10} \\ 0 \end{bmatrix} \tag{2.29}$$

$$\boldsymbol{\alpha}_{ref} = \begin{bmatrix} \alpha_1 \\ \alpha_2 \end{bmatrix}_{ref} = (\tilde{\mathbf{S}}^* - \boldsymbol{\rho}) \begin{bmatrix} e^{-j2k\ell} a'_{10} \\ 0 \end{bmatrix} - \begin{bmatrix} a'_{10} \\ 0 \end{bmatrix} \tag{2.30}$$

Assume now that the stream consists of N well separated identical photons a_{10} per second. Of these let N_{tr} be transmitted to the load and N_{ref} be reflected back. The total expenditure of photon supplements in the overall routing process is the sum of the constituents of (2.29) and (2.30). Dividing by N and adding we find:

$$\frac{1}{N}\left(\sum_{tr} \boldsymbol{\alpha}_{tr} + \sum_{ref} \boldsymbol{\alpha}_{ref}\right) = (\tilde{\mathbf{S}}^* - \boldsymbol{\rho}) \begin{bmatrix} \frac{N_{ref}}{N} b'_1 \\ \frac{N_{tr}}{N} b'_2 \end{bmatrix} - \begin{bmatrix} \frac{1}{N}\sum_N a'_{10} \\ 0 \end{bmatrix} \tag{2.31}$$

If N becomes a very large number, the expressions on the right must be interpreted as macroscopic averages, according to the following:

$$a_{10,macro} = \frac{1}{N}\sum_N a'_{10} \qquad b_{1,macro} = \frac{N_{ref}}{N} b'_1 \qquad b_{2,macro} = \frac{1}{N}\sum_N a'_{10} \tag{2.32}$$

Here, $a_{10,macro}$ is the time average photon flux a_{10} into the configuration, and $b_{1,macro}$ and $b_{2,macro}$ are the time average classical transmitted and reflected parts shown as b_1 and b_2 in Fig. 2.1c. Accordingly, the right hand side of (2.31) must be identified with the classical *macroscopic* relation (2.22):

$$(\tilde{\mathbf{S}}^* - \boldsymbol{\rho}) \begin{bmatrix} b_{1,macro} \\ b_{2,macro} \end{bmatrix} - \begin{bmatrix} a_{10,macro} \\ 0 \end{bmatrix} = 0 \tag{2.33}$$

which shows that the left hand side of (2.31) vanishes.

$$\lim_{N\to\infty} \frac{1}{N}\left(\sum_{tr} \boldsymbol{\alpha}_{tr} + \sum_{ref} \boldsymbol{\alpha}_{ref}\right) = 0 \tag{2.34}$$

In the neoclassical theory this relation is a special case of a very fundamental theorem which in the following chapters is formulated in more general terms:

- *The routing of individual photons to full absorption or to full reflection is orchestrated by the environmental photon doublet energy fluctuations (PDE) so that the time average of the doublet energy supplement vanishes, a condition that also is the criterion for the time average ratio of transmission and reflection to equal the macroscopic ratio.*

This stipulation is quite general, holding regardless of the specifics of the configurations, expressed for instance by their scattering matrix **S** and reflection matrix $\boldsymbol{\rho}$. The physics behind these results is a central theme in the following chapters. At the present time let us just point out that the results are in complete agreement with the formal relations (1.1)-(1.3) of Chapter 1. The routing of the photons is organized by the environmental photon doublet energy fluctuations (PDE). A particular incident single photon, faced with two options—being fully absorbed or fully reflected—is directed to the terminal minimizing the *total* time average PDE.

Before leaving the discussion of photon routing let us present the actual explicit supplements $\boldsymbol{\alpha}_{tr}$ and $\boldsymbol{\alpha}_{ref}$ performing the task of transmitting and reflecting the photon, respectively. In obtaining these from (2.29) and (2.30) we need the scattering matrix **S** and the reflection matrix $\boldsymbol{\rho}$, which for the configurations in Fig. 2.1 are specified by

$$\mathbf{S} = \begin{bmatrix} 0 & e^{-jk\ell} \\ e^{-jk\ell} & 0 \end{bmatrix} \qquad \boldsymbol{\rho} = \begin{bmatrix} 0 & 0 \\ 0 & \rho_2 \end{bmatrix} \tag{2.35}$$

The composite matrix $\tilde{\mathbf{S}}^* - \boldsymbol{\rho}$ is then given by

$$\tilde{\mathbf{S}}^* - \boldsymbol{\rho} = \begin{bmatrix} 0 & e^{jk\ell} \\ e^{jk\ell} & -\rho_2 \end{bmatrix} \tag{2.36}$$

Using this relation for evaluation of the doublet supplements $\boldsymbol{\alpha}_{tr}$ and $\boldsymbol{\alpha}_{ref}$ we obtain

$$\begin{aligned} \boldsymbol{\alpha}_{tr} &= \begin{bmatrix} 0 \\ -\rho_2 e^{-jk\ell} \end{bmatrix} \bar{a}_{10} \\ \boldsymbol{\alpha}_{ref} &= \begin{bmatrix} -1 \\ e^{-jk\ell} \end{bmatrix} \bar{a}_{10} \end{aligned} \tag{2.37}$$

The schematic displays in Fig. 2.5 illustrate the details of the photon routing process, the case of full transmission in Fig. 2.5a, and full reflection in Fig. 2.5b. In the case of full transmission the only supplement is $\alpha_2 = -a_{10}\,\rho_2 \exp(-jk\ell)$, which is intuitively understood as *cancellation* of the classically reflected part traveling in the backward direction. In the diagram of Fig. 2.4 this supplementary wave component belongs to the second quadrant. Its negative power adds to the forward component b_2, exactly compensating for the reflected component $\rho_2 b_2$ of quadrant four.

The case of full reflection shown in Fig. 2.5b is a bit less obvious because both components of $\boldsymbol{\alpha}_{refl}$ are nonzero. The supplement $\alpha_1 = -\bar{a}_{10}$ at the virtual terminal 1 causes cancelation of the incident photon a_{10}. The supplement $\alpha_2 = a_{10}\exp(-jk\ell)$ at terminal 2 propagates backward towards terminal 1, adds another phase shift $k\ell$, and finally ends up at the virtual terminal 1 as the reflected component $b_1 = a_{10}\exp(-j2k\ell)$. The position of the virtual terminal 1 is quite arbitrary and could just as well coincide with terminal 2, which would perhaps provide a better mental association of the reflection process. In both situations the *apparent* behavior of the incident photon is indicated by gradient shade arrows. The incident photon travels straight to the plate and is either fully absorbed or fully reflected.

It is remarkable that this behavior is exactly what one would observe experimentally in some properly arranged equivalent quantum experiment, the details of which are irrelevant and need not be specified. The transmission or reflection of the incident photons would be recorded by photon counters placed at the appropriate locations. The large scale ratio of transmitted to reflected photons would match exactly the macroscopic average specified in (2.4). Quantum mechanics would explain these experimental results by the arguments discussed earlier in the comments immediately following (2.4). The flavor of various quantum theoretical explanations differ, but all involve in some ways the *duality* principle in describing transmission or reflection, as well as the *correspondence* principle linking the observed average number ratio N_{tr}/N_{ref} to the macroscopic average.

2.4 PHOTON DOUBLETS—THE NEW ENTITIES

As noted earlier, the photon doublets consist of a combination of one positive energy pulse and one negative energy pulse propagating as a composite single pulse. This formulation can be brought on a formal basis by selecting appropriate combinations of respective field components from (2.7) and (2.10). Apparently, there are four combinations, two spatially localized doublets and two entangled doublets. The spatially localized positive and negative energy pulses separate into *forward*

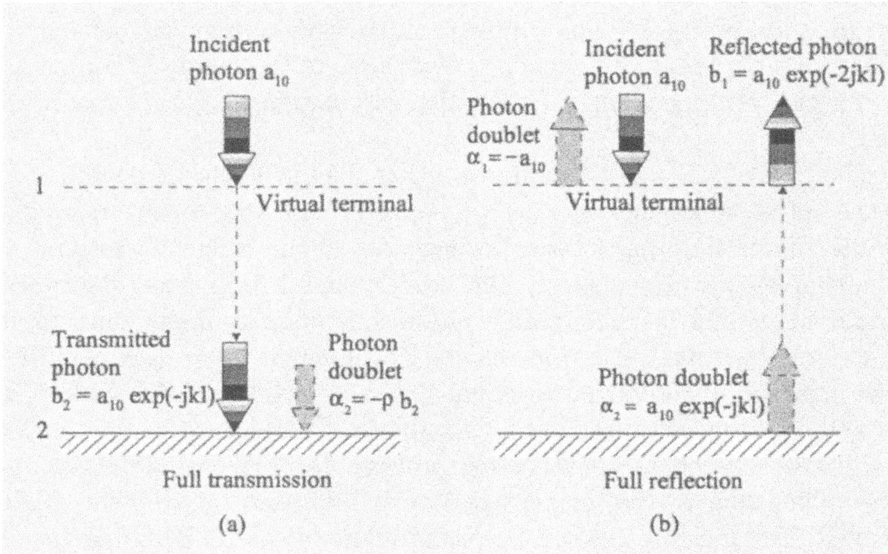

Figure 2.5. The drawings relate to the configuration of Fig. 2.1, illustrating schematically the required photon doublet supplements for routing of the incident single photon a_{10} for full transmission and full reflection. (a) Complete transmission and absorbtion in the lossy surface. (b) Complete reflection from the surface.

and *backward* propagating doublets $\hat{\mathbf{D}}_{fw}$ and $\hat{\mathbf{D}}_{bw}$:

$$\begin{bmatrix} \hat{e}(Z,T) \\ \hat{h}(Z,T) \end{bmatrix}_{loc} = \hat{\mathbf{D}}_{fw} + \hat{\mathbf{D}}_{bw} \tag{2.38}$$

where

$$\hat{\mathbf{D}}_{fw} = \begin{bmatrix} \hat{F}(Z-T) + \hat{G}(Z-T) \\ j[\hat{F}(Z-T) - \hat{G}(Z-T)] \end{bmatrix} \tag{2.39}$$

$$\hat{\mathbf{D}}_{bw} = \begin{bmatrix} \hat{F}(Z+T) + \hat{G}(Z+T) \\ j[\hat{F}(Z+T) - \hat{G}(Z+T)] \end{bmatrix} \tag{2.40}$$

The two remaining combinations of the basic fourfold set of Fig. 2.4 represent nonlocalized entangled doublets $\hat{\mathbf{D}}_a$ and $\hat{\mathbf{D}}_b$ specified by the following superpositions of (2.7) and (2.10):

$$\begin{bmatrix} \hat{e}(Z,T) \\ \hat{h}(Z,T) \end{bmatrix}_{ent} = \hat{\mathbf{D}}_a + \hat{\mathbf{D}}_b \tag{2.41}$$

where

$$\hat{\mathbf{D}}_a = \begin{bmatrix} \hat{F}(Z-T) + \hat{G}(Z+T) \\ j[\hat{F}(Z-T) - \hat{G}(Z+T)] \end{bmatrix} \tag{2.42}$$

$$\hat{\mathbf{D}}_b = \begin{bmatrix} \hat{F}(Z+T) + \hat{G}(Z-T) \\ j[\hat{F}(Z+T) - \hat{G}(Z-T)] \end{bmatrix} \tag{2.43}$$

Let us keep in mind that the basic doublet concept is referred to identically shaped photon pulses, so that $\hat{F}(Z–T)$ and $\hat{G}(Z–T)$ are of the same shape and amplitudes. But their polarizations are not necessarily the same. Their relative polarization is a significant parameter giving rise to several distinct classes of photon doublets for both the localized and entangled types. The following is a discussion of the properties of the two types of doublets beginning with the spatially localized doublets specified by (2.38)-(2.40).

2.4.1 Spatially localized photon doublets

The two photon doublets $\hat{\mathbf{D}}_{fw}$ and $\hat{\mathbf{D}}_{bw}$in (2.39) and (2.40) are illustrated schematically in Fig. 2.6, the left diagram representing the forward doublet $\hat{\mathbf{D}}_{fw}$ (45 degree line), and the right diagram the backward traveling doublet $\hat{\mathbf{D}}_{bw}$ (–45 degree line). Let us consider the forward doublet $\hat{\mathbf{D}}_{fw}$ and discuss some of its basic properties. The electric and magnetic fields are specified by

$$\begin{bmatrix} \hat{e}(Z,T) \\ \hat{h}(Z,T) \end{bmatrix}_{fw} = \begin{bmatrix} \hat{F}(Z-T) + \hat{G}(Z-T)) \\ j[\hat{F}(Z-T) - \hat{G}(Z-T))] \end{bmatrix} \tag{2.44}$$

Even if the amplitudes and shapes of $\hat{F}(Z-T)$ and $\hat{G}(Z-T)$ are the same, their *polarizations* may not be the same. Let the difference in the angles of polarization be φ_{fw} so that the two pulses are related by

$$\hat{G}(Z-T) = \hat{F}(Z-T)\, e^{j\varphi_{fw}} \tag{2.45}$$

whereby (2.44) reduces to

$$\begin{bmatrix} \hat{e}(Z,T) \\ \hat{h}(Z,T) \end{bmatrix}_{fw} = 2\,\hat{F}(Z-T)\, e^{j\varphi_{fw}/2} \begin{bmatrix} \cos(\varphi_{fw}/2) \\ \sin(\varphi_{fw}/2) \end{bmatrix} \tag{2.46}$$

The equation shows that the overall electric and magnetic fields of the doublet are *colinear* regardless of their relative polarization. This remarkable property is tied in with the fact that the net energy carried by the doublet vanishes. The ratio of electric and magnetic field is equal

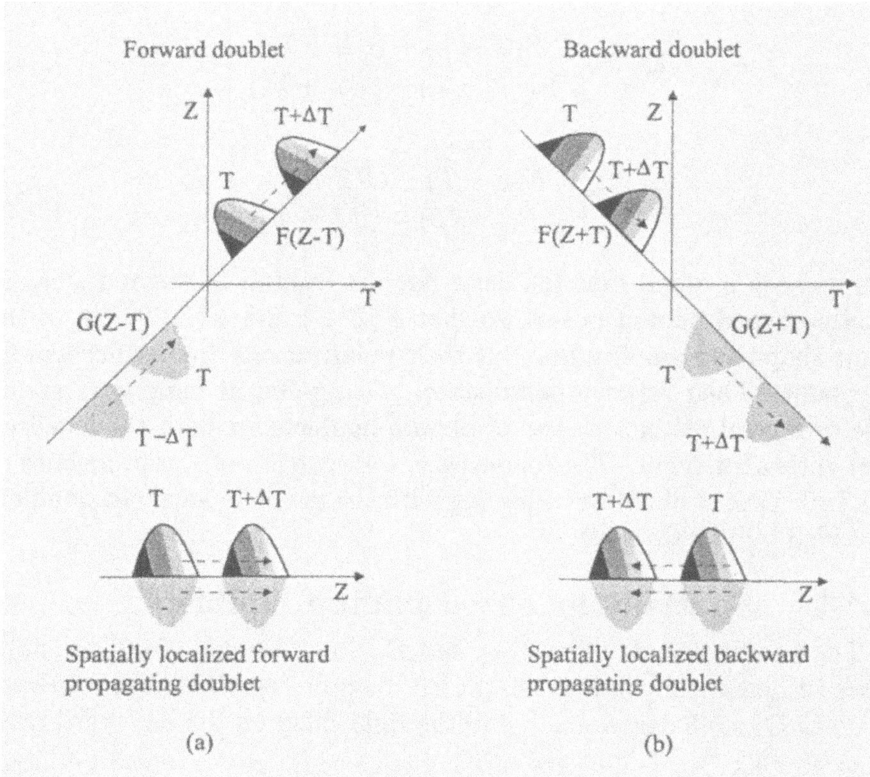

Figure 2.6. **Schematic illustration of spatially localized photon doublets. (a) Forward propagating doublet $\hat{\mathbf{D}}_{fw}$. (b) Backward propagating doublet $\hat{\mathbf{D}}_{bw}$.**

to cotan$(\varphi_{fw}/2)$, i.e., the ratio spans the range from minus infinity to infinity according to the value of the relative polarization angle φ_{fw}. These relations are illustrated in Fig. 2.7 for an arbitrary angle φ_{fw}. Two special cases are of particular significance, the *electric doublet* and the *magnetic doublet*, corresponding to $\varphi_{fw}=0$, and $\varphi_{fw}=\pi$, respectively.

Electric doublet:

$$\begin{bmatrix} \hat{e}(Z,T) \\ \hat{h}(Z,T) \end{bmatrix}_{el} = 2\hat{F}(Z-T)\begin{bmatrix} 1 \\ 0 \end{bmatrix} \tag{2.47}$$

Magnetic doublet:

$$\begin{bmatrix} \hat{e}(Z,T) \\ \hat{h}(Z,T) \end{bmatrix}_{mag} = j2\hat{F}(Z-T)\begin{bmatrix} 0 \\ 1 \end{bmatrix} \tag{2.48}$$

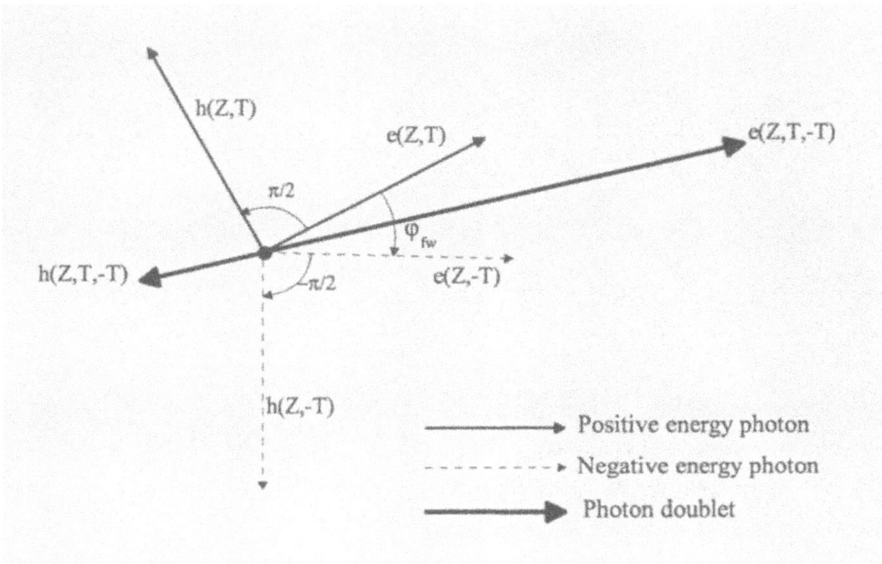

Figure 2.7. **Illustration of the colinear overall electric and magnetic fields (thick arrows) of a forward photon doublet with polarizations of the constituents differing by** φ_{fw}.

The fields of the two doublets, illustrated schematically in Fig. 2.8, can be used for expansion purposes of any doublet because they are orthogonal.

$$\left[\begin{array}{cc} \hat{e}(Z,T)^* & \hat{h}(Z,T)^* \end{array}\right]_{el} \left[\begin{array}{c} \hat{e}(Z,T) \\ \hat{h}(Z,T) \end{array}\right]_{mag} = 0 \tag{2.49}$$

Equation (2.46) and the sketches in Fig. 2.7 and Fig. 2.8 refer to the forward doublet. Quite similar relations apply to the backward doublet, the only difference being a reversal of the magnetic field relative to the electric field. If the polarizations of the backward components differ by φ_{bw}, they are related by:

$$\hat{F}(Z+T)) = \hat{G}(Z+T)\, e^{j\varphi_{bw}} \tag{2.50}$$

yielding the following equation:

$$\left[\begin{array}{c} \hat{e}(Z,T) \\ \hat{h}(Z,T) \end{array}\right]_{bw} = 2\hat{F}(Z+T)e^{j\varphi_{bw}/2} \left[\begin{array}{c} \cos(\varphi_{bw}/2) \\ -\sin(\varphi_{bw}/2) \end{array}\right] \tag{2.51}$$

The equation demonstrates the expected colinearity of electric and magnetic fields also for the backward propagating doublet. The colinear

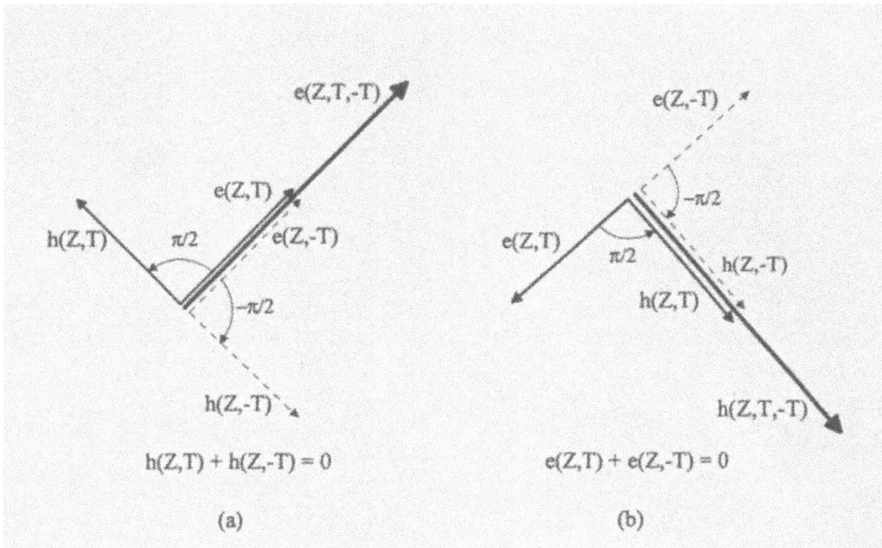

Figure 2.8. **Figure showing the electric and magnetic fields of two special photon doublets: (a)** ***Electric doublet*** **with φ_{fw}=0. The magnetic fields cancel. (b)** ***Magnetic doublet*** **with $\varphi_{fw}=\pi$. The electric fields cancel.**

electric and magnetic fields of doublets are just manifestations of the fact that their power density and energy density vanish. The vanishing of power density S_z is readily verified from (2.9) applied to (2.46) and (2.51). The results are

$$S_z(Z,T)_{fw} = -\frac{1}{Z_0}\mathrm{Im}\left[\hat{e}(Z,T)_{fw}\,\hat{e}(Z,T)^*_{fw}\tan(\frac{\varphi_{fw}}{2})\right] = 0 \qquad (2.52)$$

$$S_z(Z,T)_{bw} = -\frac{1}{Z_0}\mathrm{Im}\left[\hat{e}(Z,T)_{bw}\,\hat{e}(Z,T)^*_{bw}\tan(\frac{\varphi_{bw}}{2})\right] = 0 \qquad (2.53)$$

According to the general relation between power density S_z and energy density w expressed by (2.15) the energy carried by a doublet must also be zero.

$$w(Z,T)_{fw} = w(Z,T)_{bw} = 0 \qquad (2.54)$$

The two constituents of the spatially localized doublets of Fig. 2.6 occupy exactly the same region in space so that the doublets propagate as one single entity irrespective of the particular functional variations of the two constituents, provided these are the same. The properties of the electric and magnetic doublets are certainly quite peculiar. The electric doublet represents an arbitrary pulse of electric field, having zero net energy, propagating with the velocity of light. Similarly, the magnetic field doublet is a propagating pulse of pure magnetic field and zero energy. In the general doublet of Fig. 2.7 the electric and magnetic fields are both nonzero, but they are *colinear* so that the net energy vanishes also in this general case.

2.4.2 Entangled photon doublets

The two remaining combinations of the fourfold basic set of Fig. 2.4 are nonlocalized entangled doublets specified by (2.41)-(2.43). The schematic diagrams in Fig. 2.9 show the two types of entangled doublets represented by the two terms $\hat{\mathbf{D}}_a$ and $\hat{\mathbf{D}}_b$. The first type $\hat{\mathbf{D}}_a$, in the left display of the figure, represents a doublet which presumably is excited spontaneously in some localized region of space. The excitation does not require any energy expenditure; it has the character of a *creation* process. Subsequent to excitation the two constituents of the doublet travel apart in opposite directions. The entangled doublet $\hat{\mathbf{D}}_b$, shown in the right display of Fig. 2.9, represents a situation where the two doublet constituents appear from some unspecified regions of space, traveling *towards* each other. At the meeting point they become completely extinguished, a phenomenon that can be interpreted as an *annihilation* process.

2.4.3 The CPT theorem for photon doublets

Throughout this chapter the description of plane waves of photons and photon doublets has consistently referred these to the *positive* time direction. The pulses are observed in real time, meaning that the positive and negative energy constituents are specified by positive and negative frequencies ω and $-\omega$, respectively. In the realm of a bitemporal theory this is more a convenience rather than a necessity. In some of the experimental situations described in the following chapters it is sometimes more illuminating to describe the doublets in the negative time domain. This transformation is analogous to the famous *CPT* theorem (Charge, Parity, Time) which is a consequence of the symmetry of the microscopic laws of physics. The laws remain invariant if the particle is swapped with its antiparticle, the coordinates converted into the mirror image, and the direction of time reversed. In the diagram of Fig. 2.4 this operation corresponds to mirror operations with regard to the Z-axis and the T-axis which is the same as mirroring with respect to the origin. Restricting the mirror operation to just one of the two axes appears

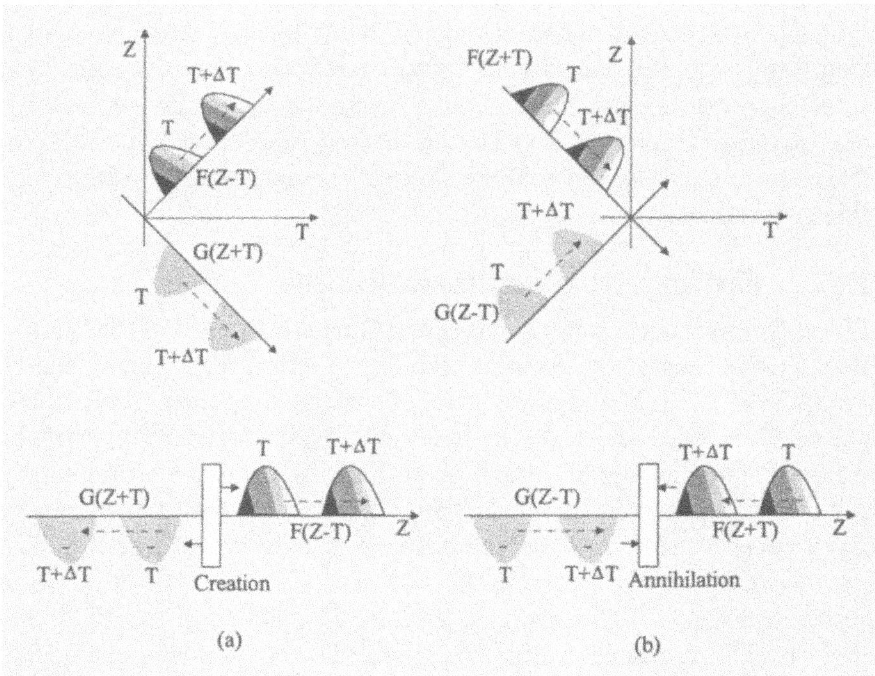

Figure 2.9. **Illustration of the two types of *entangled* photon dublets. (a) The doublet $\hat{\mathbf{D}}_a$ is created spontaneously in some region of space without any expenditure of energy, whereupon the two constituents propagate in opposite directions along the Z-axis. (b) In the doublet of type $\hat{\mathbf{D}}_b$ the two constituents travel in opposite directions *towards* each other, annihilating each other at the meeting point.**

possible, such as the following:

$$\text{photon} \rightarrow \text{antiphoton} \qquad T \rightarrow -T \qquad v \rightarrow -v \tag{2.55}$$

where the direction along Z is maintained. Letting $T \rightarrow -T$ implies a simple sign reversal of the velocity. As a demonstration of the transformation (2.55) the display in Fig. 2.10 shows how the entangled doublets in Fig. 2.9 appear in the reversed time domain. In the bitemporal neoclassical theory the two representations are formally completely equivalent, and the conceptual extensions from positive time to negative time must be accepted as realistic propositions. We shall indeed find that such transformations are quite useful in some of the experimental configurations to be discussed in the forthcoming chapters, in particular those representing delayed choice experiments on single photons.

At the end of this discussion it is tempting to compare the concepts of antiphotons and photon doublets with familiar phenomena in elementary particle physics, sharing analogue properties with the doublets. The first phenomenon that comes to mind is the electron and its antiparticle, the positron. Particle physics describes the positron in two equivalent ways: either as an electron with regular negative charge moving in the *negative* time direction—or as an equivalent electron with opposite, i.e., positive charge moving in the *positive* time direction. At no place is the similarity demonstrated more convincingly than in high energy particle colliders, of which The Stanford Linear Accelerator (SLAC) is a typical example. In this installation pulses of positrons are first produced by high energy electrons bombarding a metallic surface. This phenomenon is conceptually similar to antiphotons being generated as a necessary admixture in the matching process of the incident photon beam impinging on the surface of Fig. 2.1. In the SLAC collider extremely narrow high density pulses of electrons and positrons are then injected into a diskloaded cylindrical waveguide supporting a TM waveguide mode fed from the outside by high power pulsed klystrons. By a clever arrangement the two charge constituents of opposite signs are subject to continuous acceleration in the same direction by riding on the opposite crests of the forward traveling waveguide mode, approaching the speed of light. Apparently, this unison travel of pulses of electrons and the antiparticles—the positrons—in the positive time direction is conceptually similar to the spatially localized photon doublets of Fig. 2.6a. The doublet has zero overall energy, the union of electron and positron has zero net charge.

A second example from particle physics that comes to mind is the similarities of doublets with *neutrinos*, these mysterious particles of copious number penetrating the universe. Both entities propagate with the velocity of light, have zero or negligible mass, i.e., energy, both exist in several distinct flairs, and both show negligible interaction with the environment. In spite of the perhaps surprising number of similarities it would be a too bold and unfounded proposition that they are identical.

2.5 CONCLUDING REMARKS

From classical electromagnetic theory, limited to macroscopic type pulses in regular time, it certainly is difficult to visualize the doublet pulses. They simply do not exist in the context of classical theory; they represent entities strictly confined to the bitemporal domain. In particular, the existence of electromagnetic disturbances having nonzero electric and magnetic fields, but carrying zero energy with them, is a completely new and at first glance bizarre concept. Nevertheless, if the

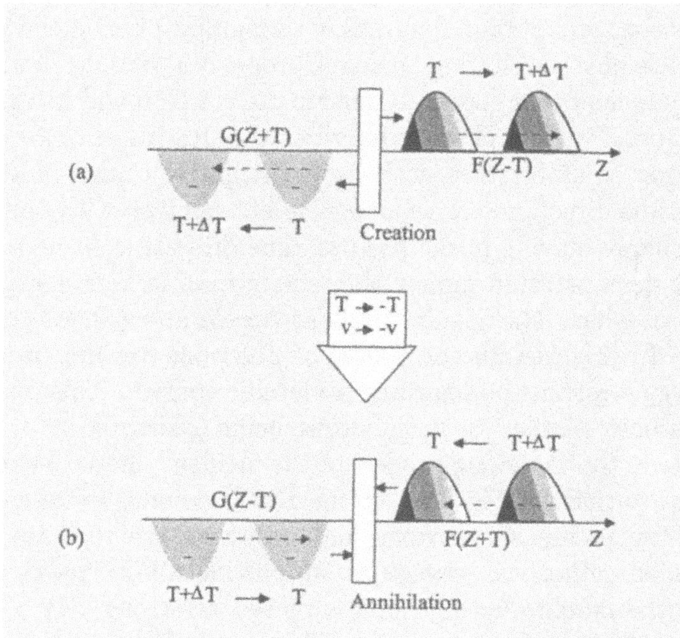

Figure 2.10. **Schematic illustration of the transformation of a photon doublet from real time to the negative time domain. The two entangled doublets in (a) and (b) are entirely equivalent.**

bitemporal hypothesis is accepted, the photon doublets must be accepted as being physically real as well.

These properties, being entirely foreign in classical electromagnetism, are likely to cause wariness amongst readers familiar with electromagnetic theory. Some readers might feel compelled to outright reject these unfamiliar concepts, perhaps arguing that nonzero fields are always associated with a corresponding nonzero energy density w which in free space is specified by the usual formula:

$$w = \frac{1}{2}(\epsilon_0 E^2 + \mu_0 H^2) \tag{2.56}$$

How is this formula to be reconciled with the concept of doublets having zero energy in spite of the fact that their fields are nonzero? The answer to this question is that (2.56) is a special case applicable to macroscopic electromagnetism only; the relation does not apply in the neoclassical formulation. In assessing these subtle differences it is well to keep in mind that a general procedure for determining energy density must be based on the excitation process itself, relating the buildup of energy to the required expenditure of power influx into the relevant volume of space. And for the plane wave doublets the result of such a process is zero energy density, as shown in (2.54). In the macroscopic classical case the same process leads to the well known formula (2.56). Both formulae are correct in the model environments for which they are defined. For the comfort of the concerned readers these questions, relating to the build up of bitemporal electromagnetic energy, are discussed more thoroughly in the next chapter.

Nevertheless, it is legitimate to ask about the physical reason for the zero energy density of the doublets. By now it should be pretty obvious that the zero energy is a consequence of balance of energy of the two constituent photons of the doublet, the regular photon carrying positive energy, and the antiphoton carrying an equal amount of negative energy. In the general neoclassical theory presented in the next chapter we shall carry the doublet concept further and demonstrate that doublet energy is not exactly zero but fluctuating about a *time average* of zero.

In summing up this chapter it is appropriate to ask about the physical significance of the new set of plane wave solutions of Maxwell's equations. Are they simply formal solutions following from the hypothesis of microscopic bidirectional time, or do they have physical reality? The question can not be decided from any logical arguments, but rather from empirical facts. In the introduction it was argued that the extra solutions are indeed physically real, a conclusion based on the successful application of the theory to numerous experimental situations exhibiting phenomena that are difficult to explain on an *objective* basis without incurring bidirectional time. The examples are not restricted to single photon behavior in the quantum domain, but are abundant in macroscopic systems such as regular microwave and low frequency circuits. The latter statement may come as a surprise to electromagnetic and electronic engineers used to firmly believe in the completeness and consistency of classical electromagnetism and its low frequency adaptations. And this is certainly true if one is not concerned with *individual* photon behavior in the circuits. Referring again to the formal relation (1.1) in Chapter 1 it is correct within the confines of the theory $(em)_{classical}$. The neoclassical electromagnetic theory has another facet that is completely absent in the $(em)_{classical}$ theory, namely the behavior of individual photons making up the overall macroscopic $(em)_{classical}$ field. It is the contention of the neoclassical theory that the photon is a wave type entity satisfying Maxwell's equations under all conceivable circumstances, with the added stipulation of photon energy integrity and bidirectional time. These premisses imply that single photon analysis of circuits and networks beyond that of classical network theory requires the supplementary modes of negative energy photons, i.e., antiphotons and the composite entities photon doublets.

From the generalizations undertaken in the next chapters the neoclassical theory comes forward as a possible merger between classical macroscopic electromagnetism and microscopic photonic theory, i.e., quantum theory of photons. At this stage the statement is not much more than a conjecture; however, the conjecture is supported by a substantial number of demonstrations presented in the remaining part of the book, adding considerable credence to the statement. It should also be emphasized that the neoclassical theory does not distinguish between single photon behavior in macroscopic circuits and in experimental setups in the quantum domain. The theory is equally applicable in the macroworld and in the microworld.

Chapter 3

NEOCLASSICAL ELECTROMAGNETICS

'The truth of a single proposition derives not from its "correspondence" with reality, but from being embedded in a true theoretical system. The truth of such a comprehensive system, in turn, derives from its generality and the logical simplicity of its foundations.'

Albert Einstein

3.1 DEFINITION OF THE GENERALIZED CIRCUIT

The simple demonstrations presented in the last chapter were intended as an introduction to the supplementary field solutions arising from extending Maxwell's equations into the bitemporal domain. Although the examples were limited to some very special everyday configurations that could be analyzed by simple mathematics, the results served well to illustrate some of the basic new features concerning single photon behavior in the neoclassical theory. But the analysis was not general enough for the many configurations and cases discussed in the forthcoming chapters.

This restriction is lifted in the present chapter which presents a fairly general theory of electromagnetics in bidirectional time, serving as the formal basis of the neoclassical theory. From this viewpoint classical electromagnetics is a subset of the bitemporal theory. The emphasis is on generality so that the theory may be applied to all special configurations and cases treated in subsequent chapters. There are a few but insignificant restrictions on generality to be discussed shortly. The reference to the theory as a *circuit theory* is not to be taken literally. In the context of the theory at hand a circuit is not just some geometrical embodiment of various metallic and dielectric configurations but rather a more general abstract concept defined as some arbitrary region of space with specified physical parameters of the embedded electromagnetic medium. Figure 3.1 illustrates a surface segment of such a generalized circuit consisting of an arbitrary enclosure of space with volume V and surface S. The interior is supposed to contain an electromagnetic medium that is linear, stationary, lossless and reciprocal, but not necessarily homogeneous and isotropic. In particular, the enclosure can contain nonuniform components of various kinds such as lossless inductors, capacitors, and dielectric members such as optical lenses. The linear and stationary conditions imply a set of restrictions ruling out the occurrence of nonlinear and parametric-type frequency conversions inside the confines of the circuit volume. Such effects are notoriously absent in the study cases treated in the book and, furthermore, their inclusion would add considerable mathematical complexity without contributing much to basic understanding. The stated conditions allow us to operate in the frequency domain, specifying all field variables as functions of one single frequency. The assumptions of losslessness and reciprocity are not undue restrictions in that they do not seriously affect any of the configurations treated in subsequent chapters.

The electromagnetic fields inside the volume V satisfy the regular Maxwell's equations:

$$\begin{aligned} \nabla \times \vec{E}(\vec{r},\omega) &= -j\omega\mu(\vec{r})\,\vec{H}(\vec{r},\omega) \\ \nabla \times \vec{H}(\vec{r},\omega) &= j\omega\varepsilon(\vec{r})\,\vec{E}(\vec{r},\omega) + \vec{i}(\vec{r},\omega) \end{aligned} \tag{3.1}$$

where ω is the angular frequency and $\vec{E}(\vec{r},\omega)$ and $\vec{H}(\vec{r},\omega)$ are complex amplitudes of the electric and magnetic fields defined by

$$\vec{E}(\vec{r},t) = \mathrm{Re}\left[\vec{E}(\vec{r},\omega)e^{j\omega t}\right] \qquad \vec{H}(\vec{r},t) = \mathrm{Re}\left[\vec{H}(\vec{r},\omega)e^{j\omega t}\right] \tag{3.2}$$

The current density $\vec{i}(\vec{r},\omega)$ in the second equation of (3.1) is partly due to the electric field $\vec{E}(\vec{r},\omega)$ and partly to some unspecified extraneous current $\vec{i}_g(\vec{r},\omega)$ imposed on the medium inside the enclosure. Ruling out relativistic effects, the following relation is an adequate representation of the current density $\vec{i}(\vec{r},\omega)$:

$$\vec{i}(\vec{r},\omega) = \int_{V_0} \mathbf{G}(\vec{r},\vec{r}_0,\omega)\,\vec{E}(\vec{r}_0,\omega)\,d\vec{r}_0 + \vec{i}_g(\vec{r},\omega) \tag{3.3}$$

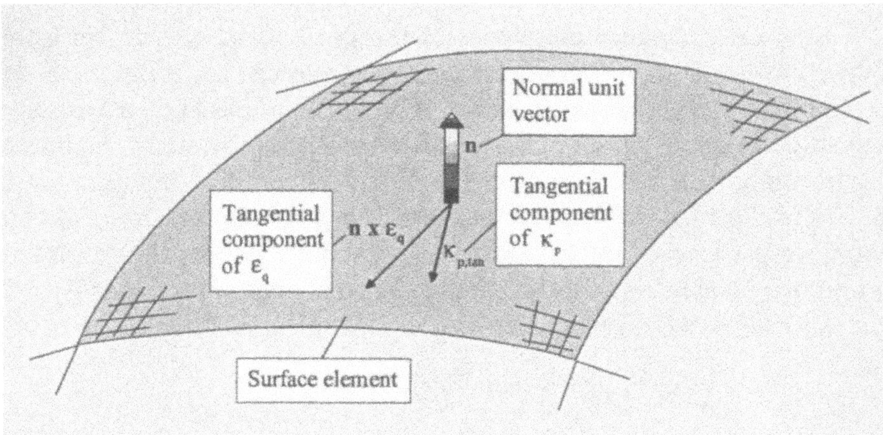

Figure 3.1. **The generalized electromagnetic circuit is defined by an arbitrary region of space enclosed by the surface S. The illustration shows a small surface segment with definitions of surface fields.**

The absence of dependence of $\vec{i}(\vec{r},\omega)$ on the magnetic field $\vec{H}(\vec{r},\omega)$ implies that the model excludes the special cases of *translated* media, such as electron beams, as well as configurations with a constant magnetic field $\vec{H}_0$ in addition to the time varying component. The integral represents what is usually referred to as the constitutive equation for the medium at hand, with the 3×3 Green matrix $\mathbf{G}(\vec{r},\vec{r}_0,\omega)$ interpreted as generalized admittance per unit volume. The noted restrictions on reciprocity and losslessness implies that the Green matrix satisfies

$$\mathbf{G}(\vec{r},\vec{r}_0,\omega) = -\tilde{\mathbf{G}}^*(\vec{r}_0,\vec{r},\omega) \tag{3.4}$$

where the tilde signifies the transposed matrix. The additional term $\vec{i}_g(\vec{r},\omega)$ in (3.3) is an extraneous driving current representing some imposed current component entirely independent of the medium properties. The usual interpretation of this *active* component is a distribution of constant current generators arising from physical causes that can not be identified as being part of the medium specifications. In the form of (3.3) the current density is expressed in adequate generality to cover most circuits of interest, at least those dealt with in this book. The generality maintained in the definition of circuits is essential for maintaining a corresponding generality of the associated *circuit theory,* with assurance that the theory is applicable to the variety of configurations analyzed in the subsequent chapters.

3.2 GENERALIZED CIRCUIT EQUATION

The electromagnetic properties of the generalized circuit are established by means of a reciprocity procedure described in detail in Appendix A at the end of this chapter. The theory is based on a formalism expressed in terms of surface field distributions rather than volumetric distributions. In this procedure the circuit is specified uniquely by its generalized *admittance matrix* $\mathbf{Y}(\omega)$ relating a set of generalized voltages and currents located on the surface S. The first step of the reciprocity procedure establishes a set of N orthonormal vector functions $\vec{\epsilon}_q(\vec{r},\omega)$ and $\vec{\kappa}_q(\vec{r},\omega)$ satisfying Maxwell's equations in the volume V.

$$\left.\begin{aligned}\nabla \times \vec{\epsilon}_q(\vec{r},\omega) &= -j\omega\mu(\vec{r})\ \vec{\kappa}_q(\vec{r},\omega)\\ \nabla \times \vec{\kappa}_q(\vec{r},\omega) &= j\omega\varepsilon(\vec{r})\ \vec{\epsilon}_q(\vec{r},\omega)\end{aligned}\right\} q = 1, 2, ...N \tag{3.5}$$

As shown in the Appendix, the N solutions are selected so that the corresponding tangential surface vectors $\vec{\psi}_q(\vec{s},\omega) = \vec{n}(\vec{s}) \times \vec{\epsilon}_q(\vec{s},\omega)$ located on the closed surface S form an orthonormal vector set for expansion purposes. The N surface vector components $\vec{\psi}_q(\vec{s},\omega)$ are conveniently lumped together in a column vector $\vec{\boldsymbol{\psi}}(\vec{s},\omega)$.

$$\vec{\boldsymbol{\psi}}(\vec{s},\omega) = col\,[\vec{n}(\vec{s}) \times \vec{\epsilon}_q(\vec{s},\omega)] \tag{3.6}$$

Similarly, the surface vectors $\vec{\kappa}_q(\vec{s},\omega)$ are lumped together in the column vector $\vec{\boldsymbol{\kappa}}(\vec{s},\omega)_{\mathrm{tan}}$ containing the *tangential* components of $\vec{\kappa}_q(\vec{s},\omega)$.

$$\vec{\boldsymbol{\kappa}}(\vec{s},\omega)_{\mathrm{tan}} = col\,[\vec{\kappa}_q(\vec{s},\omega)_{\mathrm{tan}}] \tag{3.7}$$

The column vectors $\vec{\boldsymbol{\psi}}(\vec{s},\omega)$ and $\vec{\boldsymbol{\kappa}}(\vec{s},\omega)_{\mathrm{tan}}$ represent normalized tangential electric and magnetic fields on the surface S. All the N vector functions in the column $\vec{\boldsymbol{\psi}}(\vec{s},\omega)$ are *independent* solutions of Maxwell's equations in the volume V. They are all orthonormal, with the overall set of conditions expressed as the following $N \times N$ matrix orthonormality relation:

$$\int_S \vec{\boldsymbol{\psi}}(\vec{s},\omega) \cdot \widetilde{\vec{\boldsymbol{\psi}}}(\vec{s},\omega)^*\, dS = \mathbf{1} \tag{3.8}$$

where the tilde signifies the transposé, and the right hand side is the unity matrix.

For a specified expansion set $\vec{\boldsymbol{\psi}}(\vec{s},\omega)$ the general circuit is described uniquely by its *admittance matrix* $\mathbf{Y}(\omega)$ representing its electromagnetic properties. The admittance matrix is defined by the implicit relation:

$$\vec{\boldsymbol{\kappa}}(\vec{s},\omega)_{\mathrm{tan}} = \mathbf{Y}(\omega)\, \vec{\boldsymbol{\psi}}(\vec{s},\omega) \tag{3.9}$$

If the circuit somehow is excited, the *actual* physical tangential electric and magnetic fields $\vec{n}(\vec{s}) \times \vec{E}(\vec{s},\omega)$ and $\vec{H}(\vec{s},\omega)_{\text{tan}}$ can be expanded in terms of the supposedly complete orthonormal expansion set $\vec{\psi}(\vec{s},\omega)$.

$$\vec{n}(\vec{s}) \times \vec{E}(\vec{s},\omega) = \widetilde{\vec{\psi}}(\vec{s},\omega)\mathbf{V}(\omega) \qquad \vec{H}(\vec{s},\omega)_{\text{tan}} = \widetilde{\vec{\psi}}(\vec{s},\omega)\mathbf{I}(\omega) \tag{3.10}$$

The N- dimensional expansion columns $\mathbf{V}(\omega)$ and $\mathbf{I}(\omega)$ appearing in the equations are interpreted as generalized voltage and current column vectors. These are readily obtained from the orthogonality condition (3.8) through inversions of the two equations in (3.10), a procedure resulting in

$$\mathbf{V}(\omega) = \int_S [\vec{n}(\vec{s}) \times \vec{E}(\vec{s},\omega)] \cdot \vec{\psi}(\vec{s},\omega)^* \, dS \tag{3.11}$$

$$\mathbf{I}(\omega) = \int_S \vec{H}(\vec{s},\omega)_{\text{tan}} \cdot \vec{\psi}(\vec{s},\omega)^* \, dS \tag{3.12}$$

If the electric and magnetic fields $\vec{E}(\vec{s},\omega)$ and $\vec{H}(\vec{s},\omega)$ are *arbitrary* functions of the surface coordinate vector $\vec{s}$, the expansions (3.10) require a mathematically *complete* set of expansion functions $\vec{\psi}(\vec{s},\omega)$, with $N \to \infty$. In most practical circuits the field distributions are not that general, but rather restricted by geometrical constraints of the circuit configuration, such as waveguide input or output terminals confining the fields to essentially the cross section of the terminals. In such cases the dimensional number N is not infinite but rather determined by the degrees of freedom allowed by the geometrical configuration.

With the established general definitions of voltage $\mathbf{V}(\omega)$, current $\mathbf{I}(\omega)$, and circuit admittance $\mathbf{Y}(\omega)$, the generalized circuit equation derived in Appendix A to this chapter has the simple form

$$\mathbf{Y}(\omega)\,\mathbf{V}(\omega) - \mathbf{I}(\omega) = \mathbf{I}_g(\omega) \tag{3.13}$$

where the right hand driving term $\mathbf{I}_g(\omega)$ is given by the volume integral

$$\mathbf{I}_g(\omega) = \int_V \vec{i}_g(\vec{r},\omega) \cdot \vec{\epsilon}(\vec{r},\omega) \, dV \tag{3.14}$$

The generalized voltage and current $\mathbf{V}(\omega)$ and $\mathbf{I}(\omega)$ are N-dimensional column vectors in the abstract N- dimensional function space, and so is the right hand driving current $\mathbf{I}_g(\omega)$. The admittance matrix $\mathbf{Y}(\omega)$ is an

$N \times N$ matrix in the same space. The assumption of losslessness of the medium implies that the admittance matrix is Skew Hermitian.

$$\mathbf{Y}(\omega) + \tilde{\mathbf{Y}}(\omega)^* = \mathbf{0} \tag{3.15}$$

which applies regardless of the particulars of the generalized circuit.

3.2.1 Virtual terminals of the generalized circuit

The N voltages and currents represented by the columns $\mathbf{V}(\omega)$ and $\mathbf{I}(\omega)$ can be associated with a corresponding set of N *virtual,* or *abstract* terminals located on the surface S. In view of the fact that the orthogonal vector expansion set $\vec{\psi}(\vec{s},\omega)$ in (3.6) is not unique and can be defined in an infinite number of ways the abstract terminal set N as well as $\mathbf{V}(\omega)$, $\mathbf{I}(\omega)$, $\mathbf{I}_g(\omega)$, and $\mathbf{Y}(\omega)$ are not unique. In the general case the number N is infinite, but in circuits with geometrical constraints N is finite. The admittance matrix $\mathbf{Y}(\omega)$ of the lossless circuit always satisfies the symmetry relation (3.15) but is otherwise specified by the electromagnetic properties inside the circuit volume *and* the set of selected virtual terminals, i.e., the expansion vectors $\vec{\psi}(\vec{s},\omega)$ and $\vec{\kappa}(\vec{s},\omega)$. In the general theory presented here the terminals are abstract in the sense that they do not necessarily represent separated localized areas on the surface S, although most practical circuits belong to this category, essentially all that are discussed in the remainder of the book.

What is the physical significance of the right hand column vector $\mathbf{I}_g(\omega)$ in (3.13)? The column can be interpreted as a set of N constant current generators located at the N virtual terminals on the surface S, with currents specified by the integral (3.14). The equation specifies the constant current generator strength as a weighted integral of the extraneous current density distribution $\vec{i}_g(\vec{r},\omega)$ over the volume V. The weight factors are the components of the normalized electric expansion vector $\vec{e}(\vec{r},\omega)$ of the circuit. Readers familiar with microwave electronics will recognize the equation as a more general formulation of *Ramo's Theorem* [39], applied extensively in microwave electronic systems and tube designs for calculation of induced current caused by a modulated electron beam. In the subsequent discussions throughout the book the driving term $\mathbf{I}_g(\omega)$ does not play a major role; it represents merely an additional option appearing in certain discussions of conceptual nature, such as excitation of the environmental doublet field (PDE) by noise fluctuations (see 3.3.3) or by single electrons (see 8.2.1).

3.2.2 Stored energy in the generalized circuit

The stored energy in the generalized circuit is evaluated from Poynting's theorem, relating the electromagnetic flux vector—the Poynting vector $\vec{E}(\vec{r},t) \times \vec{H}(\vec{r},t)$—and the stored energy density $w(\vec{r},t)$.

$$\nabla \cdot [\vec{E}(\vec{r},t) \times \vec{H}(\vec{r},t)] + \frac{\partial w(\vec{r},t)}{\partial t} = 0 \tag{3.16}$$

In this equation all variables are real quantities. By means of Gauss's theorem integration over the volume V converts the equation to

$$\frac{\partial W(t)}{\partial t} + \int\limits_S [\vec{E}(\vec{s},t) \times \vec{H}(\vec{s},t)] \cdot \vec{n}(\vec{s})\, dS = 0 \tag{3.17}$$

which is the same as

$$\frac{\partial W(t)}{\partial t} + \int\limits_S [\vec{n}(\vec{s}) \times \vec{E}(\vec{s},t)] \cdot \vec{H}(\vec{s},t)_{\mathrm{tan}}\; dS = 0 \tag{3.18}$$

In these equations $W(t)$ is the total instantaneous stored energy inside the volume V expressed as a function of time.

$$W(t) = \int\limits_V w(\vec{r},t)\, dV \tag{3.19}$$

As shown in the Appendix at the end of the chapter the surface integral in (3.18) can be expressed by the generalized voltage and current vectors $\mathbf{V}(t)$ and $\mathbf{I}(t)$ using the equivalent formulations of (3.10) with respect to time rather than frequency, a procedure converting the differential equation to:

$$\frac{\partial W(t)}{\partial t} + \tilde{\mathbf{V}}(t)\, \mathbf{I}(t) = 0 \tag{3.20}$$

The change in stored energy from time $t_1 = -\infty$ to the present time t is obtained by integration.

$$W(t) = -\int\limits_{t_1=-\infty}^{t} \tilde{\mathbf{V}}(t_1)\, \mathbf{I}(t_1)\, dt_1 \tag{3.21}$$

The formula serves as the basis for the subsequent generalization of stored energy in electromagnetic circuits to the bitemporal domain.

3.2.3 Stored energy at positive frequencies

It will be assumed that the circuit has been filled with electromagnetic energy through a process in which the fields grow exponentially caused by some external small excitation. The details of the source causing the excitation need not be specified. The voltage and current vectors $\mathbf{V}(t)$ and $\mathbf{I}(t)$ can be expressed by their complex amplitudes $\mathbf{V}(\omega)$ and $\mathbf{I}(\omega)$, where ω is now a complex frequency specified by

$$\omega = \omega_r + j\omega_i \tag{3.22}$$

The imaginary frequency ω_i is presumably a very small *negative* number so that the exponential growth can be arbitrarily small. The stored energy at time t (the present) is then specified by

$$W(t) = -\lim_{\omega_i \to 0} \int_{t_1=-\infty}^{t} \tilde{\mathbf{V}}(t_1)\,\mathbf{I}(t_1)\,dt_1 \tag{3.23}$$

The physical process of slow exponential growth of the field amplitude from the beginning of time $t_1 = -\infty$ to the present time t is illustrated in Fig. 3.2 with a greatly compressed time scale.The amplitude growth is infinitely slow so that the final amplitudes are indistinguishable from a pure sinusoid at the time of observation. Under these presumptions let us evaluate the integral (3.23) by expressing $\mathbf{V}(t)$ and $\mathbf{I}(t)$ in terms of their complex amplitudes $\mathbf{V}(\omega)$ and $\mathbf{I}(\omega) = \mathbf{Y}(\omega)\mathbf{V}(\omega)$.

$$\mathbf{V}(t) = \mathrm{Re}\left[\mathbf{V}(\omega)\, e^{j\omega_r t}\right] e^{-\omega_i t} \qquad \mathbf{I}(t) = \mathrm{Re}\left[\mathbf{I}(\omega)\, e^{j\omega_r t}\right] e^{-\omega_i t} \tag{3.24}$$

The outcome of the integration procedure is as follows. The stored energy $W(t)$ consists of a time varying component $W(t)_\sim$ and a time average (constant) component $\langle W \rangle$:

$$W(t) = W(t)_\sim + \langle W \rangle \tag{3.25}$$

where

$$W(t)_\sim = \frac{1}{4}\,\mathrm{Re}\left\{\tilde{\mathbf{V}}(\omega_r)\frac{\mathbf{Y}(\omega_r)}{j\omega_r}\mathbf{V}(\omega_r)\, e^{2j\omega_r t}\right\} \tag{3.26}$$

$$\langle W \rangle = -\lim_{\omega_i \to 0}\left\{\frac{1}{4\omega_i}\,\mathrm{Re}\left[\tilde{\mathbf{V}}(\omega)^* \mathbf{I}(\omega)\right]\right\} \tag{3.27}$$

The time varying part $W(t)_\sim$ at frequency $2\omega_r$, specified by (3.26), is in its final form, with ω_r being the real part of the frequency ω. Equation (3.27), specifying the time average part $\langle W \rangle$ as a function of the complex frequency ω, is processed further by noting that complex amplitudes $\mathbf{V}(\omega)$ and $\mathbf{I}(\omega)$ are analytical functions of ω, permitting first

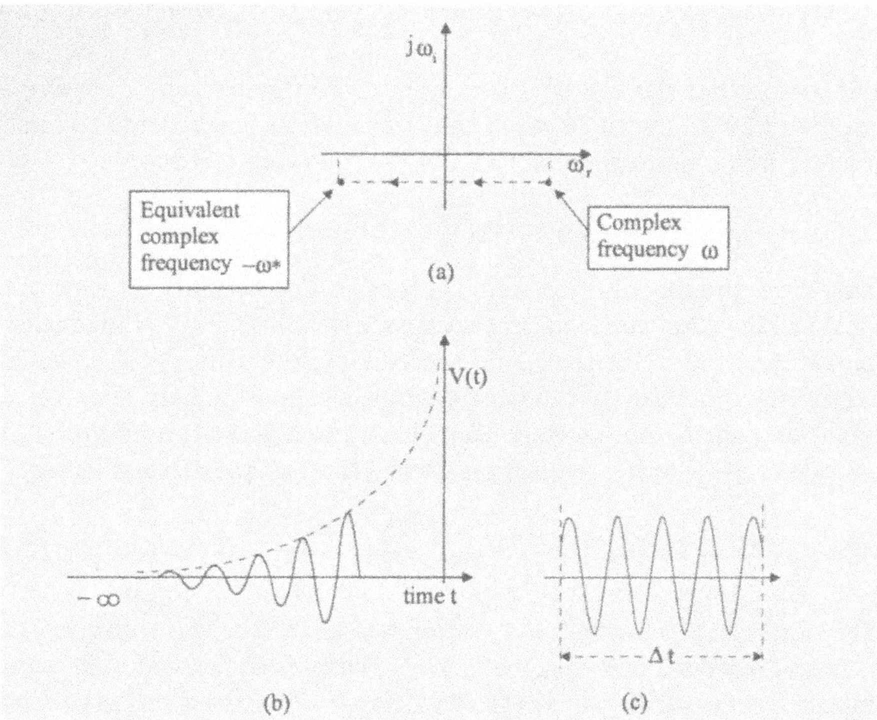

Figure 3.2. **Illustration of the process of filling a general circuit with electromagntic energy at the positive frequency ω_r. (a) Complex frequency plane. (b) Compressed scale of slowly growing fields from $t = -\infty$ to the present time. (c) Observance of constant amplitude over a *finite* interval Δt.**

order analytical continuations into the complex frequency domain. To first order:

$$\mathbf{V}(\omega) = \mathbf{V}(\omega_r) + j\omega_i \frac{\partial \mathbf{V}(\omega_r)}{\partial \omega_r} \qquad \mathbf{I}(\omega) = \mathbf{I}(\omega_r) + j\omega_i \frac{\partial \mathbf{I}(\omega_r)}{\partial \omega_r} \tag{3.28}$$

With these expansions (3.27) takes the form

$$\langle W \rangle = -\lim_{\omega_i \to 0} \left\{ \frac{1}{4\omega_i} \mathrm{Re}\left[\Delta_1 + \Delta_2 + \Delta_3\right] \right\} \tag{3.29}$$

where the three terms Δ_1, Δ_2, and Δ_3 are given by

$$\Delta_1 = \tilde{\mathbf{V}}(\omega_r)^* \mathbf{I}(\omega_r) \tag{3.30}$$

$$\Delta_2 = -\omega_i \tilde{\mathbf{V}}(\omega_r)^* \frac{\partial}{\partial \omega_r} \left(\frac{\mathbf{Y}(\omega_r)}{j} \right) \mathbf{V}(\omega_r) \tag{3.31}$$

$$\Delta_3 = j\omega_i \left[\tilde{\mathbf{V}}(\omega_r)\tilde{\mathbf{Y}}(\omega_r)\frac{\partial \mathbf{V}^*(\omega_r)}{\partial \omega_r} + \tilde{\mathbf{V}}(\omega_r)^*\mathbf{Y}(\omega_r)\frac{\partial \mathbf{V}(\omega_r)}{\partial \omega_r} \right] \tag{3.32}$$

Let us consider the contributions from these terms to the time average energy (3.29). First, it is noted that the analysis is restricted to lossless circuits which quite generally satisfy the following relation:

$$P_{loss} = \mathrm{Re}\,[\tilde{\mathbf{V}}(\omega_r)^*\,\mathbf{I}(\omega_r)] = 0$$

Therefore, the contribution from Δ_1 is zero. Furthermore, in view of the fact that the admittance matrix satisfies $\mathbf{Y}(\omega_r) = -\tilde{\mathbf{Y}}(\omega_r)^*$, the bracketed expression in (3.32) is real. And accordingly, the term Δ_3 is imaginary, contributing nothing to the time average energy. The only nonzero contribution comes from the term Δ_2 which, upon substitution into (3.29), provides the following expression for the time average stored energy:

$$\langle W \rangle = \langle W(\omega_r) \rangle = \frac{1}{4}\tilde{\mathbf{V}}(\omega_r)^*\,\frac{\partial}{\partial \omega_r}\left(\frac{\mathbf{Y}(\omega_r)}{j}\right)\mathbf{V}(\omega_r) \tag{3.33}$$

The admittance matrix $\mathbf{Y}(\omega_r)$ of lossless circuits is pure imaginary so that the expression is real, which of course is expected. The superimposed time variations $W(t)_\sim$ associated with the slow exponential growth and occurring at twice the frequency ω_r were already specified in (3.26).

The infinitely slow growing exponential excitation process is a mathematical convenience which is not expected to bear much resemblance to actual physical processes. But it can be assured that the energy must be independent of the details of excitation. Otherwise there would exist some cycling process allowing net energy to be extracted over some period of time, in violation of energy conservation principles. Therefore, the time average energy $\langle W \rangle$ is uniquely specified by (3.33) regardless of the details of the excitation process. The same argument does not apply to the superimposed energy fluctuations. The expression for $W(t)_\sim$ specified in (3.26) is applicable only to the specific exponential excitation procedure. Although other excitation methods might produce different energy fluctuations, this point is of minor significance because $W(t)_\sim$ is not used explicitly in the forthcoming presentations.

Up to this point the analysis of stored energy follows classical electromagnetic theory in expressing the stored energy by the properties of the admittance matrix, notably its derivative with respect to the positive real frequency ω_r. Let us now find out what happens for negative frequencies.

3.2.4 Stored energy at negative frequencies

Let us first be specific by what is meant by negative frequencies. If the time varying voltage is $v(t)$, then the Fourier decomposition is given by

$$v(t) = \int_{\omega} V(\omega) e^{j\omega t}\, d\omega \tag{3.34}$$

where ω is the frequency. Because $v(t)$ is a real quantity the following manipulation renders $v(t)$ invariant:

$$v(t)^* = v(t) = \int_{\omega} V(\omega)^* e^{-j\omega t} d\omega = \int_{\omega^*} V(-\omega^*)^* e^{j\omega^* t} d\omega^* \tag{3.35}$$

Comparison with (3.34) shows that

$$V(\omega) = V(-\omega^*)^* \tag{3.36}$$

The transformation $\omega \to -\omega^*$ is shown in the complex frequency plane of Fig. 3.2a. It merely transforms the complex amplitude $V(\omega)$ to its complex conjugate. If ω is real, then $-\omega^* = -\omega$, leaving the apparent impression that negative frequencies are superfluous in electromagnetic theory. But the difference is more subtle, being related to a conceptual change of the direction of time, which is an allowed operation in the neoclassical theory. This point is demonstrated by the following exposition. The definition (3.34) is still valid if we let $t \to -t$ in which case the equation transforms to

$$v(-t) = \int_{\omega} V(\omega)\, e^{j\omega(-t)}\, d\omega = \int_{\omega} V(-\omega)\, e^{j\omega t}\, d\omega \tag{3.37}$$

The equation shows that the Fourier transform of $v(-t)$ is equal to $V(-\omega)$ which is different from $V(-\omega^*)^*$ in (3.36). The transformation $\omega \to -\omega$ is shown in the complex frequency plane of Fig. 3.3a. The difference between the two transforms in Fig. 3.2a and Fig. 3.3a is important for analytical continuations from the real frequency axis into the complex frequency domain, being in opposite direction for $v(t)$ and $v(-t)$. This particular point is essential for determination of stored energy at negative frequencies.

Equation (3.37) also tells us that the calculations can be done in two equivalent ways: In the first Fourier integral $v(-t)$ is expressed by a positive frequency ω but time is progressing in the *negative* direction. In the last integral the frequency is negative, but time is progressing in the *positive* direction. This statement implies that the cases involving *negative* time are described by a simple frequency sign reversal and retaining *positive* time. We shall make extended use of this property in the subsequent chapters. In particular, the stored energy for negative frequency is obtained directly from (3.26) and (3.33) by letting $\omega_r \to -\omega_r$. The corresponding admittance matrix $\mathbf{Y}(-\omega_r)$ is determined from (3.15) which shows that $\mathbf{Y}(\omega)$ is Skew Hermitian. At real frequencies ω_r this property implies that

$$\mathbf{Y}(-\omega_r) = -\mathbf{Y}(\omega_r) \tag{3.38}$$

With these substitutions the time average stored energy becomes

$$\langle W(-\omega_r)\rangle = -\frac{1}{4}\tilde{\mathbf{V}}(-\omega_r)^* \frac{\partial}{\partial\omega_r}\left(\frac{\mathbf{Y}(\omega_r)}{j}\right)\mathbf{V}(-\omega_r) \tag{3.39}$$

whereas the periodic component is given by

$$W(-\omega_r)_\sim = \frac{1}{4}\,\mathrm{Re}\left[\tilde{\mathbf{V}}(-\omega_r)\frac{\mathbf{Y}(\omega_r)}{j\omega_r}\mathbf{V}(-\omega_r)\,e^{-2j\omega_r t}\right] \tag{3.40}$$

The formula (3.39) shows that the stored energy, i.e., the time average energy is negative. How this is to be understood conceptually is illustrated in Fig. 3.3b. This figure illustrates a scenario in which the field amplitudes grow exponentially with an extremely slow growth rate from time $t = \infty$ in the negative time direction. At the finite time t (the present) the time variation is indistinguishable from a pure sinusoid at frequency ω_r. By evaluation of the integrated power influx to the circuit during the entire process, using the same mathematical formulation as in (3.20) to (3.33), it is found that the total expenditure of power is *negative.* In the process of being excited the circuit is emitting *positive* power to the surroundings, a process equivalent to the growth of negative energy $\langle W(-\omega_r)\rangle$ as described by (3.39). It is of course realized that causality is violated in the indicated process, and circuit theorists would be quick to tell us that. But the present deliberations are put forward with very good conscience because they are consistent with the á priori assumption of bidirectional time and causal violation in phenomena dealing with single events in the microcosm.

The fact that the stored energy is negative in the negative frequency domain prompts the obvious question whether a superposition of electromagnetic fields of positive and negative frequency might result in net zero time average energy. And in fact, we shall find this to be true.

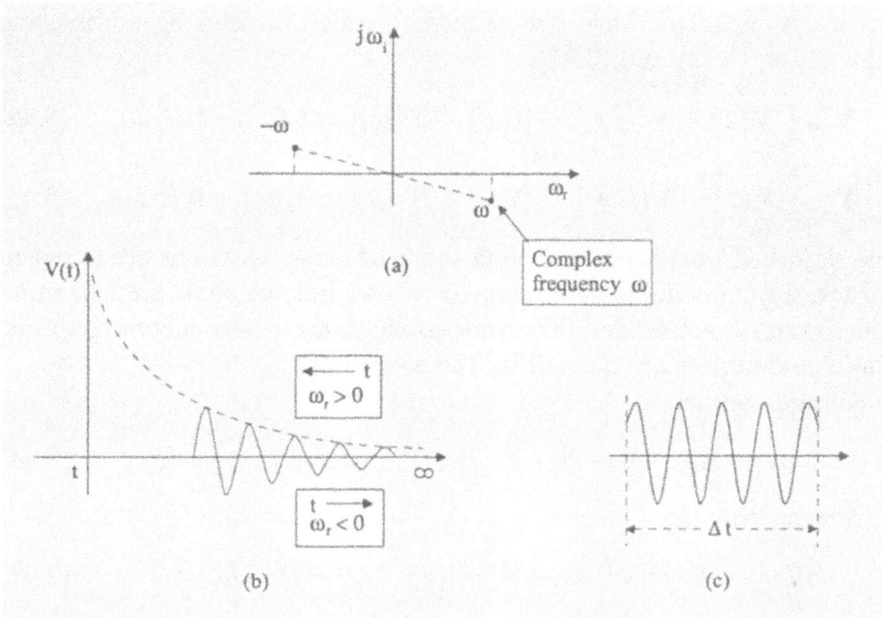

Figure 3.3. Process of negative energy storage in a general circuit at *negative* frequency. (**a**) Complex frequency plane. (**b**) Illustration of the two alternative procedures for evaluation of stored energy: Time t in the *negative* direction, $\omega_r > 0$; time t in the *positive* direction, $\omega_r < 0$. (**c**) Observance of constant amplitude over a *finite* time interval Δt.

3.3 GENERAL DEFINITION OF PHOTON DOUBLETS

The concept of photon doublets was introduced in Chapter 2 for the special case of plane wave pulse propagation in bidirectional time. The photon doublet was there defined as a composite entity consisting of a regular photon and an antiphoton so that the overall time average energy vanishes. The general theory developed in this chapter allows us to generalize the concept of photon doublets beyond plane waves. As a preamble to this generalization let us first determine the form of the circuit equation in the negative frequency domain, by transforming the regular circuit equation (3.13) to negative frequencies, making use of the Skew Hermitian properties of the admittance matrix $\mathbf{Y}(\omega)$ expressed by (3.38). Omitting subscripts r in the real frequency ω_r the result of this transformation is

$$\mathbf{Y}(\omega)\,\mathbf{V}(-\omega) + \mathbf{I}(-\omega) = -\mathbf{I}_g(-\omega) \tag{3.41}$$

The equation forms the basis of circuit equations for photon doublets which, as expected, have two different forms obtained by adding and subtracting (3.13) and (3.41).

$$\mathbf{Y}(\omega)\left[\mathbf{V}(\omega)+\mathbf{V}(-\omega)\right]-\left[\mathbf{I}(\omega)-\mathbf{I}(-\omega)\right]=\mathbf{I}_g(\omega)-\mathbf{I}_g(-\omega) \qquad (3.42)$$

$$\mathbf{Y}(\omega)\left[\mathbf{V}(\omega)-\mathbf{V}(-\omega)\right]-\left[\mathbf{I}(\omega)+\mathbf{I}(-\omega)\right]=\mathbf{I}_g(\omega)+\mathbf{I}_g(-\omega) \qquad (3.43)$$

The right hand extraneous driving terms of these equations are unessential for the immediate discussion to follow, but we shall keep in mind that the terms are needed in certain applications in subsequent chapters. The true doublets are defined by the equations:

Voltage doublet:

$$\mathbf{V}(\omega,-\omega)_- = \mathbf{V}(-\omega)-\mathbf{V}(\omega)=\mathbf{0} \qquad \mathbf{I}(\omega)+\mathbf{I}(-\omega)=0 \qquad (3.44)$$

Current doublet:

$$\mathbf{V}(\omega,-\omega)_+ = \mathbf{V}(-\omega)+\mathbf{V}(\omega)=\mathbf{0} \qquad \mathbf{I}(-\omega)-\mathbf{I}(\omega)=\mathbf{0} \qquad (3.45)$$

For these doublets the left hand sides of (3.42) and (3.43) are identically zero. The specific conditions described by these equations have the character of doublet resonances, a concept to which we shall return in a moment.

3.3.1 Stored energy of doublets

The positive and negative energy components, referring to different frequencies ω and $-\omega$, are orthogonal and can be added directly. Thus, the total time average stored energy is simply the sum of the constituents (3.33) and (3.39).

$$\langle W(\omega,-\omega)\rangle = \langle W(\omega)\rangle + \langle W(-\omega)\rangle \qquad (3.46)$$

The sum can be expressed in the form

$$\langle W(\omega,-\omega)\rangle = \frac{1}{4}\mathrm{Re}\left\{\tilde{\mathbf{V}}(\omega,-\omega)^*_+\left[\frac{\partial}{\partial\omega}\left(\frac{\mathbf{Y}(\omega)}{j}\right)\right]\mathbf{V}(\omega,-\omega)_-\right\} \qquad (3.47)$$

The expression (3.47) is proved by direct expansion, noting that the admittance $\mathbf{Y}(\omega)$ of lossless circuits is imaginary so that $\mathbf{Y}(\omega)/j$ is real.

It is readily verified that the expression reduces properly to $\langle W(\omega)\rangle$ and $\langle W(-\omega)\rangle$ in the special cases that either $\mathbf{V}(-\omega)=0$, or $\mathbf{V}(\omega)=0$. It is also observed that the time average energy of the doublets defined in (3.44) and (3.45) vanish identically.

$$\langle W(\omega,-\omega)\rangle_{doublet} = 0 \qquad (3.48)$$

The corresponding time varying part of stored energy is evaluated to

$$W(\omega,-\omega)_{\sim}=\frac{1}{4}\mathrm{Re}\left\{\tilde{\mathbf{V}}(\omega,-\omega)_a\left[\frac{\mathbf{Y}(\omega)}{j\omega}\right]\mathbf{V}(\omega,-\omega)_a\,e^{2j\omega t}\right\} \tag{3.49}$$

where

$$\mathbf{V}(\omega,-\omega)_a=\mathbf{V}(\omega)+\mathbf{V}(-\omega)^* \tag{3.50}$$

The generalized doublets (3.44) and (3.45) apply to any conceivable configuration of circuits. Conceptually, they are the generalized equivalents of the plane wave doublets discussed in Chapter 2. The voltage doublet represents a pure electric field configuration with the magnetic field being zero. The current doublet is the dual of the voltage doublet, specifying a nonzero magnetic field configuration with no electric fields. The two conditions can be interpreted as resonances in bidirectional time, a concept that we shall return to shortly. The energy of these resonances oscillates between positive and negative values according to (3.49), maintaining zero time average energy. The periodic oscillation of energy around zero is obtained from (3.49) by insertion of the resonance conditions (3.44) and (3.45) with the following results:

Voltage resonance; the net current is zero:

$$[W(\omega)_{\sim}]_{voltage}=\mathrm{Re}\,[\tilde{\mathbf{V}}(\omega)]\frac{\mathbf{Y}(\omega)}{j\omega}\,\mathrm{Re}\,[\mathbf{V}(\omega)]\,\cos(2\omega t) \tag{3.51}$$

Current resonance; the net voltage is zero:

$$[W(\omega)_{\sim}]_{current}=-\,\mathrm{Im}\,[\tilde{\mathbf{V}}(\omega)]\frac{\mathbf{Y}(\omega)}{j\omega}\,\mathrm{Im}\,[\mathbf{V}(\omega)]\,\cos(2\omega t) \tag{3.52}$$

It appears that the excitation *phase* determines the fluctuation magnitudes through the terms $\mathrm{Re}[\mathbf{V}(\omega)]$ and $\mathrm{Im}[\mathbf{V}(\omega)]$. If one of these is small the other is large. Referring to the discussions following (3.33) these formulae are expected to apply to the infinitely slow growing exponential excitation process only, and may differ in alternative excitation methods. This point is of minor importance in the forthcoming discussions where the exact knowledge of the magnitude of the energy fluctuations plays an insignificant role.

The voltage and current doublet excitations in (3.44) and (3.45) have the flair of resonances in bidirectional time. In the neoclassical theory they are the causes of the environmental photon doublet energy fluctuations, shortened *PDE*, a concept that was introduced in Chapters 1 and 2, and subject to further generalization in this chapter. The following discussion is hopefully shedding more light on this important concept.

3.3.2 Photon doublet resonances

As a preamble to the subject of doublet resonances let us first look at the *regular* resonances at positive frequencies, referred to the generalized circuit in Fig. 3.1. Without the driving term $\mathbf{I}_g(\omega)$, which is irrelevant for the resonance problem, the circuit equation (3.13) has the form

$$\mathbf{Y}(\omega)\,\mathbf{V}(\omega) = \mathbf{I}(\omega) \tag{3.53}$$

It was commented earlier that the admittance matrix $\mathbf{Y}(\omega)$ and the voltage and current columns $\mathbf{V}(\omega)$ and $\mathbf{I}(\omega)$ are not unique. But a particular selection is unique in the sense that it serves to diagonalize the system. The process of diagonalization is based on the eigenvalue relation for $\mathbf{Y}(\omega)$, which can be expressed in the following matrix form:

$$\mathbf{Y}(\omega)\mathbf{R}(\omega) = \mathbf{R}(\omega)\mathbf{Y}(\omega)_{diag} \tag{3.54}$$

In this equation the diagonal matrix $\mathbf{Y}(\omega)_{diag}$ represents the N eigenvalues, whereas the columns of the $N \times N$ matrix $\mathbf{R}(\omega)$ specify the corresponding eigenfunctions. For the presumably lossless circuit the admittance matrix $\mathbf{Y}(\omega)$ is symmetric and $\mathbf{R}(\omega)$ is unitary so that $\mathbf{R}(\omega)^{-1} = \tilde{\mathbf{R}}(\omega)$. Through this procedure the circuit equation (3.53) converts to

$$\mathbf{Y}(\omega)_{diag}\,\mathbf{V}(\omega)_0 = \mathbf{I}(\omega)_0 \tag{3.55}$$

where the transformed voltage and current columns $\mathbf{V}(\omega)_0$ and $\mathbf{I}(\omega)_0$ are related to the original ones by

$$\mathbf{V}(\omega)_0 = \tilde{\mathbf{R}}(\omega)\mathbf{V}(\omega) \qquad \mathbf{I}(\omega)_0 = \tilde{\mathbf{R}}(\omega)\mathbf{I}(\omega) \tag{3.56}$$

The voltage resonances are characterized by $\mathbf{I}(\omega)_0 = \mathbf{0}$ so that the circuit equation becomes homogeneous

$$\mathbf{Y}(\omega)_{diag}\,\mathbf{V}(\omega)_0 = \mathbf{0} \tag{3.57}$$

In order for $\mathbf{V}(\omega)_0$ to be nonzero, one or several of the eigenvalues must be zero. The qth resonance is characterized by the qth eigenvalue being zero:

$$Y(\omega_q)_{diag} = 0 \tag{3.58}$$

The corresponding voltage vector is then

$$\mathbf{V}(\omega_q)_0^q = \begin{bmatrix} 0 \\ \vdots \\ V(\omega_q)_0^q \\ \vdots \\ 0 \end{bmatrix} \tag{3.59}$$

At this particular resonance the original voltage vector $\mathbf{V}(\omega_q)$ is obtained from the relation

$$\mathbf{V}(\omega_q) = \mathbf{r}(\omega_q)^q \, V(\omega_q)_0^q \tag{3.60}$$

where $\mathbf{r}(\omega_q)^q$ is the *qth* eigenvector of the original admittance matrix, evaluated at the real resonance frequency ω_q. Depending on the specifics of the configuration, the circuit can have one or any number $N_{res} \leq N$ of distinct resonance frequencies. The condition (3.58) is conventionally referred to as a *parallel* resonance. As an auxiliary condition the occurrence of a particular resonance is contingent on providing the right open circuit boundary conditions at N virtual terminals, so that the current column $\mathbf{I}(\omega)_0$ is zero.

The *series* resonances are characterized by the voltage vector $\mathbf{V}(\omega)$ being identically zero so that

$$\mathbf{Y}(\omega)^{-1} \, \mathbf{I}(\omega) = \mathbf{Z}(\omega) \, \mathbf{I}(\omega) = \mathbf{V}(\omega) = \mathbf{0} \tag{3.61}$$

The occurrence of series resonances are formally equivalent to parallel resonances, with the *impedance* matrix $\mathbf{Z}(\omega)$ replacing the admittance matrix $\mathbf{Y}(\omega)$ and $\mathbf{I}(\omega)$ replacing $\mathbf{V}(\omega)$. The N terminals must be short circuited in order to provide the right boundary conditions for series resonance. All the conventional resonances in the presumably lossless circuit exhibit the common feature that the stored energy is nonzero, oscillating periodically between pure electric energy and magnetic energy, the sum of the two being constant. A particular resonance can be set up through influx of an equivalent amount of energy from some extraneous source such as $I_g(\omega)$ in (3.13). Once excited in the lossless circuit, the resonance continues without further power expenditure.

With this brief account of the characteristics of regular resonances, let us return to the doublet resonances and ask in what sense they differ from the regular resonances. First of all they are resonances in bidirectional time rather than being spatial resonances, with the ramifications following from this fundamental difference. The conditions for voltage and current resonance in bidirectional time can be expressed in a form similar to (3.53) and (3.61). Applying the doublet definitions (3.44) and (3.45) we obtain the following equivalents of the resonance conditions (3.57) and (3.61):

Voltage resonance in bidirectional time:

$$\mathbf{Y}(\omega) \, [\mathbf{V}(\omega) - \mathbf{V}(-\omega)] = \mathbf{0} \tag{3.62}$$

Current resonance in bidirectional time:

$$\mathbf{Z}(\omega) \, [I(\omega) - \mathbf{I}(-\omega)] = \mathbf{0} \tag{3.63}$$

By now it should be pretty clear that these *doublet* resonances are conceptually quite different from the regular resonances expressed by (3.57) and (3.61). The two homogeneous equations are satisfied regardless of the particulars of $\mathbf{Y}(\omega)$ and $\mathbf{Z}(\omega)$ because the doublet variables $\mathbf{V}(\omega)$–$\mathbf{V}(-\omega)$ or $I(\omega)$–$\mathbf{I}(-\omega)$ are identically zero at resonance. This demonstration shows that the existence of doublet resonances do not in any way depend on the particulars of the circuit, i.e., they are independent of the admittance and impedance matrices. The doublet resonances exist at all frequencies from zero and upwards in all conceivable electromagnetic circuits, including free space. In these resonances the energy fluctuates between positive and negative values with a time average energy of zero. Even if the existence of doublet resonances is not contingent on specific circuit parameters, the *spatial* distributions of all doublet field patterns are still determined by the circuit configuration simply because the appropriate circuit equations must be satisfied for the positive as well as the negative frequency components.

3.3.3 Doublet resonances—spontaneous excitation?

How are the doublet resonances excited? It is evident that all usual concepts of resonance excitation must be shed, simply because no energy is required. No net power influx to the circuit takes place in the excitation process. Is it possible that they are spontaneously excited like some kind of instability? In the procedure shown earlier the energy content of the generalized circuit was determined by a process in which the sinusoidal field amplitudes were supposed to grow infinitely slow; observed over any finite time interval they look like pure sinusoids. This procedure, which is merely a mathematical convenience, is not expected to bear much resemblance to actual physical mechanisms of excitations. However, the procedure is justified by imposing energy conservation principles. The energy content at a given frequency must be *independent* of the way the fields are excited. Otherwise, there would exist some cycling process allowing net energy to be extracted over some period of time, in apparent violation of energy conservation principles. A bit of insight on the excitation process of doublet resonances can possibly be gained from consideration of the *inhomogeneous* admittance equation (3.43) at resonance:

$$\mathbf{Y}(\omega)\left[\mathbf{V}(\omega) - \mathbf{V}(-\omega)\right] = \mathbf{I}_g(\omega, -\omega) \tag{3.64}$$

where now the frequency ω is assumed to be complex, allowing exponential growth of the amplitude, putting

$$\omega = \omega_r - j\omega_i \tag{3.65}$$

The doublet resonance condition assures that $\mathbf{V}(\omega_r) \equiv \mathbf{V}(-\omega_r)$ for all real frequencies ω_r. Accordingly, analytical continuation of the left side of (3.64) results in

$$\omega_i \frac{\partial}{\partial \omega_r}\left(\frac{\mathbf{Y}(\omega_r)}{j}\right)[\mathbf{V}(\omega_r) - \mathbf{V}(-\omega_r)] = \mathbf{I}_g(\omega_r, -\omega_r) \tag{3.66}$$

Multiplying both sides by $[\tilde{\mathbf{V}}(\omega_r)^* + \tilde{\mathbf{V}}(-\omega_r)^*]/4 = [\tilde{\mathbf{V}}(\omega_r)^*]/2$ and taking the real parts we obtain the following transformed formula:

$$\omega_i \left\langle W(\omega_r, -\omega_r)\right\rangle = \frac{1}{2}\operatorname{Re}\left[\tilde{\mathbf{V}}(\omega_r)^* \mathbf{I}_g(\omega_r, -\omega_r)\right] \tag{3.67}$$

where $\langle W(\omega_r, -\omega_r)\rangle$ is the time average doublet energy density specified earlier in (3.47). The expected growth rate ω_i is then given by

$$\omega_i = \frac{\operatorname{Re}\left[\tilde{\mathbf{V}}(\omega_r)^* \mathbf{I}_g(\omega_r, -\omega_r)\right]}{2\left\langle W(\omega_r, -\omega_r)\right\rangle} \tag{3.68}$$

The denominator $\langle W(\omega_r, -\omega_r)\rangle$ is zero because the time average energy density of doublet resonances vanishes. Therefore, the equation tells us that the *tiniest* amount of driving current $\mathbf{I}_g(\omega_r, -\omega_r)$ causes infinite growth rate ω_i. In accordance with the definition (3.14) the driving current $\mathbf{I}_g(\omega_r, -\omega_r)$ is indeed nonzero if there exists some small current source $\vec{i}_g(\vec{r}, \omega_r)$ whose scalar product with the normalized circuit field column $\vec{\epsilon}(\vec{r}, \omega_r)$ integrated over the circuit volume is nonzero. It is utterly unlikely that tiny currents from moving charge carriers such as individual electrons would be orthogonal to *all* the components of the column $\vec{\epsilon}(\vec{r}, \omega_r)$. It is much more likely that there exists some small but nevertheless nonzero coupling to most, if not all the characteristic modes of whatever circuit might be thought of, so that the corresponding components of the driving current $\mathbf{I}_g(\omega_r, -\omega_r)$ are nonzero and the corresponding growth rate ω_i infinite.

The analysis presented here leading to (3.68) is based on first order analytical continuation in the complex frequency plane, with ω_i supposedly being small. Because equation (3.68) predicting infinite ω_i certainly violates this assumption, the result should possibly be interpreted as a kind of absolute instability of explosive nature. It can just be speculated that various higher order effects and nonlinear interactions saturate the amplitudes. Regardless of the lacking details, the first order excitation of doublet resonances described here is certainly quite spectacular. It has the character of an absolute instability in bidirectional time. The amplitude growth at any frequency takes place at each space point with no need of any influx of energy. And this brings us to the tentative conclusion that the doublet resonances might be excited spontaneously, in all kinds of circuits and in free space. Not only do the resonances appear to be excited spontaneously whenever needed, but they probably *are* already excited everywhere, permeating free space as photon doublet energy (PDE) fluctuations, and being present in any configuration that can be labeled as a circuit in the widest meaning of the word.

3.3.4 Doublet spectral energy density

It is tempting to carry this line of thought even further, speculating that the set of photon doublet resonances form an essential ingredient of the vacuum state, with the doublets visualized as being in some kind of equilibrium by mutual interactions through various nonequilibrium active processes and also by nonlinear effects in material bodies. Interactions of the latter kind occur if the electromagnetic properties are either time varying or nonlinear, giving rise to the familiar frequency conversions known from classical electromagnetics. Often referred to as parametric processes, the frequency conversions are governed by the Manley Rowe relations [33] expressing conservation of energy and momentum. It is tempting to speculate further that the doublet energy fluctuations through such interactions form a stationary state, tending to redistribute themselves so that the *total rms* energy fluctuations are minimized. The *stable* equilibrium is certainly expected to be a minimum. The *rms* energy fluctuation $w(\omega)$ of a single doublet of frequency ω would be specified by an equation like (3.49). If it is assumed that the equilibrium number density at frequency ω is $n(\omega)$ then the total doublet *rms* energy fluctuation is given by the integral

$$W_{rms} = \int_{\omega_{\min}}^{\omega_{\max}} w(\omega)\, n(\omega)\, d\omega \tag{3.69}$$

Let us make the reasonable guess that the product $w(\omega)\, n(\omega)$ is proportional to $C\omega^q$ where C is some constant and the exponent q has to be determined so that the integral

$$W_{rms} = \int_{\omega_{\min}}^{\omega_{\max}} C\, \omega^q\, d\omega \tag{3.70}$$

is minimized. This happens for $q = -1$ yielding

$$(W_{rms})_{\min} = \int_{\omega_{\min}}^{\omega_{\max}} \frac{C}{\omega} \, d\omega_r = C \log \left(\frac{\omega_{\max}}{\omega_{\min}} \right) \tag{3.71}$$

The optimum spectral energy density is then specified by

$$w_{rms} = \frac{C}{\omega} \tag{3.72}$$

Compared to (3.71) all exponents differing from the optimum $q = -1$ results in orders of magnitude higher energy density, characterized by singularities of the type $(\omega_{\max})^q$ and $(\omega_{\min})^{-q}$. The optimum (3.71) still has logarithmic infinities at $\omega_{\min} = 0$ and $\omega_{\max} = \infty$, but we shall see that these are of little practical significance. Provided the initial proposition of an energy minimum is accepted, the minimization process leading to (3.71) predicts constant energy for a fixed ratio $\omega_{\max}/\omega_{\min}$. Thus, the energy is the same in the range 0.1 Hz to 1 Hz as in 1 Hz to 10 Hz, or in any other decade of frequency. In principle, the number of decades is infinite, but there are physical limitations to the frequencies $\omega_{\max}$ and $\omega_{\min}$ causing the number of decades to be finite. Assuming stationary fluctuations, the lowest frequency can be assessed making use of the *Ergodic Theorem* of statistical mechanics. According to this theorem the ensemble average and the time average are equal. This correspondence is expressed by the equation

$$\int_{\omega_{\min}}^{\omega_{\max}} w(\omega) \, n(\omega) \, d\omega = \frac{1}{2\pi} \frac{1}{T} \int_{t=0}^{T} w(t) n(t) dt \tag{3.73}$$

If we visualize the fluctuations determined by a measurement process involving the time average on the right, the maximum time interval T, i.e., the upper limit in the integral, can certainly not be longer than the age of the universe, thought to be approximately 10^{17} seconds. This upper time limit puts the lower frequency limit $\omega_{\min}$ equal to the inverse age of the universe, which then is roughly $10^{-17}\,\text{sec}^{-1}$. As far as the upper frequency is concerned let us take $\omega_{\max}$ to be exceedingly high and equal to the *Planck frequency* which is roughly $10^{43}\,\text{sec}^{-1}$. In spite of this enormous range the number of *decades* in the interval is only 60, so that the total *rms* energy density $(W_{rms})_{\min}$ in (3.71) is a meager 60 times the amount of energy contained in each frequency decade, and the total *rms* voltage is only a factor $\sqrt{60} = 7.75$ times the *rms* voltage in one decade. Therefore, in view of the limitations imposed by cosmological boundary conditions (age of the universe) and the tiniest dimensions (Planck frequency) the logarithmic infinities are of little concern. In fact, regardless of the span of the selected frequency range, if within reason, the number of decades will be finite. These simple deliberations show that the logaritmic infinities in (3.71) are merely mathematical and of no concern as far as physical realities are concerned. It must be concluded that the total *rms* energy density, i.e., the energy per unit volume, is finite and in all probability not very large.

If we accept the initial hypothesis that the total *rms* energy fluctuations should be a minimum, the inevitable conclusion is the result stated in (3.72) that the *rms* energy density is inversely proportional to frequency. It further implies that free space and all circuit configurations are exposed to fields of the doublet types, having no time average energy but superimposed energy fluctuations (*PDE*) inversely proportional to frequency. In particular, the associated voltage or current fluctuations are expected to exist across the terminals of every electronic circuit and gadget surrounding us in the daily life.

At this point circuit designers will probably object vehemently, uttering remarks such as "This is pure nonsense. We never see or measure anything like these doublet fields and fluctuations in any circuit of ours. If they existed, they would completely ruin the performance of any electronic device". But the designers can rest assured that nothing of this kind happens, for the simple reason that electromagnetic doublet fields carry no net energy, and therefore are unable to supply any energy to a measuring or registering device, such as a voltmeter placed across the terminals of a passive circuit. In short, the basic doublet fields can not be detected by an ordinary measurement process.

However, circuit designers may not be entirely correct in their assessment of never seeing *PDE* fluctuations in their circuits. On the contrary, fluctuations that very well might be traced back to this source are observed regularly in *active* and *dissipative* circuits, manifesting themselves as the familiar $1/f$ noise, a constant nuisance to circuit designers. As the name indicates the observed noise power spectrum is typically inversely proportional to frequency. In spite of a continued search over many decenniums, the basic physical mechanism causing $1/f$ noise has stubbornly refused to reveal itself [4]. The general occurrence of the phenomenon in essentially all active electronic devices points to a very fundamental physical cause. This cause could possibly be the photon doublet fluctuations discussed here because the agreement with (3.72) certainly indicates a common source. The readers taking special interest in the subject are referred to Chapter 9 which presents an extensive analysis and discussion of $1/f$ noise based on doublet fluctuations as the possible noise source, employing a more direct analytical approach. The results presented in that chapter appear to provide considerable support for the common source hypothesis.

3.3.5 Quantum mechanical ZPE fluctuations

The universal photon doublet fluctuations described in the preceding section appear to be unavoidable consequences of the neoclassical hypothesis of a bidirectional time frame in the microcosm domain, i.e., in the description of single event behavior. We have seen that this assumption leads to vacuum fluctuations of zero time average energy (PDE), differing markedly from the quantum mechanical concept of zero point energy fluctuations (ZPE). The ZPE fluctuations, described in any textbook on quantum mechanics, see for instance [30], are attributed to the fluctuations in the ground states of quantum oscillators. It is claimed to be a consequence of the uncertainty relation. According to this quantum principle the ground mode is populated with a photon half the time. This population, common to all modes, add up to the zero point energy in free space with the spectral energy density

$$w(\omega) = \frac{\hbar\omega^3}{2\pi^2 c^3} \tag{3.74}$$

which is seen to be proportional to the third power of frequency. The total energy density is obtained by integration over the frequency range from zero to some maximum frequency ω_{max}.

$$W(\omega_{\text{max}}) = \int_0^{\omega_{\text{max}}} w(\omega)\, d\omega = \frac{\hbar\omega_{\text{max}}^4}{8\pi^2 c^3} \tag{3.75}$$

In principle, the allowed frequencies have no upper limit and the integral is highly divergent. But even if the integration range is truncated by specifying some value for ω_{max}, the integral will be excessively large. Which value to pick is an open question because there is no great physical theorem predicting the value of ω_{max}, except that quantum effects smaller than a certain wavelength may presumably be altered due to gravitational effects, thereby putting an upper limit on ω_{max}. It is argued in quantum mechanics that ω_{max} should be perhaps of the order of the Planck Frequency which is roughly 10^{43} sec^{-1} . The only trouble with this upper limit is that the resulting energy density would still be excessively large, so large in fact that the universe would shrink to atomic dimensions due to gravity. In fact, almost regardless of the upper limit taken for ω_{max}, if it is within reasonable limits, the energy density would still be quite enormous, and exceeding atomic energy densities. The cosmologists tell us that the geometric structure of the universe is extremely sensitive to the value of the ZPE vacuum energy. The *cosmological constant*, which presumably is proportional to the vacuum energy density and therefore directly related to it, is an empirical parameter introduced to explain observational data of the expansion of the universe. The only trouble is that the most recent observational data on the cosmological constant puts its value at 120 orders of magnitude below the calculated vacuum energy [1],[28],[45]. This discrepancy between predictions from orthodox quantum theory and observations is one of the most perplexing quantitative puzzles in physics today. The whole concept of such high energy density is difficult to take seriously, and it is understandable that the concept of universal ZPE fluctuations is approached with reluctance in quantum physics as well as in cosmology.

In addition to the energy catastrophe associated with the ZPE fluctuations there is another just as serious problem tied in with the circumstance that the corresponding rms electromagnetic *fields* would add up to enormous strengths. Quantum theory naively treats the ZPE field as a uniform distribution pervading all space including material bodies. But the electromagnetic fields are photons obeying Maxwell's equations, otherwise the fields would have to be visualized as some kind of abstractions quite foreign to regular photons. But ideas in this direction are not seriously suggested. It must therefore be maintained that the ZPE fields are indeed electromagnetic, adding up to extreme *rms* intensity, which is entirely unrealistic from the point of view of elementary electromagnetic *boundary* conditions. Interaction with charge carriers in conducting objects in the universe would cause current responses of extreme strength and corresponding power dissipation. In fact, the whole universe would blow up in one big flash.

Attempts to circumvent the problem with the excessive energy density are many. In quantum electrodynamics (QED) a technique known as renormalization, by which the infinity is cleverly absorbed or compensated by other infinities, is well known. A problem with the renormalization concepts is their inability to point to specific physical processes causing the compensation. It is argued, sometimes, that the fields are not real, but rather virtual, representing merely quantum possibilities rather than certainties. But a more common attitude amongst quantum physicists is to simply explain the problem away with arguments that at best are highly questionable. It is insisted that there is no great embarrassment in the existence of infinite zero point energy for the simple reason that no measurements have been devised in which the response would be proportional to the infinite zero point energy, because practical experiments respond to *changes* in the total energy. In order to be detected, nonuniformities must somehow be created, or so the argument goes. This assertion has somehow penetrated the views of the physics community, as testified by quotes in a large number of books and publications, too numerous to be cited. The argument is in itself a highly questionable proposition because the ZPE fields are supposedly not different from the regular electromagnetic radiation surrounding us, and therefore would be detected by a variety of electronic apparatus such as radios, television sets, and radar installations and, by all means it would be detected by the retina of the human eye, converting the incident optical ZPE photon energy to electrophysiological responses in the brain. How that would affect our well being is anybody's guess. Without exception all of these receivers, and other conceivable detectors as well, respond directly to the flux density of the incoming radiation, whether regular or of the ZPE kind, with responses directly proportional to the radiation *strength* and not to the radiation *gradient* as claimed in these assertions. In short, it does not take a special kind of detector to register the ZPE radiation fields.

It is somewhat surprising that arguments as those above, in obvious disregard of elementary electromagnetic boundary conditions, are put forward in textbooks and publications. It is difficult to see such propositions as anything else than some kind of psychological trauma. The typical quantum physicist, if he is an orthodox follower of Bohr and Heisenberg, is really in a bind because the concept of infinite zero point energy follows from or at least is associated with Heisenberg's uncertainty relation, surely one of the cornerstones of quantum physics. The physicist might find himself torn between two conflicting and mutually incompatible choices. He must accept the infinite zero point energy, otherwise he would implicitly have to abandon the uncertainty relation. He can not find himself to reject Heisenberg but would still like to be on terms with Maxwell by somehow explaining away the electromagnetic implications of the high zero point energy by arguments of the kind discussed above.

The concept of the ZPE fluctuations is a direct consequence of the basic quantum mechanical postulates, specifically Heisenberg's uncertainty relation. The questions raised concerning the improbability of these fluctuations are bound to somehow affect our view of the uncertainty relation. If the concept of ZPE energy is in for a major revision, then the uncertainty equation must be viewed with a great deal of suspicion. Readers interested in a further discussion on this subject are referred to Chapter 14.

In the neoclassical theory the problem of the energy density catastrophe of the *ZPE* concept is nonexistent. In this theory the corresponding *PDE* fields are also thought to pervade all space, but the associated energy problem is alleviated due to the cancellation of positive and negative energy components. The only detectable fields appear to be the energy *fluctuations* around the zero energy equilibrium. It is an interesting speculation that the energy cancellation in this theory might be considered a sort of renormalization, but differing from the abstract mathematical renormalization concepts in quantum physics by its physically based negative energy components. The discussion on the subject of *PDE* fluctuations, based on a hypothesis of minimization of energy, predicted the $1/\omega$ spectral density variation shown in (3.72). Although the total energy density was shown to have a logaritmic infinity at $\omega = 0$ and $\omega = \infty$, these limits were shown to be of little consequence due to cosmological and other physical constraints limiting both the lower and upper frequencies to finite values, so that the fluctuation energy as well as the associated fields are expected to be finite and in all probability quite small.

3.4 THE SCATTERING FORMULATION OF CIRCUIT RESPONSE

The general theory of electromagnetism in bidirectional time developed in this chapter was formulated in terms of the admittance matrix $\mathbf{Y}(\omega)$ of the generalized circuit, or the equivalent impedance matrix $\mathbf{Z}(\omega) = \mathbf{Y}(\omega)^{-1}$. The *scattering* description represents an alternative and often more illuminating way of specifying circuit response. This formulation is used in several of the configurations and experiments analyzed in the remainder of the book.

As a preamble to a general scatter formulation for the bitemporal domain, let us first establish the interconnection between the two formulations in the regular time frame, based on regular electromagnetics. The generalized voltage and current column vectors $\mathbf{V}(\omega)$ and $\mathbf{I}(\omega)$ are replaced by the new variables $\mathbf{a}(\omega)$ and $\mathbf{b}(\omega)$ through the linear transformations

$$\begin{aligned} \mathbf{V}(\omega) &= \mathbf{Y}_0^{-1/2}[\mathbf{a}(\omega) + \mathbf{b}(\omega)] \\ \mathbf{I}(\omega) &= \mathbf{Y}_0^{1/2}[\mathbf{a}(\omega) - \mathbf{b}(\omega)] \end{aligned} \tag{3.76}$$

where we have introduced the real diagonal admittance matrix $\mathbf{Y}_0$, to be defined shortly by the virtual terminal configuration of the actual circuit. The physical dimension of $\mathbf{a}(\omega)$ and $\mathbf{b}(\omega)$ is the square root of power. The two equations taken together with the admittance relation

(3.53) results in the following familiar scatter formulation:

$$\mathbf{b}(\omega) = \mathbf{S}(\omega)\,\mathbf{a}(\omega) \tag{3.77}$$

where the *scattering matrix* $\mathbf{S}(\omega)$, expressed by the admittance matrix $\mathbf{Y}(\omega)$ and the terminal matrix $\mathbf{Y}_0$, is given by

$$\mathbf{S}(\omega) = \mathbf{Y}_0^{1/2}[\mathbf{Y}_0 + \mathbf{Y}(\omega)]^{-1}[\mathbf{Y}_0 - \mathbf{Y}(\omega)]\mathbf{Y}_0^{-1/2} \tag{3.78}$$

The N-dimensional column vectors $\mathbf{a}(\omega)$ and $\mathbf{b}(\omega)$ can be interpreted as *influx* and *outflux* vectors associated with the N abstract terminals, specifying a set of N inward traveling wave components and a set of N outward traveling components, respectively. The net power dissipated in the circuit is obtained from (3.76)

$$P = \frac{1}{2}\,\mathrm{Re}\left[\tilde{\mathbf{V}}(\omega)^*\,\mathbf{I}(\omega)\right] = \frac{1}{2}\left[\tilde{\mathbf{a}}(\omega)^*\mathbf{a}(\omega) - \tilde{\mathbf{b}}(\omega)^*\mathbf{b}(\omega)\right] \tag{3.79}$$

In the lossless circuit treated here, the power P is zero so that

$$\tilde{\mathbf{a}}(\omega)^*\mathbf{a}(\omega) - \tilde{\mathbf{b}}(\omega)^*\mathbf{b}(\omega) = \mathbf{0} \tag{3.80}$$

In view of (3.77) this relation shows that

$$\tilde{\mathbf{S}}(\omega)^*\mathbf{S}(\omega) = \mathbf{1} \tag{3.81}$$

demonstrating that the scattering matrix of the generalized lossless circuit is unitary. The elements of the diagonal matrix $\mathbf{Y}_0$ represent the characteristic admittances of the virtual transmission lines feeding the circuit with the in- and outgoing wave components $\mathbf{a}(\omega)$ and $\mathbf{b}(\omega)$. In the present generalized circuit $\mathbf{Y}_0$ can be specified at will. However, in practical circuits, for instance in the microwave range, the terminals are well defined physical transmission lines or waveguides in which case the elements of the diagonal $N \times N$ matrix $\mathbf{Y}_0$ are the characteristic waveguide admittances. During the following analysis, in order to avoid unessential mathematical complexities, $\mathbf{Y}_0$ is assumed to be frequency independent.

3.4.1 Stored energy in the scattering formulation

In the scattering description the stored energy is evaluated from the corresponding admittance formulation (3.33) by expressing the admittance matrix $\mathbf{Y}(\omega)$ in terms of the scattering matrix $\mathbf{S}(\omega)$ through rearrangement of (3.78)

$$\mathbf{Y}(\omega) = \mathbf{Y}_0^{1/2}\,[\mathbf{1} - \mathbf{S}(\omega)]\;[\mathbf{1} + \mathbf{S}(\omega)]^{-1}\,\mathbf{Y}_0^{1/2} \tag{3.82}$$

The stored energy in the admittance formulation (3.33) is expressed by the derivative $\partial \mathbf{Y}(\omega)/\partial\omega$. In evaluating the derivative from (3.82) the following identity is needed:

$$\frac{d}{d\omega}[\mathbf{S}(\omega)^{-1}] = -\mathbf{S}(\omega)^{-1}\frac{\partial \mathbf{S}(\omega)}{\partial\omega}\mathbf{S}(\omega)^{-1} \tag{3.83}$$

which results in

$$\frac{\partial \mathbf{Y}(\omega)}{\partial\omega} = -2\mathbf{Y}_0^{1/2}[\mathbf{1}+\mathbf{S}(\omega)]^{-1}\frac{\partial \mathbf{S}(\omega)}{\partial\omega}[\mathbf{1}+\mathbf{S}(\omega)]^{-1}\mathbf{Y}_0^{1/2} \tag{3.84}$$

The voltage vectors $\mathbf{V}(\omega)$ and $\tilde{\mathbf{V}}(\omega)^*$ can be expressed in terms of the scatter variables $\mathbf{a}(\omega)$ and $\tilde{\mathbf{b}}(\omega)^*$ through (3.76) and (3.77).

$$\begin{aligned}[\mathbf{1}+\mathbf{S}(\omega)]^{-1}\mathbf{Y}_0^{1/2}\mathbf{V}(\omega) &= \mathbf{a}(\omega)\\ \tilde{\mathbf{V}}(\omega)\mathbf{Y}_0^{1/2}[\mathbf{1}+\mathbf{S}(\omega)]^{-1} &= \tilde{\mathbf{b}}(\omega)^*\end{aligned} \tag{3.85}$$

By insertion into (3.33) the following relation is obtained:

$$\langle W(\omega)\rangle = -\frac{1}{2}\tilde{\mathbf{b}}(\omega)^*\frac{\partial}{\partial\omega}\left(\frac{\mathbf{S}(\omega)}{j}\right)\mathbf{a}(\omega) \tag{3.86}$$

which is the general scatter formulation for stored energy at positive frequencies. It is noted that the formula is independent of the particular selection of the terminal matrix $\mathbf{Y}_0$.

3.4.2 Scattering relations for negative frequencies

The scatter relations (3.76) to (3.86) apply to regular positive frequencies. For negative frequencies the corresponding scattering matrix is obtained by direct substitution of $-\omega$ in (3.78).

$$\mathbf{S}(-\omega) = \mathbf{Y}_0^{1/2}[\mathbf{Y}_0+\mathbf{Y}(-\omega)]^{-1}[\mathbf{Y}_0-\mathbf{Y}(-\omega)]\mathbf{Y}_0^{-1/2} \tag{3.87}$$

Inserting $\mathbf{Y}(-\omega) = -\mathbf{Y}(\omega)$ and inverting the equation we obtain

$$\mathbf{S}(-\omega)^{-1} = \mathbf{Y}_0^{-1/2}[\mathbf{Y}(\omega)+\mathbf{Y}_0]^{-1}[\mathbf{Y}(\omega)-\mathbf{Y}_0]\mathbf{Y}_0^{1/2} \tag{3.88}$$

Comparison with (3.78) shows that the scattering matrix for negative frequencies is the inverse of the regular macroscopic scattering matrix.

$$\mathbf{S}(-\omega) = \mathbf{S}(\omega)^{-1} \tag{3.89}$$

Thus, for negative frequencies the scattering relation (3.77) transforms to

$$\mathbf{b}(-\omega) = \mathbf{S}(-\omega)\,\mathbf{a}(-\omega) = \mathbf{S}(\omega)^{-1}\mathbf{a}(-\omega) \tag{3.90}$$

which inverted results in the following scatter formula:

$$\mathbf{a}(-\omega) = \mathbf{S}(\omega)\,\mathbf{b}(-\omega) \tag{3.91}$$

Thus, for negative frequencies the roles of the influx and outflux components are interchanged. By means of the identity (3.83) the time average stored energy for negative frequencies is obtained directly from (3.86) by substituting $-\omega$ for ω.

$$\langle W(-\omega)\rangle = -\frac{1}{2}\tilde{\mathbf{a}}(-\omega)^* \frac{\partial}{\partial\omega}\left(\frac{\mathbf{S}(\omega)}{j}\right)\mathbf{b}(-\omega) \tag{3.92}$$

In the following section the established scatter representations of stored energy at positive and negative frequencies are combined to the more interesting scatter relations for photon doublets.

3.4.3 Doublet scattering relations and resonances

The appropriate scatter relation for photon doublets is obtained by adding and subtracting (3.91) and (3.77).

$$\mathbf{b}(\omega) \pm \mathbf{a}(-\omega) = \mathbf{S}(\omega)[\,\mathbf{a}(\omega) \pm \mathbf{b}(-\omega)] \tag{3.93}$$

The composite entities $\mathbf{a}(\omega) \pm \mathbf{b}(-\omega)$ and $\mathbf{b}(\omega) \pm \mathbf{a}(-\omega)$ are interpreted as generalized *input* doublets and *output* doublets, respectively. The doublet *resonances* are special cases for which the input and output vectors are both zero.

$$\begin{aligned} \mathbf{b}(\omega) \pm \mathbf{a}(-\omega) &= \mathbf{0} \\ \mathbf{a}(\omega) \pm \mathbf{b}(-\omega) &= \mathbf{0} \end{aligned} \tag{3.94}$$

These conditions specify valid solutions of the overall field equations with *nonzero* $\mathbf{a}(\omega)$, $\mathbf{b}(\omega)$, $\mathbf{a}(-\omega)$, and $\mathbf{b}(-\omega)$. Thus, even if the *net* input and output vectors are zero according to (3.94), their positive and negative frequency constituents are nonzero, exciting the interior circuit with doublet type field components carrying no time average energy. The upper signs in (3.94) correspond to the voltage doublet in (3.44), whereas the lower signs are associated with the current doublet (3.45), statements that are readily verified from (3.76) by forming $\mathbf{V}(\omega)+\mathbf{V}(-\omega)$ and $\mathbf{I}(\omega)+\mathbf{I}(-\omega)$ and comparing with (3.94).

It is physically appealing to think of the doublet resonances as describing a continuous excitation process taking place in all conceivable circuits, including free space, through interaction with the *PDE* fields pervading the entire universe. In this scenario the doublet fields are engaged in a process of establishing a stationary doublet sea of fluctuations having zero time average energy and *rms* spectral density inversely proportional to frequency, in accordance with the proposition made earlier.

In the next few chapters we shall elaborate on *single* photon behavior using the general results of the present chapter. Chapter 4 investigates the possible occurrence of bitemporal effects in electronic media. The basic issue in Chapters 5 and 6 is the *routing* of photons to specific independent terminals, a process that is fundamental for understanding photon energy integrity. Chapter 5 is an application of the general theory to the photon duality problem in quantum physics—the notion that a photon sometimes has the costume of a true wave satisfying Maxwell's equations and sometimes behaves as a particle. In this chapter we shall demonstrate that application of the present generalized circuit theory in bitemporal frame resolves the problem of photon transition from a wavelike to a particlelike entity in a consistent manner without recurring to the concept of duality. In Chapter 6 the general theory of the present chapter is applied to the important class of circuits having well defined and physically separate terminals. Such configurations represent a large group of everyday electronic circuits as well as special circuits employed in quantum mechanical experiments on single photons. We shall meet with a number of those in the later chapters such as the double slit experiment for photons and electrons, the Mach-Zehnder interferometer, the *EPR* configuration, and others, all well known and familiar in quantum physics.

APPENDIX A TO CHAPTER 3

A.1 FIELD EQUATIONS

Referring to the generalized circuit in Fig. 3.1, represented by an arbitrary enclosure of space of volume V and surface S, the electromagnetic fields inside the volume V satisfy Maxwell's equations:

$$\nabla \times \vec{E}(\vec{r},\omega) = -j\omega\mu(\vec{r})\,\vec{H}(\vec{r},\omega)$$

$$\nabla \times \vec{H}(\vec{r},\omega) = j\omega\varepsilon(\vec{r})\,\vec{E}(\vec{r},\omega) + \vec{\imath}(\vec{r},\omega) \tag{A.1}$$

$$\vec{\imath}(\vec{r},\omega) = \int_V \mathbf{G}(\vec{r},\vec{r}_0,\omega)\,\vec{E}(\vec{r}_0,\omega)d\vec{r}_0 + \vec{\imath}_g(\vec{r},\omega)$$

In the last equation, specifying current density $\vec{i}(\vec{r},\omega)$, the field induced current density is specified by the volume integral, whereas the term $\vec{i}_g(\vec{r},\omega)$ represents some extraneous current density which presumably is independent of the electromagnetic properties inside the volume V.

A.2 RECIPROCITY FORMULATIONS FOR THE GENERALIZED CIRCUIT

Let us consider two different field distributions at the same frequency ω, both satisfying (A.1). Denoting these by subscripts p and q, and omitting the variables $\vec{r}$ and ω, the fields ϵ_p, κ_p and ϵ_q, κ_q are specified by

$$\begin{aligned} \nabla \times \vec{\epsilon}_p &= -j\omega\mu\vec{\kappa}_p \qquad & \nabla \times \vec{\kappa}_p &= j\omega\varepsilon\vec{\epsilon}_p + \vec{j}_p \\ \nabla \times \vec{\epsilon}_q^{\,*} &= j\omega\mu\vec{\kappa}_q^{\,*} \qquad & \nabla \times \vec{\kappa}_q^{\,*} &= -j\omega\varepsilon\vec{\epsilon}_q^{\,*} + \vec{j}_q^{\,*} \end{aligned} \tag{A.2}$$

where the two last equations are shown in their complex conjugate forms. The reciprocity relations are obtained by multiplying the four equations scalarly by respectively $\vec{\kappa}_q^{\,*}$, $\vec{\epsilon}_q^{\,*}$, $\vec{\kappa}_p$, and $\vec{\epsilon}_p$, and taking appropriate linear combinations of the resulting equations, a procedure yielding the final result:

$$\nabla \cdot (\vec{\epsilon}_p \times \vec{\kappa}_q^{\,*} + \vec{\epsilon}_q^{\,*} \times \vec{\kappa}_p) = -\vec{j}_q^{\,*} \cdot \vec{\epsilon}_p - \vec{j}_p \cdot \vec{\epsilon}_q^{\,*} \tag{A.3}$$

By integration over the volume V, and use of Gauss's theorem, the equation is converted to

$$F_{p,q} = \int_S [\vec{\epsilon}_p(\vec{s}) \times \vec{\kappa}_q(\vec{s})^* + \vec{\epsilon}_q(\vec{s})^* \times \vec{\kappa}_p(\vec{s})] \cdot \vec{n}(\vec{s})\, dS = \int_V (\vec{j}_q^{\,*} \cdot \vec{\epsilon}_p + \vec{j}_p \cdot \vec{\epsilon}_q^{\,*})\, dV \tag{A.4}$$

The first objective is to establish a set of orthogonal expansion vectors. For these vectors the extraneous imposed current densities are zero, so that the current components are specified by

$$\vec{j}_q(\vec{r})^* = \int_{V_0} \mathbf{G}_q(\vec{r},\vec{r}_0)^* \,\vec{\epsilon}_q(\vec{r}_0)^*\, d\vec{r}_0 \qquad \vec{j}_p(\vec{r}) = \int_{V_0} \mathbf{G}_p(\vec{r},\vec{r}_0)\,\vec{\epsilon}_p(\vec{r}_0)\, d\vec{r}_0 \tag{A.5}$$

Substitution into the second integral of (A.4) results in the relation

$$F_{p,q} = -\int_V \int_{V_0} \left[\, \vec{\epsilon}_p(\vec{r})\, \mathbf{G}_q(\vec{r},\vec{r}_0)^* \,\vec{\epsilon}_q(\vec{r}_0)^* + \vec{\epsilon}_q(\vec{r})^*\, \mathbf{G}_p(\vec{r},\vec{r}_0)\,\vec{\epsilon}_p(\vec{r}_0) \right] d\vec{r}\, d\vec{r}_0 \tag{A.6}$$

In the first scalar product let us interchange integration variables $\vec{r}$ and $\vec{r}_0$ and transpose the result, a procedure leaving it invariant. These operations result in the following:

$$F_{p,q} = -\int_V \int_{V_0} \vec{e}_q(\vec{r})^* \left[\mathbf{G}_p(\vec{r},\vec{r}_0) + \tilde{\mathbf{G}}_q(\vec{r}_0,\vec{r})^* \right] \vec{e}_p(\vec{r}_0)\, d\vec{r}\, d\vec{r}_0 \quad \text{(A.7)}$$

The electromagnetic media considered in this text are reciprocal and lossless, a condition implying that

$$\mathbf{G}_p(\vec{r},\vec{r}_0) = -\tilde{\mathbf{G}}_q(\vec{r}_0,\vec{r})^*, \qquad \text{or} \quad F_{p,q} = 0 \quad \text{(A.8)}$$

so that (A.4) reduces to

$$\int_S \left[\vec{e}_p(\vec{s}) \times \vec{\kappa}_q(\vec{s})^* + \vec{e}_q(\vec{s})^* \times \vec{\kappa}_p(\vec{s}) \right] \cdot \vec{n}\,(\vec{s})\, dS = 0 \quad \text{(A.9)}$$

where $p, q = 1, 2...N$, with N being the number of independent solutions. The equation is recognized as a standard form of reciprocity relation in classical electromagnetic theory. The relation is to be processed a bit further, by establishing a set of orthogonal vector functions serving as the base for expansion of the *actual* physical electromagnetic fields. Permutation of the triple scalar products in (A.9) yields a more convenient and relevant form of the reciprocity theorem. It is readily verified that

$$[\vec{e}_p(\vec{s}) \times \vec{\kappa}_q(\vec{s})^*] \cdot \vec{n}(\vec{s}) = \vec{\kappa}_q(\vec{s})^* \cdot [\vec{n}(\vec{s}) \times \vec{e}_p(\vec{s})] = \vec{\kappa}_q(\vec{s})^*_{\tan} \cdot [\vec{n}(\vec{s}) \times \vec{e}_p(\vec{s})] \quad \text{(A.10)}$$

and similarly for the second product. The last part of (A.10) asserts that the only contribution from $\vec{\kappa}_q(\vec{s})^*$ to the triple product is the component $\vec{\kappa}_q(\vec{s})^*_{\tan}$ which lies in the tangential plane, but not in general pointing in the direction of $\vec{n}(\vec{s}) \times \vec{e}_p(\vec{s})$. With these modifications the reciprocity relation takes the form

$$\int_S \left\{ \vec{\kappa}_q(\vec{s})^*_{\tan} \cdot [\vec{n}(\vec{s}) \times \vec{e}_p(\vec{s})] + \vec{\kappa}_p(\vec{s})_{\tan} \cdot [\vec{n}(\vec{s}) \times \vec{e}_q(\vec{s})^*] \right\} dS = 0 \quad \text{(A.11)}$$

where $p, q = 1, 2, ...N$. Let the spatial vector components be lumped together in column vectors $\vec{\psi}(\vec{s})$ and $\vec{\kappa}(\vec{s})$.

$$\vec{\psi}(\vec{s}) = \mathrm{col}\,[\vec{n}(\vec{s}) \times \vec{e}_q(\vec{s})] \qquad \vec{\kappa}(\vec{s}) = \mathrm{col}\,[\vec{\kappa}_q(\vec{s})_{\tan}] \quad \text{(A.12)}$$

We shall also need the column

$$\vec{\epsilon}(\vec{r}) = \mathrm{col}\,[\vec{\epsilon}_q(\vec{r})] \tag{A.13}$$

With these definitions the reciprocity relation takes the compact form

$$\int_S \left[\vec{\kappa}(\vec{s})^* \cdot \tilde{\vec{\psi}}(\vec{s}) + \vec{\psi}(\vec{s})^* \cdot \tilde{\vec{\kappa}}(\vec{s})\right] dS = \mathbf{0} \tag{A.14}$$

A.2.1 Generalized admittance matrix

The two sets of vector functions contained in the column vectors $\vec{\psi}(\vec{s})$ and $\vec{\kappa}(\vec{s})$ lie in the tangential plane S. Because both sets presumably are complete, the two are not independent. The set $\vec{\psi}(\vec{s})$ will be designated as the basic set for expansion purposes. It is always possible to specify a denumerable set of independent solutions $\vec{\psi}(\vec{s})$ of Maxwell's equations in the volume V. Such a selection is by no means unique, and can be accomplished in N different ways. For any set the selection is normalized to satisfy the following orthogonality condition:

$$\int_S \vec{\psi}(\vec{s}) \cdot \tilde{\vec{\psi}}(\vec{s})^*\, dS = \mathbf{1} \tag{A.15}$$

where the right hand term is the unity matrix. The normalization condition implies that the expansion set $\vec{\psi}(\vec{s})$ is normalized to dimension $1/\ell$. As noted above the set $\vec{\kappa}(\vec{s})$ is not independent and can be expressed by a linear combination of the set $\vec{\psi}(\vec{s})$.

$$\vec{\kappa}(\vec{s}) = \mathbf{Y}\, \vec{\psi}(\vec{s}) \tag{A.16}$$

where the expansion terms are contained in the $N \times N$ matrix $\mathbf{Y}$. We shall show shortly that $\mathbf{Y}$ is the *admittance matrix* of the generalized circuit. By insertion of (A.16) into (A.14) the reciprocity relation takes the form

$$\mathbf{Y} \int_S \vec{\psi}(\vec{s})^* \cdot \tilde{\vec{\psi}}(\vec{s})\, dS + \int_S \vec{\psi}(\vec{s}) \cdot \tilde{\vec{\psi}}(\vec{s})^* dS\ \tilde{\mathbf{Y}}^* = \mathbf{0} \tag{A.17}$$

In view of the orthogonality condition (A.15) the equation reduces to

$$\mathbf{Y} = -\tilde{\mathbf{Y}}^* \tag{A.18}$$

The admittance matrix $\mathbf{Y}$ is *Skew Hermitian,* a property that in electromagnetic theory is characteristic of lossless circuits. The additional specification of reciprocity implies that the admittance matrix is *symmetric*, i.e., $\mathbf{Y} = \tilde{\mathbf{Y}}$. This condition taken together with (A.18) shows that $\mathbf{Y}$ is pure imaginary.

$$\mathbf{Y} = -\mathbf{Y}^* \tag{A.19}$$

The next section shows how the set of N orthogonal vector functions represented by the elements of $\vec{\psi}(\vec{s})$ serve the purpose of expanding any actual field solutions in the generalized circuit. In principle, the number N of independent vector functions is infinite.

A.2.2 Generalized voltage and current representation

The procedure involves development of a reciprocity relation quite similar to the analysis leading to (A.14), with the *actual* electromagnetic fields replacing set p in (A.2):

$$\vec{\epsilon}_p(\vec{r}) \to \vec{E}(\vec{r}) \qquad \vec{\kappa}_p(\vec{r}) \to \vec{H}(\vec{r}) \tag{A.20}$$

The driving term in the last equation of (A.1) has the extra term $\vec{i}_g(\vec{r})$ which was absent for the expansion vectors, but is retained in the field equations. The multiplication procedure following (A.2) and the subsequent integration produce the additional term

$$F_q = \int_V \vec{i}_g(\vec{r}) \cdot \vec{\epsilon}_q(\vec{r})\, d\vec{r} \tag{A.21}$$

Instead of (A.11) the reciprocity relation takes the form

$$\int_S \left[\vec{\kappa}(\vec{s})^*_{\tan} \cdot (\vec{n}(\vec{s}) \times \vec{E}(\vec{s})) + \vec{H}(\vec{s})_{\tan} \cdot \vec{\psi}(\vec{s})^* \right] dS = -\int_V \vec{i}_g(\vec{r}) \cdot \vec{\epsilon}(\vec{r})\, d\vec{r} \tag{A.22}$$

where the normalized field column $\vec{\epsilon}(\vec{r})$ is defined in (A.13). The next step is to expand the electric and magnetic fields $\vec{n}(\vec{s}) \times \vec{E}(\vec{s})$ and $\vec{H}(\vec{s})$ in terms of the presumably complete set $\vec{\psi}(\vec{s})$.

$$\vec{n}(\vec{s}) \times \vec{E}(\vec{s}) = \tilde{\vec{\psi}}(\vec{s})\, \mathbf{V} \qquad \vec{H}(\vec{s})_{\tan} = \tilde{\vec{\psi}}(\vec{s})\, \mathbf{I} \tag{A.23}$$

where the elements in the column vectors $\mathbf{V}$ and $\mathbf{I}$ represent generalized voltages and currents. By substitution into (A.22) and use of (A.15), (A.16), and (A.19) the following simple equation is obtained:

$$\mathbf{Y}\,\mathbf{V} - \mathbf{I} = \mathbf{I}_g \tag{A.24}$$

The driving current $\mathbf{I}_g$, referred to the N virtual terminals, is specified by the column vector

$$\mathbf{I}_g = \int_V \vec{i}_g(\vec{r}) \cdot \vec{\epsilon}(\vec{r})\, d\vec{r} \tag{A.25}$$

The components of the driving current $\mathbf{I}_g$ are weighed averages of the extraneous constant current generators $\vec{i}_g(\vec{r})$ located in the interior of the volume V. In view of the orthonormality condition (A.15) the inverted forms of (A.23) are

$$\mathbf{V} = \int_S \vec{\psi}(\vec{s})^* \cdot [\vec{n}(\vec{s}) \times \vec{E}(\vec{s})]\, dS \qquad \mathbf{I} = \int_S \vec{\psi}(\vec{s})^* \cdot \vec{H}(\vec{s})_{\tan}\, dS \tag{A.26}$$

which specify the generalized voltage and current vectors for given tangential field distributions $\vec{n}(\vec{s}) \times \vec{E}(\vec{s})$ and $\vec{H}(\vec{s})_{\tan}$ on the surface S.

A.2.3 Transformation of the circuit to different bases

The set of vector functions $\vec{\psi}(\vec{s})$ forming the orthonormal base for the field expansions is not unique, but is rather one set out of N possible sets, where N may be infinite. Assuming $\vec{\psi}(\vec{s})$ to be given, then any new set $\vec{\phi}(\vec{s})$ is formed by expansion in terms of the presumably complete old set $\vec{\psi}(\vec{s})$.

$$\vec{\phi}(\vec{s}) = \tilde{\mathbf{T}}^* \vec{\psi}(\vec{s}) \tag{A.27}$$

where $\tilde{\mathbf{T}}$ is an $N \times N$ transformation matrix. With a specified new base $\vec{\phi}(\vec{s})$ the appropriate transformation is obtained by inversion of the equation. Post-multiplying by $\widetilde{\vec{\psi}^*}(\vec{s})$ and integrating over S, using the orthonormality condition (A.15), we obtain

$$\tilde{\mathbf{T}}^* = \int_S \vec{\phi}(\vec{s}) \cdot \widetilde{\vec{\psi}}(\vec{s})^*\, dS \tag{A.28}$$

Because the old set satisfies Maxwell's equations so does the new set. The normalization condition (A.15) imposed for the new set gives the equation

$$\int_S \vec{\phi}(\vec{s}) \cdot \widetilde{\vec{\phi}}(\vec{s})^*\, dS = \mathbf{1} = \tilde{\mathbf{T}}^* \int_S \vec{\psi}(\vec{s}) \cdot \widetilde{\vec{\psi}}(\vec{s})^*\, \mathbf{T} dS = \tilde{\mathbf{T}}^* \mathbf{T} \tag{A.29}$$

which shows that the transformation matrix $\mathbf{T}$ is *unitary*.

$$\tilde{\mathbf{T}}^* \mathbf{T} = \mathbf{1} \tag{A.30}$$

Except for the unitary condition the transformation matrix has no restrictions. Selection of a particular matrix is just one out of N possible **T**-matrices. The general circuit equation in the form of (A.24) retains validity, with the admittance matrix **Y** and normalized field vectors $\vec{\psi}(\vec{s})$, as well as **V**, **I**, and $\mathbf{I}_g$ relating to the new base.

Chapter 4

ARE ELECTRON MEDIA BITEMPORAL?

'We must dare to think 'unthinkable' thoughts. We must learn to explore all the options and possibilities that confront us in a complex and rapidly changing world. We must learn to welcome and not fear the voices of dissent. We must dare to think about 'unthinkable' things because when things become unthinkable, thinking stops and action becomes mindless.'

J. William Fulbright

4.1 DOUBLET INTERACTIONS IN COLLISIONLESS PLASMAS

The introductory theory of photon doublets presented in the last chapters was restricted to free space with no charge carriers involved. The behavior of photons in material media is not a dominant issue in the experimental configurations discussed in the subsequent chapters. Therefore, the present venture into material media is a modest and somewhat speculative effort to investigate the possible existence of bitemporal *charge carriers* and, if they exist, to find out if they show conceptual similarities to photon doublets. The discussion has a rather limited scope, being restricted to the expected carrier responses to photon doublets in collisionless plasmas and in conducting media, i.e., metallic conductors. Based on the bitemporal hypothesis as the foundation of this investigation we shall find that the free electrons in a plasma appear to interact

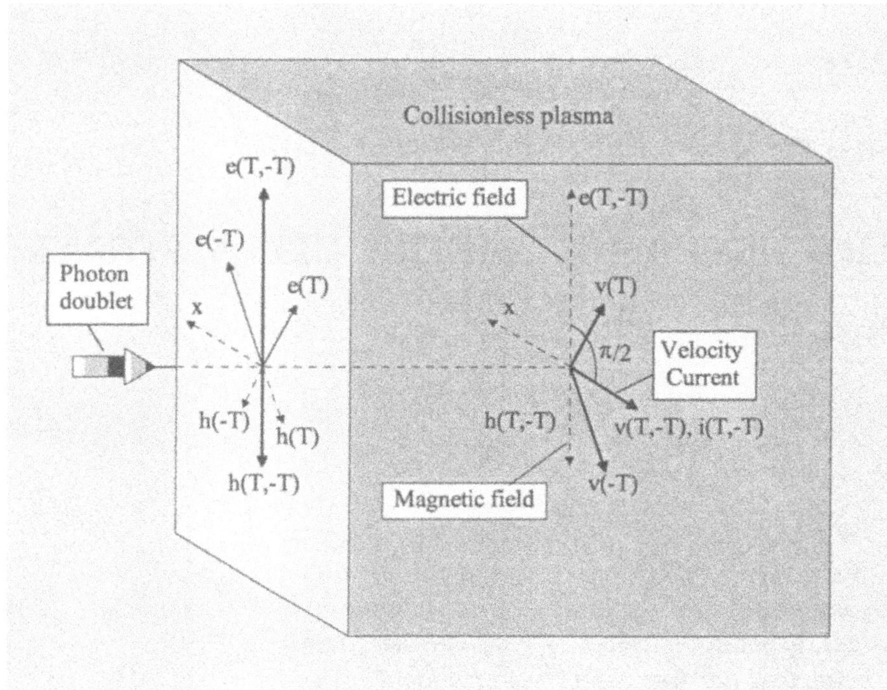

Figure 4.1. **Collisionless plasma excited by a photon doublet of the general kind, showing expected bitemporal carrier response with velocity and current perpendicular to the electric field.**

with the doublets in a manner quite different from their usual behavior based on classical electromagnetic theory. The difference is due to the composite electric field of the photon doublet, consisting of positive as well as negative time constituents. In the same theory applied to metallic conduction we shall also find that the extra degree of freedom introduced through the bitemporal model gives rise to a prediction of additional conduction modes not present in regular conduction.

4.1.1 Plasma response to general photon doublets

The enclosure illustrated schematically in Fig. 4.1, containing a collisionless plasma, is presumably excited with a plane wave photon doublet of the general kind, shown earlier by the schematic drawing in Fig. 2.7, Chapter 2, but now propagating through the plasma interior rather than in free space. The figure shows the electronic response obtained from the following simple analysis. In the positive time frame a particular electron of charge $-q$ and mass m is subject to a force $-q\,\hat{e}(T)$ due to the electric field component $\hat{e}(T)$. The corresponding exchange of momentum $m\hat{v}(T)$ in this regular interaction is specified by the differential equation

$$m\frac{d\,\hat{v}(T)}{dT} = -q\,\hat{e}(T) \tag{4.1}$$

where the variables are complex quantities. If the bitemporal hypothesis is accepted, the electrons must also respond to the field constituent $\hat{e}(-T)$ belonging to the negative time domain. The momentum exchange in the negative time frame is obtained from (4.1) by letting $T \rightarrow -T$.

$$m\frac{d\,\hat{v}(-T)}{d(-T)} = -q\,\hat{e}(-T) \tag{4.2}$$

Observed in the *positive time* frame the equation is equivalent to

$$m\frac{d\,\hat{v}(-T)}{dT} = q\,\hat{e}(-T) \tag{4.3}$$

which differs from (4.1) by the sign of the force term. The overall momentum exchange for the composite pair of electrons, observed in positive time, is then specified by

$$m\frac{d}{dT}[\hat{v}(T) + \hat{v}(-T)] = m\frac{d\,\hat{v}(T,-T)}{dT} = -q\,[\hat{e}(T) - \hat{e}(-T)] \tag{4.4}$$

which can be written

$$m\frac{d\,\hat{v}(+)}{dT} = -q\,\hat{e}(-) \tag{4.5}$$

where we have defined

$$\hat{v}(+) = \hat{v}(T) + \hat{v}(-T) \qquad \hat{e}(-) = \hat{e}(T) - \hat{e}(-T) \tag{4.6}$$

For later use let us also specify

$$\hat{v}(-) = \hat{v}(T) - \hat{v}(-T) \qquad \hat{e}(+) = \hat{e}(T) + \hat{e}(-T) \tag{4.7}$$

The *total* velocity is $\hat{v}(+)$ and the *total* electric field $\hat{e}(+)$. An important property of the electron pair follows through scalar multiplication of both sides of (4.5) with $\hat{e}(+)$. Because the photon doublet is characterized by $|\hat{e}(T)| = |\hat{e}(-T)|$ the procedure results in

$$m\,\hat{e}(+)\frac{d\,\hat{v}(+)}{dT} = -j\,2q\,|\hat{e}(T)|^2 \sin(2\varphi) \tag{4.8}$$

where φ is the angle formed by $\hat{e}(T)$ with the x-axis. Because the product is imaginary, the total velocity $\hat{v}(+)$ imposed from the photon doublet to the electron pair is perpendicular to the electric field, and also perpendicular to the magnetic field. The corresponding current component is given by

$$\hat{\imath}(+) = -q\,\hat{v}(+) \tag{4.9}$$

In view of (4.8) the imparted energy to the electron pair is zero because

$$\mathrm{Re}\,[\hat{\imath}(+)\,\hat{e}(+)] = 0 \tag{4.10}$$

The formulae (4.8)-(4.10) derived for doublets of the general kind hold also for the special electric and magnetic doublets discussed in the next few sections. The absence of energy transfer is of course expected because the photon doublet does not posses any energy to supply to the carriers. But still the electrons have gained velocity and current *perpendicular* to the electric field. The electron pair can be interpreted as a new kind of charge carrier with a total inseparable charge of $-2q$. In analogy with photon doublets we shall designate the name *electron doublet* for the composite carrier. Together with the photon doublet the electron doublet shares the common property of having zero net energy and also zero net momentum. In spite of having zero energy both kinds of doublets have nonvanishing physical attributes. How is the concept of nonvanishing velocity and current to be reconciled with the absence of any energy and momentum? Apparently, the explanation is linked with the bitemporal nature of the electron doublet, only to be understood in the context of the neoclassical theory. The particular electron interacting with the negative time field component $\hat{e}(-T)$ carries negative energy, thereby compensating the positive energy of the other constituent. The negative energy might be understood by the following reasoning: In the process of being accelerated in the negative time direction the electron gains *positive* energy in the process, exactly in the same way as the regular electron. But observed in *positive time,* i.e., in the reversed time direction, the electron appears to loose a corresponding amount of energy. This apparent energy *loss* is interpreted as an equivalent *gain* of negative energy. The interpretation is quite analogous to the corresponding case of photon doublets described by Fig. 3.3 in Chapter 3.

The sketch in Fig. 4.1 referred to a general photon doublet with colinear nonzero components of both electric and magnetic fields. The special cases of plasma interaction with pure electric and pure magnetic doublets are perhaps of more interest because they lead to carrier responses defining electron pairs conceptually similar to these pure photon doublets. The discussion in the next few sections is addressing this subject. For more information on the basic properties of the electric and magnetic photon doublets the reader is referred to Chapters 2 and 3.

4.1.2 Plasma response to electric and magnetic doublets

The electric doublets are characterized by no net magnetic field. The two electric field constituents $\hat{e}(T)$ and $\hat{e}(-T)$ add up to the overall nonzero composite field $\hat{e}(+)$. It follows, from the general equations (4.1)-(4.10), applied to the case of electric doublets, that the two velocity components $\hat{v}(T)$ and $\hat{v}(-T)$ of the electron pair point in opposite directions, i.e., $\hat{v}(+) = 0$. The details of the process are illustrated in Fig. 4.2. This *antisymmetric electron doublet* has similar antisymmetric bitemporal current components adding up to no net current, meaning that $\hat{\imath}(+) = 0$. But the antisymmetric current $\hat{\imath}(-)$ *is* different from zero. To the photon doublet of the electric type the plasma looks very much like an insulator, but a very special insulator responding with a peculiar kind of antisymmetric current excitation.

The magnetic doublet is characterized by nonzero magnetic field but no net electric field so that $\hat{e}(+) = 0$. Its interaction with a collisionless plasma is illustrated schematically by Fig. 4.3. The electric field components $\hat{e}(T)$ and $\hat{e}(-T)$ point in opposite directions, cancelling each other. It is found, from the general analysis (4.1)-(4.9), that the corresponding excitation of the electron pair is *symmetric,* with the two velocities $\hat{v}(T)$ and $\hat{v}(-T)$ adding up to a net velocity $\hat{v}(+)$ perpendicular to the magnetic field of the photon doublet. Thus, this *symmetric electron doublet* is characterized by nonzero velocity $\hat{v}(+)$ and associated net current $\hat{\imath}(+) = -q\,\hat{v}(+)$. Again, no energy is expended in the excitation process. The general relation (4.10) is satisfied in view of $\hat{e}(+)$ being zero for the magnetic doublet.

The discussions in this section on carrier response to photon doublets serve the sole purpose of introducing the concept of electron doublets. The presented results are contingent on acceptance of bitemporal effects in material media. Under these presumptions the probable existence of bitemporal carrier responses in collisionless plasmas caused by interaction with photon doublets has been demonstrated. No attempts were made to describe the dynamics of the full coupled system, i.e., the overall interaction between the photon doublets and the carrier doublets over an extended length and time.

It is an interesting observation that the symmetric electron doublet appears to have much of the same characteristics as the 'Cooper electron pair' supposed to be responsible for low temperature superconductivity. The Cooper pair supposedly interacts with a special phonon vibration of the lattice producing the coherent large scale phenomenon of superconductivity. In both cases the active carriers consist of two electrons in a composite form behaving as a single entity. If the electron pairs are in-

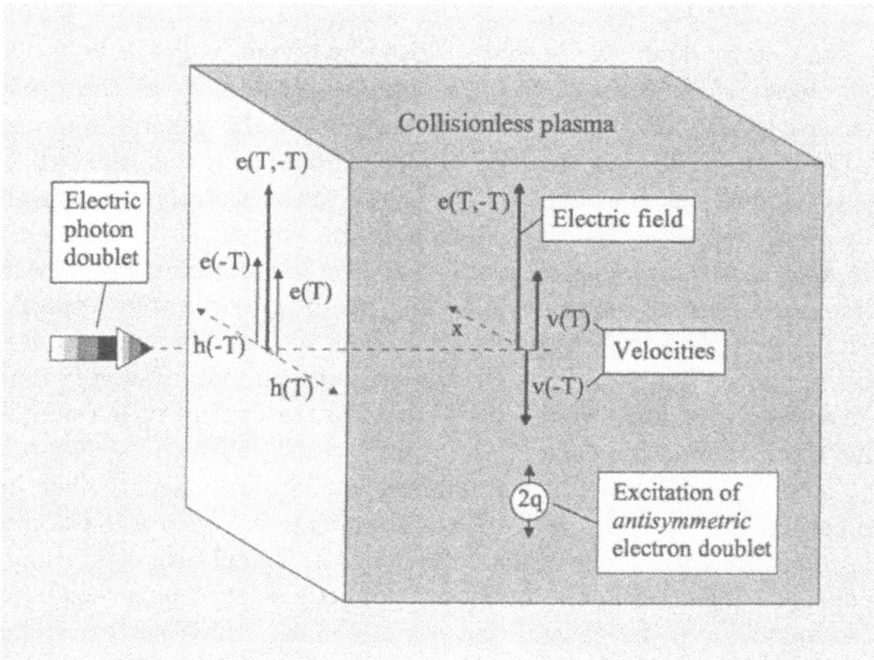

Figure 4.2. **Collisionless plasma excited by a photon doublet of the *electric* type. The response of the electronic medium is specified by *antisymmetric* electron doublets characterized by zero net velocity and current.**

deed the same, a proposition that at the present time is pure speculation, it might link superconductivity to bitemporal effects and involvement of photon doublets as part of the overall phenomenon of superconductivity. Continuing this speculation, the absence of an adequate theoretical explanation (as of 2000) of high temperature superconductivity might just point to bitemporal effects as an overlooked ingredient. The following section, dealing with photon doublets in metallic conductors, might throw some light on this interesting, but admittedly speculative proposition.

4.2 BITEMPORAL COLLISION SCATTERING IN CONDUCTORS

The issue addressed in this section concerns possible bitemporal effects in collision dominated media such as metallic conductors. The treatment is an effort to expand the results from the collisionless plasma to include collision phenomena due to lattice scattering. The discussion

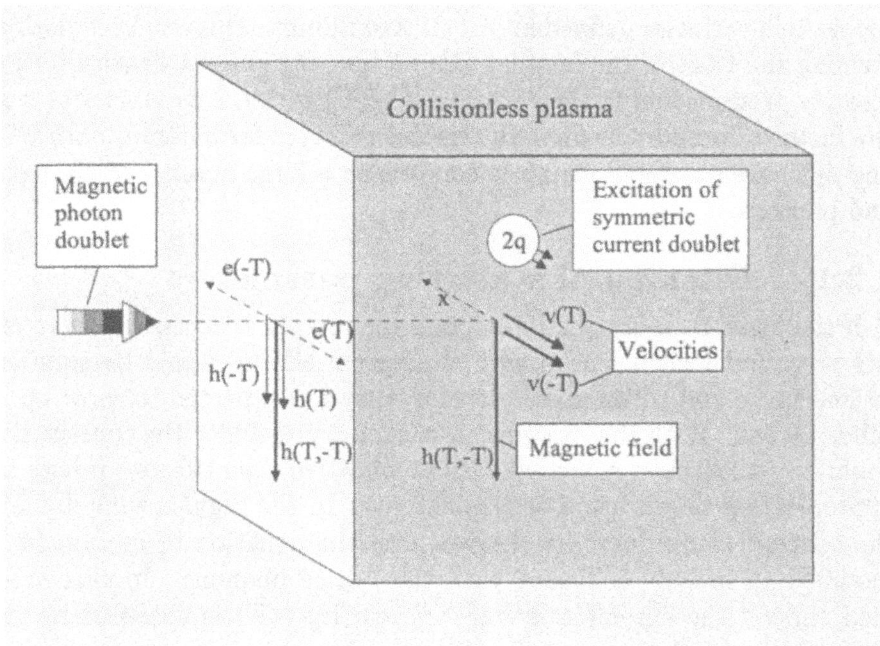

Figure 4.3. **Collisionless plasma excited by a photon doublet of the *magnetic kind*. The response of the electronic medium is specified by *symmetric* electron doublets characterized by finite net velocity and current.**

has a limited scope, being restricted to DC conduction, with scattering handled on a purely macroscopic basis through a generalization of metallic conductivity to the bitemporal domain.

Let us first specify the normalization of electromagnetic variables employed throughout.

$$T = ct \qquad \hat{h}(T) = Z_0 \hat{H}(T) \qquad \hat{\imath}(T) = Z_0 \hat{\imath}(T)_0 \tag{4.11}$$

Here T is the normalized time, with $c=1/\sqrt{\mu_0\varepsilon_0}$ being the velocity of light, $\hat{h}(T)$ is the normalized magnetic field, and $\hat{\imath}(T)$ is the normalized current density, with $Z_0=\sqrt{\mu_0/\varepsilon_0}$ representing the impedance of free space.

The macroscopic material parameter referred to as conductivity defines the constitutive relation between current density and electric field in Maxwell's equations. In the regular time frame the constitutive equation has the simple form:

$$\hat{\imath}(T) = \sigma\hat{e}(T) \tag{4.12}$$

where $\sigma = Z_0\sigma_0$ is the normalized conductivity appropriate to the macroscopic field variables defined in (4.11). According to this empiric relation, forming the basis of the familiar *Ohm's* law, the current density $\hat{\imath}(T)$ is directly proportional to the electric field $\hat{e}(T)$ with the conductivity σ as the factor of proportionality. In this macroscopic formulation both $\hat{\imath}(T)$ and $\hat{e}(T)$ are ensemble variables comprising a large number of electrons and photons.

4.2.1 Bitemporal scattering parameters

If the general thesis of a bitemporal microcosm is accepted, there exists no sound logical reason why bitemporal effects should be confined to free space and collisionless plasmas; they are expected to show up in solids as well. With the recognition of such a possibility the constitutive equation (4.12) must somehow include negative time effects through an appropriate generalization of conductivity. In the regular time domain the concept of conductivity derives from the equation of motion (4.1) modified to include collisions with the lattice phonons. In this modified version the ensemble average of velocity $\hat{v}(T)$ is specified by the differential equation

$$\frac{d\,\hat{v}(T)}{dT} + \frac{\hat{v}(T)}{\tau_+} = -\frac{q}{m}\,\hat{e}(T) \tag{4.13}$$

The added term containing the normalized *dielectric relaxation time* τ_+ accounts for macroscopic scattering. In the absence of electric fields the velocity $\hat{v}(T)$ decays exponentially according to

$$\frac{d\,\hat{v}(T)}{dT} + \frac{\hat{v}(T)}{\tau_+} = 0 \tag{4.14}$$

which has the exponential solution

$$\hat{v}(T) = \hat{v}(T_0)e^{-\frac{T-T_0}{\tau_+}} \tag{4.15}$$

where $T > T_0$. In the negative time domain we expect the corresponding equation to have the following form:

$$\frac{d\,\hat{v}(-T)}{d(-T)} + \frac{\hat{v}(-T)}{\tau_-} = -\frac{q}{m}\,\hat{e}(-T) \tag{4.16}$$

With $\hat{e}(-T) = 0$ the corresponding relaxation process in the direction of negative time is specified by

$$\frac{d\,\hat{v}(-T)}{d(-T)} + \frac{\hat{v}(-T)}{\tau_-} = 0 \tag{4.17}$$

which is solved by

$$\hat{v}(-T) = \hat{v}(-T_1)e^{-\frac{-(T-T_1)}{\tau_-}} \tag{4.18}$$

Because time presumably progresses in the negative direction so that $T < T_1$, the difference $-(T-T_1)$ is positive, showing that the relaxation term τ_- must be negative in order for $\hat{v}(-T)$ to decay in the *negative* time direction. The expected scattering relations for the two velocity components $\hat{v}(T)$ and $\hat{v}(-T)$ are illustrated in the sketches of Fig. 4.4 showing how the regular decay of $\hat{v}(-T)$ in negative time is equivalent with an exponential growth of $\hat{v}(-T)$ in the positive time direction. However, this should not be construed as meaning that there exist overall carrier responses growing with respect to time. The composite scattering is expected to show a predominance of regular positive time scattering. Only in the limiting case when the two constituents are equal do we expect to find special effects manifesting themselves as a lossless state. This limiting condition is discussed more fully shortly.

The reader might have observed that cross scattering between the velocity components $\hat{v}(T)$ and $\hat{v}(-T)$ in (4.13) and (4.16) is assumed to be nonexistent. Very little can be said about the possible occurrence of such cross scattering phenomena in addition to regular scattering, but it is unlikely that they represent a dominating part of the overall scattering processes. For this reason it was found legitimate to put this question aside and focus attention on the regular scattering terms.

4.2.2 Scatter dominated regime

In the scatter dominated regimes the inertia terms in (4.13) and (4.16) are negligible compared to the scatter terms so that the equations are well approximated by their scatter dominated representations.

$$\begin{aligned} \hat{v}(T) &= -\frac{q}{m}\tau_+ \hat{e}(T) \\ \hat{v}(-T) &= -\frac{q}{m}\tau_- \hat{e}(-T) \end{aligned} \tag{4.19}$$

Because the relaxation term τ_- is negative, the corresponding mobility, and therefore the conductivity, must be negative. Accordingly, we shall take the generalized constitutive relation to have the following tentative form:

$$\begin{bmatrix} \hat{\imath}(T) \\ \hat{\imath}(-T) \end{bmatrix} = \begin{bmatrix} \sigma_1 & 0 \\ 0 & -\sigma_2 \end{bmatrix} \begin{bmatrix} \hat{e}(T) \\ \hat{e}(-T) \end{bmatrix} \tag{4.20}$$

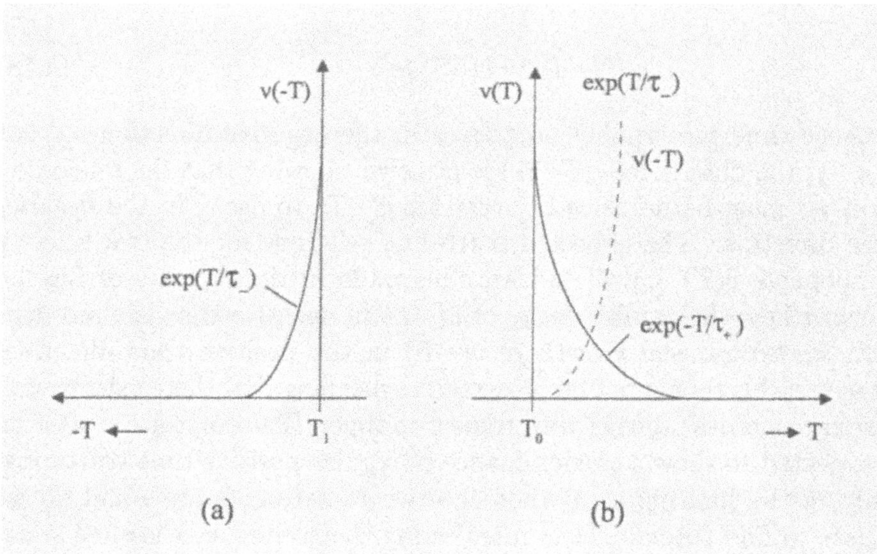

Figure 4.4. **Sketch illustrating collision scattering of positive and negative time components. (a) Decay of $\hat{v}(-T)$ in the direction of negative time. (b) Solid curve: Regular decay of $\hat{v}(T)$ in positive time. Dotted curve: Equivalent *growth* of $\hat{v}(-T)$ in the positive time direction.**

where the conductivities σ_1 and σ_2 are given by

$$\sigma_1 = \frac{q^2\tau_+}{m} \qquad \sigma_2 = -\frac{q^2\tau_-}{m} \tag{4.21}$$

Note that the definitions of currents and conductivities are normalized, or averaged, to single charge carriers rather than the full population.

In the formulation (4.20) the simple scalar conductivity specified by (4.12) is generalized to a diagonal 2×2 conductivity matrix with elements σ_1 and $-\sigma_2$ representing the scattering processes that conceivably take place in the bitemporal domain. The positive elements σ_1 and σ_2 are self scattering terms, relating directly to the relaxation times τ_+ and τ_-, and therefore referring to regular time and negative time, respectively. The assumed absence of any cross scattering between positive and negative time variables is reflected by vanishing off-diagonal elements.

In the analysis to follow the relative magnitudes of σ_1 and σ_2 turn out to be an important parameter in deciding the details of the bitemporal conduction processes in the conductor. At this point we shall merely demonstrate how a difference between σ_1 and σ_2 manifests itself in the equivalent constitutive relation expressed in terms of *doublet* ensemble variables. Noting that

$$\hat{\imath}(+) = \hat{\imath}(T) + \hat{\imath}(-T) \qquad \hat{\imath}(-) = \hat{\imath}(T) - \hat{\imath}(-T) \tag{4.22}$$

the constitutive equation (4.20) transforms to

$$\begin{bmatrix} \hat{\imath}(+) \\ \hat{\imath}(-) \end{bmatrix} = \begin{bmatrix} (\sigma_1 - \sigma_2)/2 & (\sigma_1 + \sigma_2)/2 \\ (\sigma_1 + \sigma_2)/2 & (\sigma_1 - \sigma_2)/2 \end{bmatrix} \begin{bmatrix} \hat{e}(+) \\ \hat{e}(-) \end{bmatrix} \tag{4.23}$$

The cross coupling term $(\sigma_1 - \sigma_2)/2$ dismantles the symmetry that was characteristic of the collisionless plasma discussed in the first part of the chapter. In the plasma the symmetric current doublet $\hat{\imath}(+)$ was excited by the electric doublet $\hat{e}(-)$, and the antisymmetric current doublet $\hat{\imath}(-)$ by the magnetic doublet $\hat{e}(+)$, with no cross coupling between them. Exactly the same kind of responses are expected in the conductor if $\sigma_1 = \sigma_2$, i.e., when the positive and negative conductivities are the same. In this special case the conductor appears to behave just like a collisionless plasma. Apparently, this particular condition represents a state of losslessness with positive and negative energy contributions compensating each other. We shall indeed find that these at first glance peculiar effects show up in the subsequent treatment of *DC* conduction.

Further support for the tentative form of the scatter relation (4.20) is obtained from evaluation of the expended power loss p:

$$p = \frac{1}{2} \mathrm{Re} \left\{ \begin{bmatrix} \hat{e}(T)^* & \hat{e}(-T)^* \end{bmatrix} \begin{bmatrix} \sigma_1 & 0 \\ 0 & -\sigma_2 \end{bmatrix} \begin{bmatrix} \hat{e}(T) \\ \hat{e}(-T) \end{bmatrix} \right\} \tag{4.24}$$

which results in

$$p = \frac{1}{2}[\sigma_1 \hat{e}(T)^* \hat{e}(T) - \sigma_2 \hat{e}(-T)^* \hat{e}(-T)] \tag{4.25}$$

The formula, expressing the dissipated power as a difference between contributions from the positive and negative time components, is in agreement with the bitemporal concepts developed in Chapters 2 and 3, thereby providing some support of the selected form of the scatter relation. In what follows we shall tentatively accept the bitemporal conductivity formulation (4.20) and the doublet equivalent (4.23), and proceed to investigate how this proposition is expected to affect conduction phenomena in a metallic conductor at *DC*, i.e., at zero frequency. We shall indeed find that the generalization of conductivity to include bitemporal effects predicts dramatic and unexpected modifications in the overall conduction process. That these show similarities with superconducting phenomena makes the investigations even more exciting.

4.3 BITEMPORAL DC CONDUCTION IN A CIRCULAR WIRE

The following discussion of *DC* conduction is based on a cylindrical model represented by a circular metallic wire, which is conceptually similar to the plane models of Figs. 4.1-4.3, but offer certain advantages in the interpretation of the overall mode structure. For positive as well as negative time components the *DC* fields of the cylindrical model of Fig. 4.5 are specified by Maxwell's two curl equations.

$$\begin{aligned} \nabla \times \vec{e}(r,\theta,T) &= 0 \\ \nabla \times \vec{h}(r,\theta,T) &= \vec{i}(r,\theta,T) \end{aligned} \tag{4.26}$$

where time T can be positive as well as negative, the two ranges being basically independent. The wire model is assumed to be infinitely long so that the fields have translational invariance with respect to the longitudinal coordinate Z. The electric field distribution is of the general form

$$\vec{e}(r,\theta,T) = \sum_{m=-\infty}^{\infty} \vec{e}(r,T)_m \begin{matrix} \cos(m\theta) \\ \sin(m\theta) \end{matrix} \tag{4.27}$$

with similar relations applying to $\vec{h}(r,\theta,T)$ and $\vec{i}(r,\theta,T)$. The higher order solutions beyond $m = \pm 1$ is of little interest, so that we shall consider only the two lowest angular modes: the *symmetric* mode $m = 0$ and the *antisymmetric* mode $m = \pm 1$. As we shall see, both modes exhibit unusual phenomena specific for the present bitemporal model of *DC* conduction.

4.3.1 The symmetric conduction mode

In the symmetric solution corresponding to $m = 0$ there are no field variations with respect to the angle θ. Expressed in cylindrical coordinates the first curl equation in (4.26) is solved by $\vec{e}(r,T) = \vec{a}_z \hat{e}_z(T) =$ constant, and similarly for $\vec{e}(r,-T)$. Because the current components $\vec{i}(r,T)$ and $\vec{i}(r,-T)$ are related to the electric fields by the general conductivity relation (4.20) the currents are also constant over the cross section, pointing in the Z-direction. Inside the cylindrical conductor the angular magnetic fields $\hat{h}_\theta(T)$ and $\hat{h}_\theta(-T)$ are found to be proportional to the radius r, on the outside the fields decay as $1/r$. Inside the wire the overall solution of (4.26) is conveniently expressed in the following

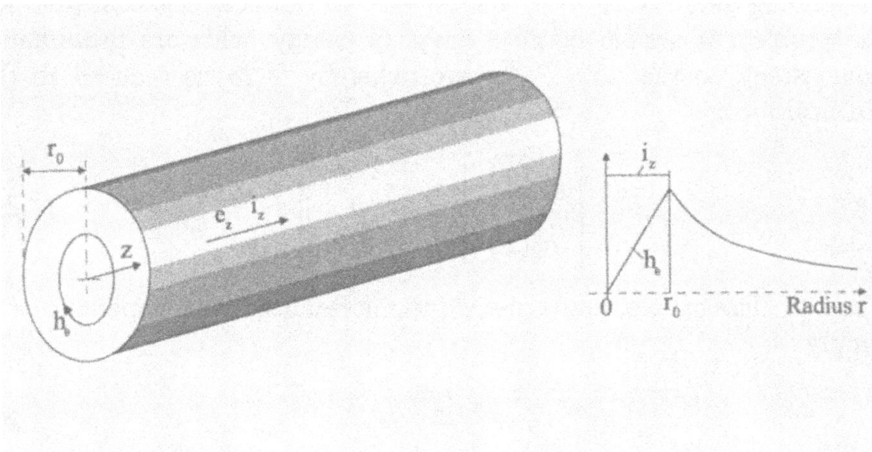

Figure 4.5. Basic configuration of the circular metallic conductor analyzed in the text. Labeling of fields and currents refer to positive time constituents in the symmetric mode $m=0$. The current density i_z and electric field e_z are constant over the cross section. Inside the conductor the angular magnetic field $h_\theta(r)=i_z r/2$; on the outside $h_\theta(r)=I/(2\pi r)$, where the current $I=\pi r_0^2 i_z$.

column:

$$\begin{bmatrix} \hat{e}_z(T) \\ \hat{\imath}_z(T) \\ \hat{h}_\theta(T) \\ \hat{e}_z(-T) \\ \hat{\imath}_z(-T) \\ \hat{h}_\theta(-T) \end{bmatrix} = \begin{bmatrix} 1 & 0 \\ \sigma_1 & 0 \\ \sigma_1 r/2 & 0 \\ 0 & 1 \\ 0 & -\sigma_2 \\ 0 & -\sigma_2 r/2 \end{bmatrix} \begin{bmatrix} \hat{A}(T) \\ \hat{A}(-T) \end{bmatrix} \tag{4.28}$$

where the mode amplitudes of the $+T$ and $-T$ constituents are specified by $\hat{A}(T)$ and $\hat{A}(-T)$, respectively. In this definition the amplitudes are seen to specify the two electric fields $\hat{e}_z(T)$ and $\hat{e}_z(-T)$. The two sets $\hat{A}(T)$ and $\hat{A}(-T)$ of (4.28), are basically orthogonal, but become coupled by the external *DC* source through the excitation procedure described next.

In order to maintain a reasonably clean mode structure we can visualize the fields excited through two large diameter metallic plates placed perpendicular to the wire. In this model the longitudinal electric field is the same inside and outside the wire, and the radial electric field is zero except for end effects which are negligible. Let us assume the wire driven from a constant current *DC* source of essentially infinite inner resistance so that the current density is constant and equal to $\hat{\imath}$. With this driving condition we are faced with the problem of determining the corresponding amplitudes $\hat{A}(T)$ and $\hat{A}(-T)$. In classical electromagnetism the problem is simple because negative energy fields are presumably nonexistent, so that $\hat{A}(-T)$ is zero, whereby (4.28) is reduced to the classical form:

$$\begin{bmatrix} \hat{e}_z(T) \\ \hat{\imath}_z(T) \\ \hat{h}_\theta(T) \end{bmatrix} = \begin{bmatrix} 1/\sigma_1 \\ 1 \\ r/2 \end{bmatrix} \hat{\imath} \tag{4.29}$$

Thus, in classical electromagnetism the normalized wire impedance z is simply

$$z = \frac{\hat{e}_z(T)}{\hat{\imath}_z(T)} = \frac{1}{\sigma_1} \tag{4.30}$$

By inclusion of the bitemporal component $\hat{A}(-T)$ in addition to $\hat{A}(T)$ this classical representation is modified in a nontrivial way because the presumably constant driving current $\hat{\imath}$ can basically excite both modes $\hat{A}(T)$ and $\hat{A}(-T)$. At this point nothing can be said about the relative excitations except that the sum of the corresponding current components is equal to the driving current.

$$\hat{\imath} = \hat{\imath}_z(T) + \hat{\imath}_z(-T) = \hat{\imath}_z(+) \tag{4.31}$$

A meter inserted into the lead for monitoring the driving current $\hat{\imath}$ just tells us the sum $\hat{\imath}_z(+) = \hat{\imath}$ but absolutely nothing about the constituents $\hat{\imath}_z(T)$ and $\hat{\imath}_z(-T)$. So how can these components be determined? Let us first note that the set of equations (4.28) represents the complete and general field solutions in the symmetric conduction problem. Nevertheless it fails to provide an answer to the question posed above because the set is incomplete in the sense that the number of available equations is less than the number of variables. This point is most easily seen by expressing (4.28) in the form of a constitutive equation, essentially (4.20), or rather the equivalent doublet form (4.23).

$$\begin{bmatrix} \hat{\imath}_z(+) \\ \hat{\imath}_z(-) \end{bmatrix} = \begin{bmatrix} \sigma_- & \sigma_+ \\ \sigma_+ & \sigma_- \end{bmatrix} \begin{bmatrix} \hat{e}_z(+) \\ \hat{e}_z(-) \end{bmatrix} \tag{4.32}$$

where

$$\sigma_- = (\sigma_1 - \sigma_2)/2 \qquad \sigma_+ = (\sigma_1 + \sigma_2)/2 \tag{4.33}$$

The magnetic fields, being directly proportional to current components, are not needed and therefore omitted. Equation (4.32) contains *four* unknown variables: $\hat{\imath}_z(+), \hat{\imath}_z(-), \hat{e}_z(+)$, and $\hat{e}_z(-)$, whereas the available number of equations is just *three*—two equations from (4.32) and a third from (4.31). Introduction of the wire impedance $z = \hat{e}_z(+)/\hat{\imath}$ as an empirical parameter obtained from measurement certainly makes the overall set of equations complete, but does not solve the basic problem of determining the theoretical variation of impedance as a function of the material parameters σ_1 and σ_2, essentially the generalized bitemporal version of (4.30). The regular components $\hat{e}(+) = \hat{e}$ and $\hat{\imath}(+) = \hat{\imath}$ are *externally* observable, whereas $\hat{e}(-)$ and $\hat{\imath}(-)$ play the role of internal *idler* components which do not manifest themselves externally.

4.3.2 Dual states of the symmetric mode

From physical reasoning it is obvious that the posed problem of determining the relative excitation of positive and negative field constituents $\hat{A}(T)$ and $\hat{A}(-T)$ has a solution, but how is it to be resolved? It seems natural to look for some stationary condition specified by minimization of stored electromagnetic energy. In the present theory, dealing with bitemporal phenomena, stored energy concepts inherited from classical electromagnetism are not applicable, as demonstrated by the neoclassical circuit theory in Chapter 3. The stored energy in an arbitrary lossless circuit was specified by (3.47). For the present *DC* condition the time average stored energy $\langle W(T,-T)\rangle$ is equal to stored energy $W(T,-T)$ itself. Expressed in terms of the *DC* electric fields $\hat{e}_z(-)$ and $\hat{e}_z(+)$ the equation simplifies to

$$W(T,-T) = \frac{1}{4}\hat{e}_z(-)\,\hat{e}_z(+)\left[\lim_{\omega\to 0}\left(\frac{db(\omega)}{d\omega}\right)\right] \tag{4.34}$$

where $b(\omega)$ is the normalized susceptance of the two-terminal wire circuit. The actual value of $b(\omega)$ is of no concern for the present discussion. For more details the reader is referred to Chapter 3 where the formula was derived under the assumption of zero loss in the circuit. Although the effect of loss was not included in the procedure, we shall make the reasonable assumption that the formula (4.32) is still valid for the lossy wire conductor. The following demonstrations will hopefully provide some justification for this stipulation.

It follows from (4.34) that the bitemporal energy is minimized if either $\hat{e}_z(-)$ or $\hat{e}_z(+)$ are zero.

$$\text{Either}\quad \hat{e}_z(-) = 0 \qquad \text{or}\quad \hat{e}_z(+) = 0 \tag{4.35}$$

Both specifications correspond to $|\hat{e}_z(T)|$ being equal to $|\hat{e}_z(-T)|$, i.e., the amplitudes of the positive and negative energy electric fields are exactly the same. The existence of *two* rather than *one* minimization condition gives rise to dual states of the wire conductor. The prevailing state appears to be determined by the relative magnitudes of the conductivities σ_1 and σ_2. We shall refer to the state characterized by $\hat{e}_z(-)=0$ as the *regular state,* and the state $\hat{e}_z(+)=0$ as the *lossless state.* It is tempting to label this last state as *superconducting* because of the similarity with regular superconductivity. At the end of the chapter we shall discuss the possible identification of the lossless state with the ordinary superconducting state at low temperatures. At present, the discussion is confined to the conditions characterizing the excitation of a particular state, in particular the relations to the material parameters σ_1 and σ_2.

4.3.3 Regular conducting state

The regular state characterized by $\hat{e}_z(-)=0$, corresponding to minimization of the idler stored energy, is expected to prevail when the negative energy conductance σ_2 is less than σ_1, a condition expected to be predominant at relatively high and moderate temperatures. This is certainly true in the limiting classical case where σ_2 is zero. Thus, this state is characterized by the two conditions

$$\hat{e}_z(-)=0 \qquad\qquad \sigma_2<\sigma_1 \tag{4.36}$$

With this extra condition satisfied the overall set of equations is complete, with the solution

$$\hat{e}_z(+)=\frac{\hat{\imath}}{\sigma_-}=\frac{2\hat{\imath}}{(\sigma_1-\sigma_2)} \tag{4.37}$$

The equation specifies the wire impedance z to

$$z=\frac{\hat{e}_z\,(+)}{\hat{\imath}_z(+)}=\frac{1}{\sigma_-}=\frac{2}{(\sigma_1-\sigma_2)} \tag{4.38}$$

The factor of 2 appearing in these equations is due to the circumstance that $\hat{e}_z(+)\;=\;2\hat{e}_z(T)$. The idler current $\hat{\imath}_z(-)$, representing a passive admixture to the observable current $\hat{\imath}_z(+)$, is given by

$$\hat{\imath}_z(-)=\sigma_+\hat{e}_z(+)=\frac{1}{2}(\sigma_1+\sigma_2)\,\hat{e}_z\,(+)=\frac{\sigma_1+\sigma_2}{\sigma_1-\sigma_2}\,\hat{\imath} \tag{4.39}$$

The ratio of the two current components appears to be an important parameter in predicting the behavior of the system.

$$\frac{\hat{\imath}_z(-)}{\hat{\imath}_z(+)}=\frac{\hat{\imath}_z(-)}{\hat{\imath}}=\frac{\sigma_+}{\sigma_-}=\frac{\sigma_1+\sigma_2}{\sigma_1-\sigma_2} \tag{4.40}$$

If the bitemporal proposition leading to the present theory of *DC* conduction is accepted, the idler component $\hat{\imath}_z(-)$ is a physically real component even if it is not observable in the experimental set up. If σ_2 approaches σ_1, the ratio $\hat{\imath}_z(-)/\hat{\imath}$ becomes infinitely large and so does the corresponding magnetic field $\hat{h}_\theta(-T)$ as well as the corresponding positive and negative temporal components $\hat{\imath}_z(T)$ and $\hat{\imath}_z(-T)$, $\hat{h}_\theta(T)$, and $\hat{h}_\theta(-T)$. Thus, the specification $\sigma_2 = \sigma_1$ appears to give rise to a resonance condition. As demonstrated by (4.40) it manifests itself as a *series resonance* of the idler current $\hat{\imath}_z(-)$, which becomes infinite at the resonance point in spite of the fact that the driving field $\hat{e}_z(-)$ is zero and the driving current $\hat{\imath} = \hat{\imath}_z(+)$ is constant. Such infinities would cause catastrophic deviations from the linear state and can not be endured in a physical system which probably would run into saturation before the resonant state occurs.

4.3.4 Lossless conducting state

The system can avoid the resonance catastrophe by switching to the second state of the dual set (4.35), specified by $\hat{e}_z(+)=0$ rather than $\hat{e}_z(-)=0$. The switching might happen a bit ahead of full saturation at resonance, but we have no way of finding out the exact switching condition. Therefore, we shall take $\sigma_2=\sigma_1$ as the nominal switching stipulation with some allowance for the two being slightly different. Accordingly, after switching to the lossless state the new specifications are

$$\hat{e}_z(+) = 0 \qquad\qquad \sigma_2 = \sigma_1 \tag{4.41}$$

During the switching process between the two states the idler current $\hat{\imath}_z(-)$ changes dramatically from the very large value in (4.40) to a very small value, essentially zero if $\sigma_1 = \sigma_2$.

$$\frac{\hat{\imath}_z(-)}{\hat{\imath}} = \frac{\sigma_1 - \sigma_2}{\sigma_1 + \sigma_2} = 0 \tag{4.42}$$

Thus, the idler component $\hat{\imath}_z(-)$ is inverted in the switching process. The idler electric field $\hat{e}_z(-)$ switches from zero to

$$\hat{e}_z(-) = \frac{\hat{\imath}}{\sigma_+} = \frac{2\hat{\imath}}{\sigma_1 + \sigma_2} \tag{4.43}$$

which is finite. The *idler impedance* $z(-)$ becomes

$$z(-) = \frac{\hat{e}_z(-)}{\hat{\imath}_z(-)} = \frac{2}{\sigma_1 - \sigma_2} \rightarrow \infty \tag{4.44}$$

In this *lossless* state the idler has the character of a *parallel* resonance because $z(-)$ is infinite, but neither this impedance nor the idler variables can be observed externally. In sharp contrast to the infinite idler impedance the regular impedance z vanishes identically because the observable electric field $\hat{e}_z(+)$ is zero.

$$z = \frac{\hat{e}_z(+)}{\hat{\imath}} \equiv 0 \tag{4.45}$$

Because the stipulation $\hat{e}_z(+) \equiv 0$ is derived from an energy minimization principle, a certain insensitivity towards possible minor system perturbations is expected to be predominant. Perturbations of various kinds might cause changes in the wire susceptance $b(\omega)$ in (4.34) but would in no way change the minimization conditions derived from the same formula. This implies that the impedance is expected to be maintained at *exactly* zero, not only close to zero. Maintaining the condition $\sigma_1 = \sigma_2$ also subsequent to switching is a prerequisite for this state to remain in a lossless mode. The stipulation of losslessness follows directly from the general expression (4.25) for power loss. With $\hat{e}_z(+) = 0$ and therefore $|\hat{e}_z(T)| = |\hat{e}_z(-)|/2$ the formula reduces to

$$p = \frac{1}{8}|\hat{e}_z(-)|^2(\sigma_1 - \sigma_2) \tag{4.46}$$

which clearly demonstrates that the lossless state can not be sustained unless $\sigma_1 - \sigma_2 = 0$.

For the idler component the condition $\sigma_1 = \sigma_2$ represents a singularity, manifesting itself as a series resonance in the regular state and a parallel resonance in the lossless state. These singular phenomena are a direct consequence of the exact balance of positive and negative energy. In the lossless state the two scattered components in positive and negative time are equal and of opposite signs, balancing out to zero, so that the conductor behaves much like the collisionless plasma. Because $\hat{e}_z(+) = 0$ the net loss disappears and the doublet current pair $\hat{\imath}(+) = \hat{\imath}(T) + \hat{\imath}(-T)$ is driven by the electric field component $\hat{e}_z(-)$, in all essential respects equivalent to the plasma case illustrated in Fig. 4.3.

Once the switching to the lossless state has occurred through a gradual rapprochement of σ_2 to σ_1, for instance caused by lowering the temperature, there are thermodynamic reasons why σ_2 can never exceed σ_1. If such a condition were allowed, the power expenditure would be negative, which clearly is an inadmissible situation. The original assumptions concerning the scattering parameters expressed in the simple constitutive equation (4.20) do not account for stability in the sense of maintaining $\sigma_2 = \sigma_1$. Cross scatter effects between $\hat{v}_z(T)$ and $\hat{v}_z(-T)$ or effects of entirely different origin might contribute to this condition. Because scattering theory belongs to the area of solid state, which is not a topic of the present book, no further speculation on the subject was undertaken. Moreover, the somewhat speculative nature of the bitemporal scattering theory presented in this chapter does not justify ruminations into details of scattering processes that at best might be highly questionable.

4.3.5 The antisymmetric conduction modes

The antisymmetric modes of the cylindrical conductor correspond to putting $m = \pm 1$ in (4.27). The solutions of (4.26) corresponding to this symmetry show that the longitudinal electric fields are always identically zero.

$$\hat{e}_z(r,T) = \hat{e}_z(r,-T) = 0 \tag{4.47}$$

Therefore, the longitudinal bulk current density is also zero.

$$\hat{\imath}_z(r,T) = \hat{\imath}_z(r,-T) = 0 \tag{4.48}$$

This particular angular symmetry gives rise to two independent modes shown in Fig. 4.6. The mode shown in Fig. 4.6a is a trivial but nevertheless basic solution of the field equations. It is represented by a uniform magnetic field in some arbitrary transverse direction, with no electric fields and currents in the conductor. If the magnetic field is pointing in the y-direction the field is expressed by:

$$\hat{h}(r,T) = \hat{h}(r,T)_y = C(T)_0 \tag{4.49}$$

where $C(T)_0$ is the mode amplitude. The second mode has the more complex magnetic field structure shown in Fig. 4.6b. The magnetic field components inside the wire conductor, where $r \le r_0$, are specified by:

$$\begin{aligned} \hat{h}_\theta(r,T)_i &= C(T)\cos(\theta) \\ \hat{h}_r(r,T)_i &= C(T)\sin(\theta) \end{aligned} \tag{4.50}$$

The field components outside the wire, where $r \ge r_0$, are given by:

$$\begin{aligned} \hat{h}_\theta(r,T)_o &= -C(T)\left(\frac{r_0}{r}\right)^2\cos(\theta) \\ \hat{h}_r(r,T)_o &= C(T)\left(\frac{r_0}{r}\right)^2\sin(\theta) \end{aligned} \tag{4.51}$$

Because the outside fields drop off with radius according to $1/r^2$, the pattern is localized to the immediate wire surroundings. As shown by

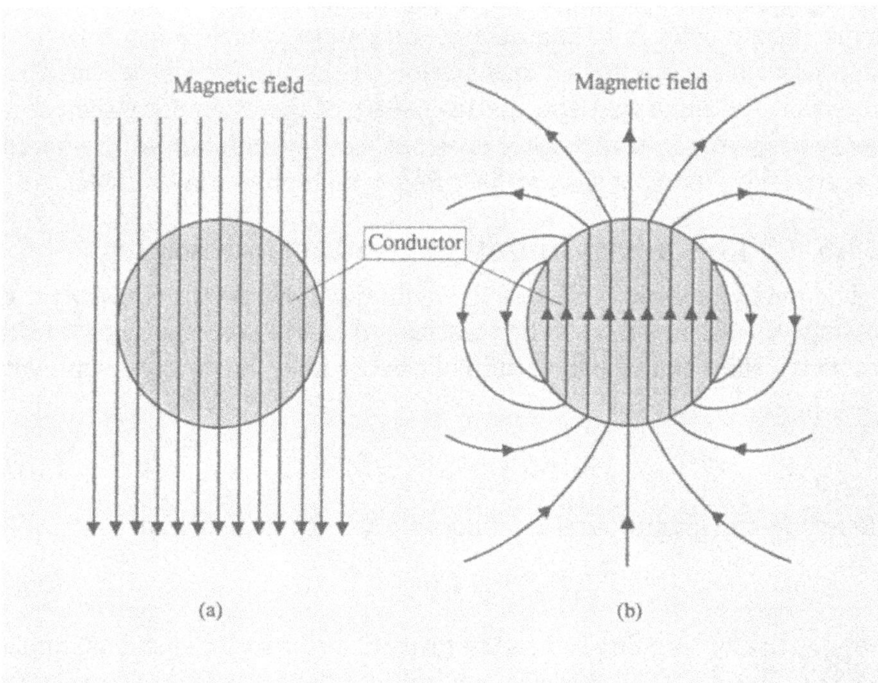

Figure 4.6. **Illustration of the two antisymmetric modes in *DC* conduction. (a) The first mode is simply a constant transverse magnetic field everywhere. (b) The second mode has a constant magnetic field *inside* the wire and a declining field outside, with a surface current excited by the magnetic field discontinuity at $r = r_0$. In the regular lossy state, when $\sigma_2 < \sigma_1$, the mode is evanescent, decaying exponentially with respect to time if somehow excited. In the lossless state, with $\sigma_2 = \sigma_1$, the mode is stationary.**

(4.48) the longitudinal *bulk* current density is identically zero, but there exists a nonzero longitudinal *surface* current $I_s(T)_z$ on the conductor surface excited by the angular magnetic field:

$$I_s(T)_z = \hat{h}_\theta(r_0, T)_o - \hat{h}_\theta(r_0, T)_i = 2C(T)\cos(\theta) \qquad (4.52)$$

The surface current, which follows directly from the second curl equation of (4.26), is a necessary ingredient of the overall boundary conditions. It is due to the discontinuity of the angular magnetic field across the surface $r = r_0$. In the regular lossy conducting state the antisymmetric mode described by equations (4.50)-(4.52) is evanescent. If the mode somehow is excited, for instance by a sudden exposure to the constant magnetic field in Fig. 4.6a, it will decay exponentially with respect to time due to the loss associated with the surface current. The evanescent process is not basic to the phenomena discussed and will not be pursued further. But the evanescent condition changes drastically after switching to the lossless state, a process expected to be similar to the switching of the symmetric mode described in the previous section. In the lossless state the antisymmetric mode persists once it has been excited. With the complete absence of electric fields the excitation can only take place by a magnetic field of the same angular symmetry, which is the pattern shown in Fig. 4.6a. Other configurations of transverse magnetic fields having different angular symmetries are all orthogonal to the $m \neq \pm 1$ field patterns and therefore not capable of exciting the lossless mode. The composite field pattern shown in Fig. 4.7 is a superposition of the two patterns in Fig. 4.6. The stored energy of the composite mode is minimized by the following relation between the amplitudes:

$$C(T) = -C(T)_0 \tag{4.53}$$

This condition renders the *internal* magnetic field exactly equal to zero. In order to excite the mode in Fig. 4.6b the mode must be in the lossless state, i.e., operating under the condition $\sigma_2 = \sigma_1$. Under these circumstances the wire refuses to let magnetic flux enter the interior. The interior is completely shielded by the surface current. In the theory of superconductivity this same phenomenon is referred to as the *Meissener* effect.

The excitation mechanisms for the symmetric $m = 0$ mode and the antisymmetric $m = \pm 1$ mode can be summarized as follows. The symmetric mode is excited by imposing a constant longitudinal *current* $\hat{\imath}_z$ of the same $m = 0$ symmetry. It exists in a dual state—the regular lossy conduction state and the lossless state. The antisymmetric mode in Fig. 4.6b is excited by an external constant *magnetic* field $\hat{h}_\theta$ of the same $m = \pm 1$ symmetry. In the lossy state the antisymmetric mode is evanescent and does not manifest itself. In the lossless state the mode persists, completely shielding the external $m = \pm 1$ magnetic driving field from the interior of the wire.

4.3.6 Comparison with the BCS superconducting theory

The similarity of the lossless state with the well known phenomenon of superconductivity is immediately apparent. The scatter components σ_1 and σ_2 are bound to depend on temperature; both are expected to increase with reduced temperature because of increased carrier mobilities. The sketch in Fig. 4.8 shows the expected qualitative impedance characteristics predicted from the present bitemporal scatter theory, where we draw on existing empirical data. Near the critical temperature T_{crit} the impedance of the regular state is known to grow with some power

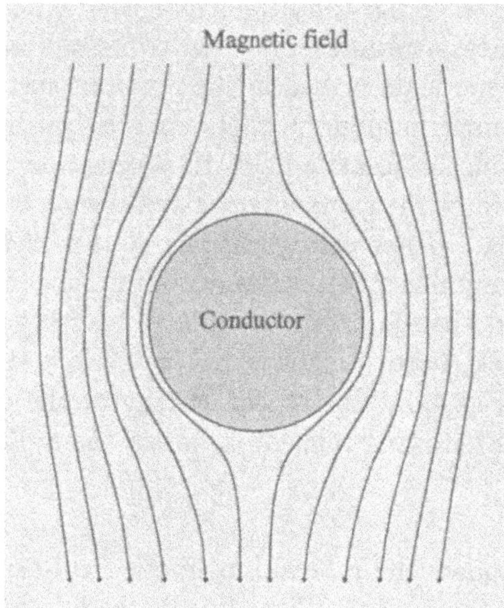

Figure 4.7. **Illustration of the excited antisymmetric lossless mode of the wire conductor, corresponding to $\sigma_2 = \sigma_1$. The field pattern is a superposition of the two basic antisymmetric modes shown in Fig. 4.6. The interior is completely shielded by the surface current flowing on the surface of the circular conductor.**

of the absolute temperature T_e. The empirical impedance variation is often specified by the relation

$$z = z_0 + cT_e^n \tag{4.54}$$

where the temperature independent constant z_0 is due to impurity scattering, and the exponent n is of the order of five. Below the critical temperature T_{crit} the impedance drops to zero. This is the superconducting state.

The theory of J. Bardeen, L. Cooper, and J. R. Schrieffer (*BCS* theory) explains superconductivity through interaction of electron pairs—the so-called Cooper pairs—with coherent large scale lattice vibrations. Except for the conceptual difference in the basic explanation of the phenomenon of superconductivity the external manifestations of the two theories are remarkably similar. In both cases the current is carried by electron pairs, the "Cooper pair" in the *BCS* theory, and the current $\hat{\imath}(+) = \hat{\imath}(T) + \hat{\imath}(-T)$ in the present theory. In both theories the total momentum of the electron pair is zero. In the present theory this is explained by one of the electrons being accelerated in the negative time direction by the field $\hat{e}(-T) = -\hat{e}(T)$, producing the velocity $\hat{v}(-T)$ in the *same* direction as $\hat{v}(T)$ but with negative momentum expenditure. Thus, the net momentum expenditure is zero (see also Fig. 4.3).

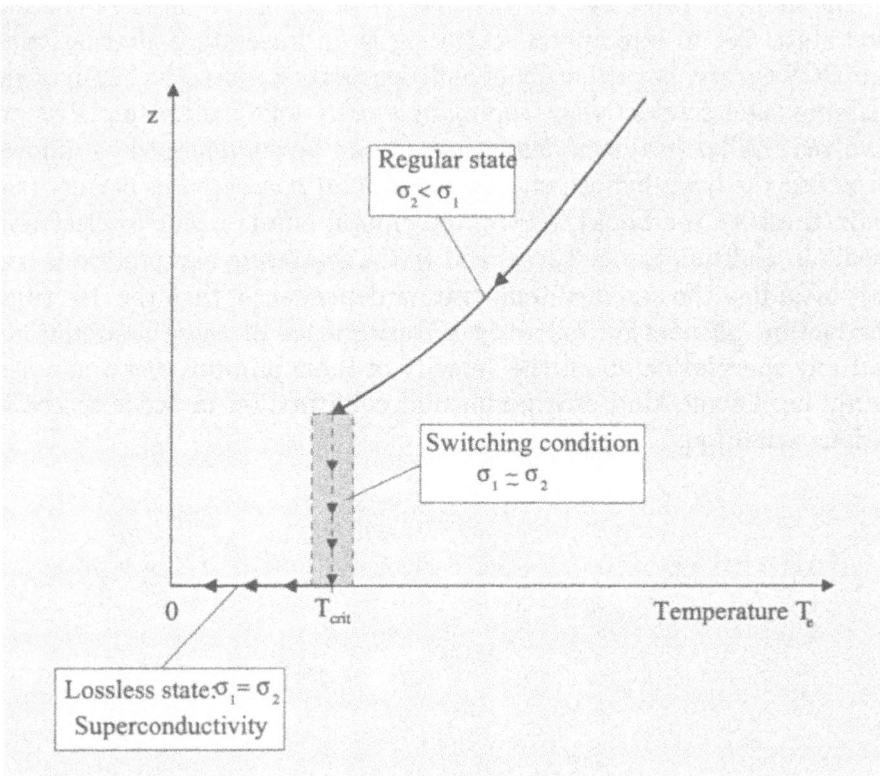

Figure 4.8. **Sketch showing the impedance variation predicted from the present scatter theory as the absolute temperature T_e is reduced towards absolute zero. The impedance variation in the *regular state* is approximate. In the *lossless state* the impedance is exactly zero, corresponding to superconductivity. The shaded region indicates uncertainty with regard to the exact switching condition $\sigma_2 \approx \sigma_1$.**

The current consensus of the solid state physics community appears to be that the *BCS* theory does not provide a fully satisfactory explanation of *high temperature* superconductivity. It is interesting to speculate whether bitemporal scattering might be involved in high temperature superconductivity. If we follow this line of thought, which is merely a conjecture, attention should be directed to a generalization of bitemporal scattering processes to the anisotropic structures of the known high temperature superconductors , effects that might possibly push the switching condition $\sigma_2 \approx \sigma_1$ towards higher temperature, i.e., increasing the critical temperature T_{crit}. Generalization of the present isotropic theory to two-dimensional conduction and scattering in the anisotropic structures of high temperature superconductors would be somewhat speculative, and far beyond the scope of this book.

The analysis presented in this chapter of a possible lossless conduction state due to bitemporal scattering is an interesting alternative to the *BCS* theory, but is hardly of sufficient rigor to justify a claim of explaining superconductivity. Too many loosely founded assumptions are involved. What has been demonstrated can be summarized as follows: Provided the basic hypothesis of a bitemporal microcosm is correct (the main thesis of the book), and *if* bitemporal effects occur in electronic media in addition to free space, and *if* the scattering assumption is correct including the assumed temperature dependence, then the described conduction phenomena follow as a consequence of these assumptions. But any speculation about the veracity of these propositions would certainly need some kind of experimental confirmation in order to create serious attention.

Chapter 5

PHOTON WAVE–PARTICLE TRANSITION

'The most incomprehensible thing about the world is that it is comprehensible.'
Albert Einstein

5.1 INTRODUCTION—THE PARADOX

One of the most puzzling phenomena in quantum physics is the apparent dual nature of photons, sometimes behaving as electromagnetic waves and sometimes as particles, and much of the early development of quantum theory was centered around this dilemma. The theoretical construct of photon duality was launched at the time of inception of quantum physics in attempts to provide some explanation of a variety of experimental facts relating to photon behavior. The most immediate example is the simple emission and absorption processes in single atoms. The atom is the pointlike *source* of photon emission as well as the pointlike *sink* of photon absorption. In its flight from source to sink the photon undergoes a magic transition from pointlike to wave costume, and then back to pointlike behavior during absorption. Your own photographic camera is an equally direct example of the apparent duality behavior. The incident beam of light from the object to be photographed consists of photons. A particular photon entering the camera lens as a true electromagnetic wave, undergoing the regular transmission through the optical system, and then approaching the photographic film, can not spread its constant energy $\hbar\omega$ amongst several silver atoms constituting the photosensitive emulsion without violating its energy integrity. It is forced to dump its energy to one single silver atom absorbing the quantum of light. The photon, in a manner of speech, is sucked into the atom, transforming itself in extremely short time from its full macroscopic cross section to atomic dimensions, a dimensional transformation of extremely high order of magnitude.

The wavelike behavior of the free photon is beyond doubt, as demonstrated by the operation of any optical instrument and, most spectacularly, by modern lasers. It is true that classical electromagnetism, through Maxwell's equations, does not predict a lower diffraction limit, but the complete absence of any kind of focusing members is certain to eliminate regular electromagnetism as a possible explanation of the photon shrinkage. Therefore, neither quantum physics, nor classical electromagnetic theory are able to provide an objective explanation for the photon behavior. Within the context of these theories it remains one of the unsolved mysteries or paradoxes of classical as well as quantum physics. The reader might have noted the phrase 'apparent duality' used in the earlier passage, which is in reference to the forthcoming exposition based on the neoclassical theory. It will be demonstrated that the wavelike to particlelike behavior is fully explained by a consistent application of the neoclassical concept of photon doublet involvement in the transition process. In this theory, which offers a realistic and objective view, the paradoxical behavior is entirely foreign with no need for the quantum mechanical principle of photon duality.

5.1.1 Neoclassical theory of photon transition

The general status of the duality principle in orthodox quantum physics prompts a general analysis of the photon transition problem viewed from the neoclassical platform. The configuration shown in Fig. 5.1, defined as an electromagnetic circuit specified by an arbitrary volume of space V enclosed by the surface S, meets the requirement of generality. The interior electromagnetic properties are specified by arbitrary distributions of dielectric and magnetic permeabilities $\varepsilon(\vec{r})$ and $\mu(\vec{r})$; otherwise the medium is specified as lossless, reciprocal, and stationary, the same restrictions that applied in the general circuit analysis in Chapter 3. The generality implies that optical instruments such as cameras, microscopes, and also the human eye with its retina are covered by the theory. The surface S is subdivided into two regions S_0 and S_L, of which the virtual open surface S_0 is closing the overall surface S and serves as an equivalent aperture providing access for the incident photon. The surface S_L of the interior wall consists of photon absorbing material such as a lossy conducting layer or a photosensitive film of silver atoms or

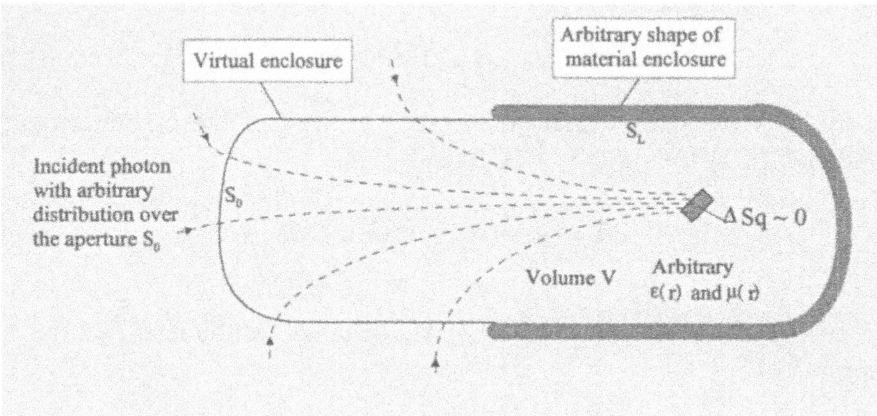

Figure 5.1. **Genereral circuit configuration used in the analysis of wave-like to particle-like photon transition. A single photon entering the enclosure from the left through the open surface S_0 as an arbitrarily shaped electromagnetic wave is either dumping all its energy at one of the miniscule surface terminals specified by ΔS_q, or being reflected back through S_0.**

any molecules that can absorb the energy of incident photons or reflect them back into free space through the aperture S_0.

The analysis in the following sections is based on the theoretical exposition in Appendix A of Chapter 3, in which we defined an infinite denumerable column of tangential vector functions $\vec{\psi}(\vec{s},\omega)$ located on the closed surface S and satisfying the orthonormality condition:

$$\int_S \vec{\psi}(\vec{s},\omega) \cdot \tilde{\vec{\psi}}(\vec{s},\omega)^* dS = \mathbf{1} \tag{5.1}$$

Referring to the discussion at the end of the mentioned appendix the set $\vec{\psi}(\vec{s},\omega)$ is not unique and can be defined in an infinite number of ways through simple unitary transformations. We shall make use of this flexibility to define a transformed set designated by $\vec{\phi}(s,\omega)$ for which the generalized voltages and currents refer to terminals represented by physically separated minuscule areas ΔS_q on the surface S. For this purpose the appropriate set of expansion functions is specified as follows:

$$\vec{\phi}(s,\omega) = col\,[\vec{\delta}_q(\vec{s} - \vec{s}_q)] \qquad q = 1, 2, ...N \tag{5.2}$$

where $\vec{\delta}_q(\vec{s}-\vec{s}_q)$ is a vector delta function. The linear transformation from some original set $\vec{\psi}(\vec{s},\omega)$ to the new set $\vec{\phi}(s,\omega)$ is specified by (A.27) of the same appendix.

$$\vec{\phi}(s,\omega) = \mathbf{T}\,\vec{\psi}(\vec{s},\omega) \tag{5.3}$$

where $\mathbf{T}$ is the transformation matrix. In view of the orthonormality condition (5.1) the matrix $\mathbf{T}$ is given by

$$\mathbf{T} = \int_S \vec{\phi}(s,\omega) \cdot \widetilde{\vec{\psi}}(\vec{s},\omega)^* \, dS \tag{5.4}$$

By means of the specification (5.2) a particular component $T_{q,j}$ can be expressed by

$$T_{q,j} = \mid \vec{\psi}(\vec{s}_q,\omega)_j \mid \cos(\theta_{q,j}) \tag{5.5}$$

where $\theta_{q,j}$ is the angle between $\vec{\delta}_q(\vec{s}-\vec{s}_q)$ and $\vec{\psi}(\vec{s}_q,\omega)_j$. The actual electric and magnetic fields were specified by (A.26), which in the present case reduce to

$$\begin{aligned} \vec{E}(\vec{s}_q,\omega)_{\mathrm{tan}} &= \sum_j V_j\,\vec{\phi}_j(s_q,\omega) = \sum_j V_j \vec{\delta}_j(\vec{s}-\vec{s}_q) = V_q \vec{\delta}_q(\vec{s}-\vec{s}_q) \\ \vec{H}(\vec{s}_q,\omega)_{\mathrm{tan}} &= \sum_j I_j\,\vec{\phi}_j(s_q,\omega) = \sum_j I_j \vec{\delta}_j(\vec{s}-\vec{s}_q) = I_q \vec{\delta}_q(\vec{s}-\vec{s}_q) \end{aligned} \tag{5.6}$$

In the transformed system the general circuit equation (A.24) is still valid.

$$\mathbf{Y}\,\mathbf{V} = \mathbf{I} \tag{5.7}$$

where the admittance matrix $\mathbf{Y}$ is appropriate for the new set of expansion functions $\vec{\phi}(s,\omega)$. By assumption, there is no internal sources in the configuration of Fig. 5.1, so that the equation lacks the right hand excitation term $\mathbf{I}_g$ appearing in the more general equation (A.24). The two equations in (5.6) can be expressed in compact vector column form:

$$\begin{aligned} \mathrm{col}[\vec{E}_i(s_q)]_{\mathrm{tan}} &= \mathrm{diag}\,[\vec{\delta}_q(\vec{s}-\vec{s}_q)]\,\mathbf{V} \\ \mathrm{col}[\vec{H}_q(s_q)]_{\mathrm{tan}} &= \mathrm{diag}\,[\vec{\delta}_q(\vec{s}-\vec{s}_q)]\,\mathbf{Y}\,\mathbf{V} \end{aligned} \tag{5.8}$$

In summing up this development, the configuration of Fig. 5.1 is described by equation (5.7) referring to a circuit with an infinite number of physically separated terminals on the closed surface S, each with a minute cross section of size ΔS approaching zero according to

$$\lim_{N\to\infty} \Delta S = \lim(S/N) = 0 \tag{5.9}$$

Each terminal is infinitely small, with the independent voltages and currents being delta type entities. The formulation is just a special case of the general transformation (A.28) in Appendix A of Chapter 3.

In order to bring the circuit equation (5.7) to an appropriate form for the specific task at hand the current column vector $\mathbf{I}$ must be expressed as the sum of two components as follows:

$$\mathbf{I} = \mathbf{I}_g + \mathbf{Y}_L \mathbf{V} \tag{5.10}$$

where the term $\mathbf{Y}_L\mathbf{V}$ represents the *passive* current flowing in the surface S_L, and the term $\mathbf{I}_g$ is the *active* driving current due to the incident photon. The diagonal matrix $\mathbf{Y}_L$, the *load matrix*, represents the lossy surface S_L in the sense that a particular component $Y_{L,q}$ is the localized load admittance at the minute terminal q. The physical loads do not have to be specified in detail; they could be associated with photosensitive atoms in a photographic film or charge carriers in a conducting surface.

Substitution of (5.10) into (5.7) transforms the equation to the form

$$(\mathbf{Y} + \mathbf{Y}_L)\,\mathbf{V} = \mathbf{I}_g \tag{5.11}$$

where the infinite set of constant current generators, i.e., the elements of $\mathbf{I}_g$ are located at the minuscule terminals on the surface S_0 and none on S_L. Equation (5.11) is perhaps best understood from the schematic diagram in Fig. 5.2 which is a familiar basic circuit equivalent in electronic circuitry.

The analysis to follow is separated into two parts, of which the first describes the macroscopic response through application of classical electromagnetics adapted to the present formulation. The second part deals with the response of single photons evaluated from the neoclassical theory.

5.2 CLASSICAL TIME AVERAGE MACROSCOPIC TRANSITION

The particular choice of expansion vectors $\vec{\phi}(s,\omega)$ defined in (5.2) as delta functions gives rise to a formulation involving an infinite number of minute but *spatially separated* virtual terminals on the enclosure S. The advantage gained by such a formulation is the ability to separate the voltage and current column vectors $\mathbf{V}$ and $\mathbf{I}$ into two subgroups defined by purely geometrical considerations decided from the two surface areas S_0 and S_L. Because each terminal has the same small area ΔS, the ratio between the total number of terminals on S_0 and S_L is simply

$$\frac{N_0}{N_L} = \frac{S_0}{S_L} \tag{5.12}$$

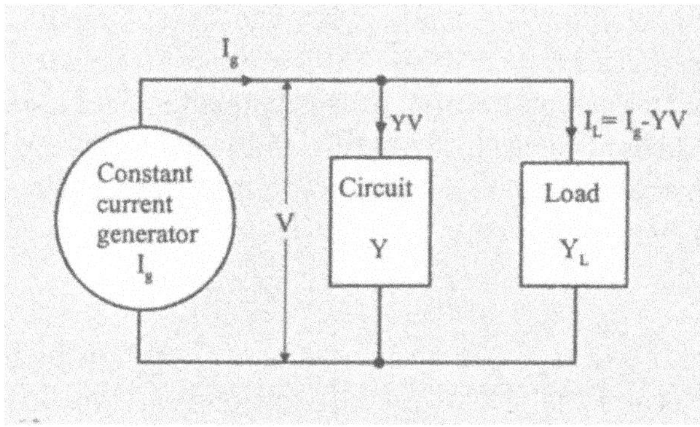

Figure 5.2. Equivalent diagram representing the circuit equiation (5.11). The lossless circuit of admittance Y is driven by the constant current generator I_g and loaded with the load admittance Y_L. The part of the current flowing in the load is specified by $I_L = I_g - YV$.

The constant current generators $\mathbf{I}_g$ are distributed amongst the terminals on surface S_0, with no generators on surface S_L. Similarly, there are no loads on the virtual surface S_0, so that the load matrix $\mathbf{Y}_L$ is nonzero on S_L only. These specifications allow us to express $\mathbf{Y}_L$ and $\mathbf{I}_g$ by their truncated forms $\mathbf{y}_L$ and $\mathbf{i}_g$ according to the relation:

$$\mathbf{Y}_L = \begin{bmatrix} 0 & 0 \\ 0 & \mathbf{y}_L \end{bmatrix} \quad \mathbf{I}_g = \begin{bmatrix} \mathbf{i}_g \\ 0 \end{bmatrix} \tag{5.13}$$

Here $\mathbf{y}_L$ is a diagonal $N_L \times N_L$ matrix and $\mathbf{i}_g$ a column vector of dimension N_0. With these definitions the circuit equation (5.11) takes the form

$$\begin{bmatrix} \mathbf{Y}_{11} & \mathbf{Y}_{12} \\ \mathbf{Y}_{21} & \mathbf{Y}_{22} + \mathbf{y}_L \end{bmatrix} \begin{bmatrix} \mathbf{V}_0 \\ \mathbf{V}_L \end{bmatrix} = \begin{bmatrix} \mathbf{i}_g \\ 0 \end{bmatrix} \tag{5.14}$$

where the dimensional ratio of the voltage subvectors $\mathbf{V}_0$ and $\mathbf{V}_L$ is equal to S_0/S_L. The equation separates into the subsystems

$$\mathbf{Y}_{11}\mathbf{V}_0 + \mathbf{Y}_{12}\mathbf{V}_L = \mathbf{i}_g \tag{5.15}$$

$$\mathbf{Y}_{21}\mathbf{V}_0 + (\mathbf{Y}_{22} + \mathbf{y}_L)\mathbf{V}_L = \mathbf{0} \tag{5.16}$$

The two equations specify the voltage column $\mathbf{V}_L$ in the form

$$\mathbf{Y}''\mathbf{V}_L = \mathbf{i}_g'' \tag{5.17}$$

where

$$\mathbf{Y}"= \mathbf{Y}_{22}+\mathbf{y}_L - \mathbf{Y}_{21}\,\mathbf{Y}_{11}^{-1}\,\mathbf{Y}_{12} \qquad \mathbf{i}_g" = -\mathbf{Y}_{21}\mathbf{Y}_{11}^{-1}\,\mathbf{i}_g \tag{5.18}$$

The relation (5.17) is an inhomogeneous linear system of dimension $N_L \times N_L$, specifying the voltage column $\mathbf{V}_L$ on the surface S_L as a function of the presumably known constant current generator column $\mathbf{i}_g$" representing the incident photon flux on the surface S_0. Because there is no reason to expect the system matrix **Y**" to be singular, the equation is readily solved for $\mathbf{V}_L$.

$$\mathbf{V}_L = (\mathbf{Y}")^{-1}\,\mathbf{i}_g" \tag{5.19}$$

The equation tells us how the incident macroscopic power distributes itself over the entire surface S_L. A particular terminal q receives the power

$$p_q = \frac{1}{2}V_{L,q}^*\,V_{L,q}\,\mathrm{Re}(y_{L,q}") = \frac{1}{2}V_{L,q}^*\,V_{L,q}\,g_{L,q} \tag{5.20}$$

The total power dissipated over the surface S_L adds up to the absorbed, or transmitted, part of the incident macroscopic power

$$P_{trans} = \frac{1}{2}\tilde{\mathbf{V}}_L^*\,\mathrm{Re}(\mathbf{Y}")\mathbf{V}_L = \frac{1}{2}\tilde{\mathbf{V}}_L^*\,\mathbf{g}_L\,\mathbf{V}_L = \frac{1}{2}\sum_{q=1}^{N_L}|\mathbf{V}_q|^2 g_{L,q} \tag{5.21}$$

where $\mathbf{g}_L$ is the real part of the matrix $\mathbf{y}_L$.

This concludes the macroscopic description of photon absorption on the material surface S_L. Let us next look at single photons in the configuration of Fig. 5.1.

5.3 WAVE TO PARTICLE TRANSITION OF SINGLE PHOTONS

Let the circuit be illuminated by N photons per second. The incident photon distribution over the input aperture specified by the surface S_0 in the drawing in Fig. 5.1 can be quite arbitrary. If the beam is spatially incoherent all the N photons may have different distributions. Following the theory developed in the last section, let a particular photon q be specified by its generator current distribution $(\mathbf{i}_g")_q$ over the aperture S_0. The *macroscopic* generator current $\mathbf{i}_g$" is simply the sum of the individual photon contributions.

$$\mathbf{i}_g" = \sum_{q=1}^{N}(\mathbf{i}_g")_q \tag{5.22}$$

If we apply the macroscopic circuit equation (5.17) to the single photon it would result in the response

$$\mathbf{Y}\text{''}\, \mathbf{V}_{L,q} = (\mathbf{i}_g\text{''})_q \tag{5.23}$$

which certainly is not valid because the equation tells us that the photon energy $\hbar\omega$ is distributed essentially over all the terminals on surface S_L. This would perhaps be acceptable if the surface conductance matrix $\mathbf{g}_L$ represented one single dissipating mechanism, with each surface point somehow coupled to all other points. But it is difficult to visualize a physical mechanism providing such a coupling. The actual absorbing entities are atoms, molecules, microcrystals, and charge carriers, all being of microscopic dimensions and not coupled strongly to adjacent members. The absence of coupling between the absorbing entities forces the photon to dump its entire energy at one atom or one charge carrier without energy fragmentation. Anything else would violate its energy integrity. Therefore, the single photon does not behave according to the macroscopic circuit equation (5.23). On the other hand the photon is an electromagnetic wave and must satisfy Maxwell's equations, and therefore also (5.23), if we apply classical electromagnetism. The resolution of this apparent contradiction, thoroughly described in Chapters 2 and 3, need not be repeated in detail. In the process the environmental photon doublets provide supplemental generator input, so that instead of (5.23) the appropriate form of the equation is

$$\mathbf{Y}\text{''}\, \mathbf{V}_{L,q} = (\mathbf{i}_g\text{''})_q + \boldsymbol{\alpha}_{q,trans} \tag{5.24}$$

where $\boldsymbol{\alpha}_{q,trans}$ is the photon doublet generator added to the regular photon generator column $(\mathbf{i}_g\text{''})_q$. Assume now that photon q is routed to terminal j on the surface S_L where it presumably dumps its full energy $\hbar\omega$ at the corresponding absorbing entity. This condition implies that the voltage vector $\mathbf{V}_{L,q}$ must have the form

$$\mathbf{V}_{L,q} = (\mathbf{V}_{L,q})^j = \begin{bmatrix} 0 \\ 0 \\ \vdots \\ V_L(j) \\ \vdots \\ 0 \end{bmatrix} \tag{5.25}$$

With this stipulation applied to (5.24) the appropriate photon doublet $(\boldsymbol{\alpha}_{q,trans})^j$ performing the task of routing photon q to terminal j is given by

$$(\boldsymbol{\alpha}_{q,trans})^j = \mathbf{Y}\text{''}\, (\mathbf{V}_{L,q})^j - (\mathbf{i}_g\text{''})_q \tag{5.26}$$

The full photon energy is dumped at terminal j because the voltages at all the other terminals are zero. It will be shown shortly that the photon travels to the particular terminal which minimizes the *cumulative* doublet expenditure. It could well be that the expenditure would be less if the photon were not absorbed on the surface S_L but rather reflected back into free space. If this happens let us determine the corresponding doublet admixture $\alpha_{q,refl}$ achieving full reflection. The criterion for complete reflection is zero voltage at *all* terminals on surface S_L. If this were not true some power would be absorbed on S_L, which is prohibitive because the reflected photon must retain its full energy. Therefore, in the full reflection case the voltage vector $\mathbf{V}_{L,q}$ must be identically zero.

$$\mathbf{V}_{L,q} = \mathbf{0} \tag{5.27}$$

The corresponding doublet admixture follows from (5.24):

$$\alpha_{q,refl} = -(\mathbf{i}_g")_q \tag{5.28}$$

Equations (5.26) and (5.28) demonstrate that appropriate doublet additives serve to either route the incident photon q to any terminal j where it is fully absorbed, or to reflect the photon back into free space.

But what is prompting the photon to select a particular terminal j for dumping is energy? Are they not all equally probable? The photon itself has no say in the matter. The selection is decided by the minimization principle discussed in earlier chapters. The doublet engagement in the routing process represents a perturbation of the environmental photon doublet energy (PDE) away from its time average energy equilibrium of zero. The particular routing of the photon is decided from the requirement of minimum doublet expenditure. The incident photon q does not enter a virgin circuit, but a circuit in which the environmental doublet fields are already perturbed by the doublet admixture already expended on the earlier photons preceding photon q. Therefore, the photon will be handled in a manner that causes a reduction of the overall accumulated doublet excitation. Thus, the distribution of photon arrivals at the surface S_L is not random, but governed by the minimization principle with an inherent memory of earlier arrivals.

5.3.1 Verification of zero time average doublet admixture

It is readily shown that the minimization principle, applied to a large number N of photons, leads to a photon distribution over the surface S_L that is exactly equal to the *macroscopic* distribution specified by (5.19).

Let $N(j)$ out of the total N photons be routed to terminal j so that

$$N_{trans} = \sum_j N(j) \tag{5.29}$$

The definition implies that $N(j)$ represents the distribution function of photons amongst the terminals. Furthermore, let N_{refl} be reflected back into free space. The total doublet supplement for all the N incident photons is obtained by summing over the transmitted and reflected photons.

$$\boldsymbol{\alpha}_{tot} = \boldsymbol{\alpha}_{trans} + \boldsymbol{\alpha}_{refl} = \sum_{q=1}^{N(j)} \sum_{j=1}^{N_{tr}} (\boldsymbol{\alpha}_{q,trans})^j + \sum_{q=1}^{N_{refl}} \boldsymbol{\alpha}_{q,refl} \tag{5.30}$$

where N_{tr} is the number of terminals on S_L. By substitution from (5.26) and (5.28) the equation becomes

$$\boldsymbol{\alpha}_{tot} = \mathbf{Y}" \begin{bmatrix} N(1)\, V_L(1) \\ N(2)\, V_L(2) \\ \vdots \\ N(j)\, V_L(j) \\ \vdots \\ N(N_{tr})\, V_L(N_{tr}) \end{bmatrix} - \sum_{j=1}^{N_{tr}} \sum_{q=1}^{N(j)} (\mathbf{i}_g")_q - \sum_{q=1}^{N_{refl}} (\mathbf{i}_g")_q \tag{5.31}$$

Apparently, the right hand sums add up to the total macroscopic generator term (5.22):

$$\sum_{j=1}^{N_{tr}} \sum_{q=1}^{N(j)} (\mathbf{i}_g")_q + \sum_{q=1}^{N_{refl}} (\mathbf{i}_g")_q = \mathbf{i}_g" \tag{5.32}$$

Substitution for $\mathbf{i}_g$" from the macroscopic circuit equation (5.17) converts (5.31) to

$$\boldsymbol{\alpha}_{tot} = \mathbf{Y}" \left\{ \begin{bmatrix} N(1)\, V_L(1) \\ N(2)\, V_L(2) \\ \vdots \\ N(j)\, V_L(j) \\ \vdots \\ N(N_{tr})\, V_L(N_{tr}) \end{bmatrix} - \mathbf{V}_L \right\} \tag{5.33}$$

where $\mathbf{V}_L$ is the *macroscopic* voltage vector. But we also know that

$$\lim_{N\to\infty} \begin{bmatrix} N(1)\,V_L(1) \\ N(2)\,V_L(2) \\ \vdots \\ N(j)\,V_L(j) \\ \vdots \\ N(N_{tr})\,V_L(N_{tr}) \end{bmatrix} = \mathbf{V}_L \tag{5.34}$$

which implies that

$$\lim_{N\to\infty} \boldsymbol{\alpha}_{tot} = \mathbf{0} \tag{5.35}$$

and also

$$\lim_{N\to\infty} \sum_{j=1}^{N_{tr}} \frac{N(j)}{N} = \frac{N_{trans}}{N} \tag{5.36}$$

These two equations confirm the general principle that the *time average* doublet energy vanishes and that the distribution of photon arrivals over the surface corresponds exactly to the macroscopic distribution. The principle of minimization is expected to apply at any time during arrival of photons. If for some reason an excessive number of photons above the equilibrium specified by (5.36) were dissipated on S_L at the expense of reflected photons, the corresponding $\boldsymbol{\alpha}_{tot}$ would be nonzero. Certainly, the next incident photon would then be reflected to improve the balance.

This concludes the proof of transition of photons from wavelike to particlelike entities under quite general conditions. It holds for any circuit-like enclosure excited through some arbitrary aperture S_0 by a stream of incident photons each having arbitrary spatial distribution of its electromagnetic wave components. An arbitrary section S_L of the interior wall consists of power absorbing material such as a lossy conducting layer or a photosensitive film of silver atoms or any molecules that can take up the photon energy. The presented analysis shows that this general circuit can be represented by an infinite number of *independent* physical terminals located on the closed surface S of the arbitrary volume enclosure constituting the circuit. Each terminal has a minuscule cross section, at least as small as any power absorbing members such as atoms or electrons. The absorption is described by the diagonal load conductance matrix $\mathbf{g}_L$ whose elements represent the *equivalent* load conductances of the absorbing members. The load conductances are arbitrary and not necessarily the same for each member. Therefore, the distribution of absorbing material over the interior is entirely arbitrary. It is remarkable that the photons are fully absorbed regardless of the particular conductance value. The process is not in any sense governed by the classical notion of *matching* the photon to the terminal load. In the neoclassical theory the supplement from environmental doublets added to the photon itself provides the matching. And contrary to the classical concept this process of single photon matching is perfect. It is not a question of say 99.9 percent match, which would be an extremely good match in any macroscopic case, but rather one hundred percent, or absolute match. Anything less would be physically inconsistent and a violation of photon energy integrity.

In conclusion, for the general configuration specified by Fig. 5.1 in the introduction, the wavelike to particlelike transition of photons appears to be fully explained by the neoclassical theory. The process describes how a photon as an electromagnetic wave of any width and shape can be squeezed to any minute cross section as small as required by the size of atomic or electron absorbers. In the neoclassical theory the quantum principle of duality is a foreign concept; it is not needed for interpretation of the apparent wavelike to particlelike behavior.

The following examples serve to illuminate the theoretical findings by presenting two widely different cases of photon wave to particle transition, illustrating the principle of photon doublet engagement in the process. The first of these is the neoclassical interpretation of the basic phenomenon of photon emission and absorption by atoms. The second is a discussion of wave to particle transition in a simple and practical configuration—the stripline—a familiar and common transmission line for microwave frequencies. The examples serve to emphasize the common aspects of photon wave to particle processes occurring in a variety of physical situations and at all frequencies, not by any means limited to optical and higher frequencies.

5.4 PHOTON EMISSION AND ABSORPTION BY ATOMS

Referring to the general configuration of Fig. 5.1 shown earlier, the incident photon entering the enclosure from the left was characterized by arbitrary lateral distribution on the virtual surface S_0. The flow pattern must be invariant with regard to reversal of the direction of power flow, in which case the minuscule absorbers on surface S_L must be identified as pointlike emitters. According to this argument the theory is applicable to photons emitted from atoms and subsequently absorbed by atoms. The relevant situation is shown schematically in Fig. 5.3a, with A being the emitting atom and B the absorbing atom, with the set of auxiliary atoms C representing alternative absorbing atoms in the environment.

It is difficult to make definite statements about the radiated wave pattern of the photon emitted from atom A, but from an electromagnetic viewpoint the most likely pattern is the lowest order dipole distribution described in any textbook on basic radiation theory. The flux density of such dipole radiation is not uniform but is still filling the entire sphere surrounding the atom with the typical dipole distribution. How can the entire photon flux of this distribution be focused to the pointlike atom B so that the full energy $\hbar\omega$ is absorbed by B? The absorption is a special example of photon routing orchestrated by the environmental photon doublets, a process fully described by the general wave to particle theory of the foregoing sections. We shall refrain from presenting the reader with the detailed analysis of this process, which involves superposition of the original dipole radiation and appropriate admixtures of environmental doublet components. Instead we shall focus on some of the more important implications resulting from the involvement of photon doublets in the basic emission-absorption process.

Quantum mechanics describes the emission of a photon as a completely *random* process. According to quantum theory there is no way of telling exactly when the atom decides to emit. The quantum randomness of emission is characterized by a nondeterministic outcome; the randomness is absolute, reflecting the basic dogma of quantum theory. In the neoclassical theory the engagement of bitemporal environmental doublets in the emission-absorption process challenges this interpretation in a fundamental way. The doublets serve to intimately couple the emitter and the absorber together so that the emission process can not be considered completely independent from the surroundings, i.e., from the absorbers and the environmental doublets. The emission-absorption process has to abide by the fundamental principle of minimization of PDE expenditure. The process takes place when the two atoms in question are in the optimum constellation with regard to the environment. More specifically, the atom emits a photon only with certain extraneous conditions satisfied, namely: the prospective absorbing atom and the emitting atom must be in such a constellation that the supplemental photon doublets maintain a balance of zero time average environmental doublet energy, or at least minimizes the energy. According to this interpretation, the emission and absorption processes are tied together in an intimate fashion. The emission does not happen unless the prospective absorber is in the right constellation for minimization of the PDE time average. The process has the character of a "handshaking" operation. The atoms must be in mutual consent between the two of them, as well

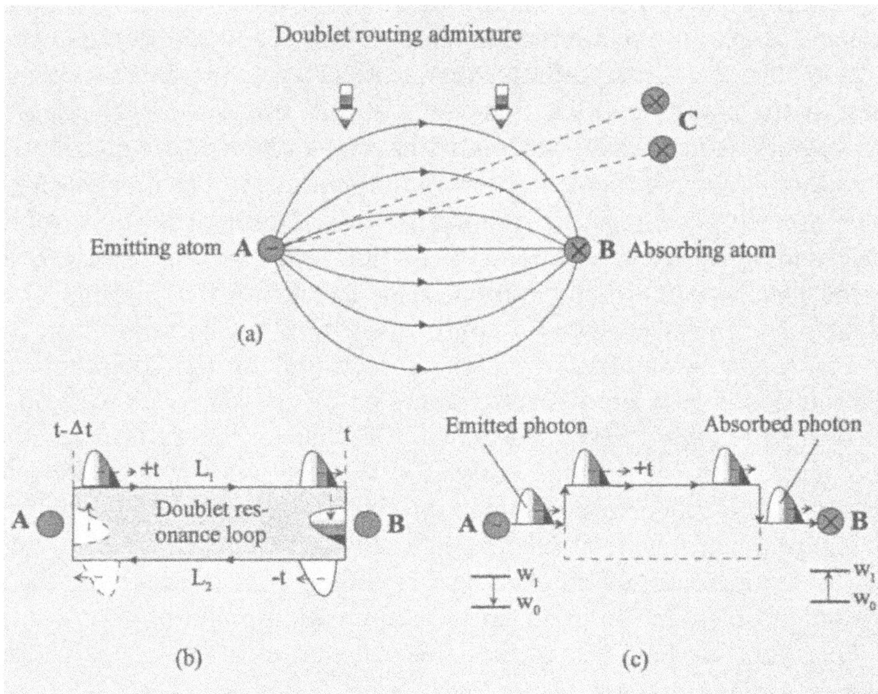

Figure 5.3. **Illustration of the emission-absorption process. (a) Atom A is the emitter, atom B the absorber. Atoms C represent some of the available auxiliary absorbers in the universe. (b) State of no emission: The two atoms are in a passive handshaking state because the overall environmental doublet energy balance is not conducive to emission. The atoms are tied together in a doublet resonance loop with time average energy equal to zero. (c) State of emission and absorption: The environmental balance of time average energy is improved through emission from atom A of a regular photon which is subsequently absorbed by atom B.**

as in agreement with the environmental sea of photon doublets. This interpretation eliminates the quantum mechanical concept of *absolute* randomness in the basic emission process. Still the process is of course random, but the random outcome is a *deterministic* phenomenon associated with *objective* randomizing factors that are unknown to us due to our ignorance of the instantaneous condition of the environmental sea of doublets governing the process. The shared responsibility of emitter and absorber is reminiscent of the Wheeler-Feynman 'Absorber Theory' [47], but their description differs from the present one in that it does not in any sense describe the wave to particle transition affected through the intermediacy of photon doublets.

The sketches in Figs. 5.3b and 5.3c represent oversimplified but nevertheless illuminating illustrations of the expected vast and very complex doublet involvement in the emission-absorption process. In these sketches the doublet set is represented by just one simple doublet, described earlier in Chapter 2 (see Fig. 2.10) as an entangled doublet pair, and in the forthcoming Chapter 6 (see Fig. 6.2) as a bitemporal doublet resonance. The resonance is characterized by a single transmission loop between the two atoms, with the positive energy photon pulse traveling in the positive time direction along path L_1, and the negative energy pulse in the negative time direction along L_2. The time average energy of the doublet resonance is zero. For more details of the resonance process the reader is referred to the citations above.

In the state of no emission, illustrated in Fig. 5.3b, the two pulses simply reflect into each other at the two atoms A and B, thereby forming a handshaking loop of bitemporal resonance. The time average energy of the resonance is zero because the photon along path L_2 has negative energy, compensating the positive energy along path L_1. The resonant loop provides information between the two atoms whether the state of the environmental sea of doublets is conducive to emission. Presumably the emission process is prohibited because the resulting change in the overall time average doublet energy would be unfavorable.

If the resulting energy change is favorable, atom A is emitting a photon which is subsequently absorbed by atom B. The process is described in the simplified sketch of Fig. 5.3c. Now being absorbed by atom B the forward traveling pulse along path L_1 fails to excite the backward traveling pulse along path L_2. In turn, the now missing reflected pulse at atom A fails to excite the forward photon pulse at time $t - \Delta t$ which therefore must be supplied through emission from atom A. As indicated in the figure, the emission and absorption processes are accompanied by corresponding jumps between atomic energy levels.

According to this picture, the emission-absorption process is a mutual event involving both atoms as well as the environmental doublet family. In the present bitemporal theory it does not make sense to talk about emission as a primary random process happening at time $t - \Delta t$ followed by absorption at some *arbitrary* atomic site at a later time t. The two processes have an equal standing. The absorption at atom B could just as well be taken as the primary event at time t followed by emission from atom A at the earlier time $t - \Delta t$. The process happens with mutual accept between emitter, absorber, and the environmental doublets. Emission takes place provided the overall constellation is conducive to minimizing the overall PDE expenditure.

5.5 PHOTON WAVE TO PARTICLE TRANSITION IN STRIPLINES

Because photons do not know any frequency restrictions, we shall investigate the wavelike to particlelike behavior in a microwave transmission line, for this purpose selecting the *stripline* as a typical example. The stripline, illustrated in Fig. 5.4, consists of two parallel metal plates of width d and spacing h. In most practical applications one plate is serving as a common ground plane serving perhaps a multitude of single lines. Provided the width d is less than half a free space wavelength, the line supports only one propagating Transverse Electric Magnetic (TEM) mode specified by their forward and backward electric and magnetic fields:

$$
\begin{aligned}
e_{y0}(Z) &= e_0^f\, e^{-jZ} + e_0^b\, e^{jZ} \\
h_{x0}(Z) &= e_0^f\, e^{-jZ} - e_0^b\, e^{jZ}
\end{aligned}
\tag{5.37}
$$

In these equations $Z = kz$ is the normalized longitudinal coordinate, where $k = \omega/c$ is the normalized frequency, and $h_{y0} = Z_0 H_{y0}$ is the normalized magnetic field with $Z_0 = \sqrt{\mu_0/\varepsilon_0}$ being the impedance of free space. Edge effects are assumed to be negligible, an assumption well justified provided the height/width ratio is small.

$$
\frac{h}{d} << 1 \tag{5.38}
$$

Under the same conditions the characteristic impedance of the line is given by

$$
Z_{line} = Z_0 \frac{h}{d} \tag{5.39}
$$

In addition to being a common component in miniaturized microwave planar circuits, the stripline is also a major component in integrated chip applications serving to interconnect the perhaps millions of tiny components in processors and memory devices. Your personal computer contains a vast number of such stripline connectors having typical widths of a fraction of a micrometer. It is true that in these applications they are not commonly referred to as striplines because they are notably of extremely short length, orders of magnitudes shorter than the free space wavelength which is 0.3 meter at a clock frequency of 1 GigaHertz. Therefore, the typical propagation characteristics are not predominant. Nevertheless, in the present discussion of single photon events they fall into the class of regular striplines, and are therefore subject to the same phenomena of wave to particle transition described in the following.

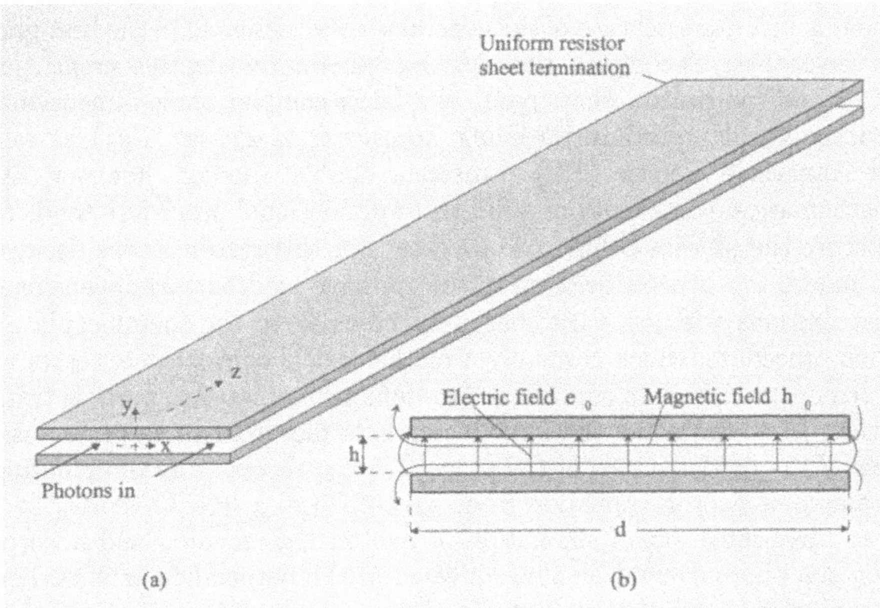

Figure 5.4. (**a**) **Basic stripline configuration. The line width** b **is less than half a free space wavelength, and the plate spacing** h **is much less than the width. (b) Sketch of the uniform field distribution of the** TEM **forward traveling mode in the main section of the line.**

As shown in the figure the transmission line is terminated with a uniform resistor sheet of some arbitrary impedance Z_L per square. The impedance is presumably different from the free space impedance Z_0, so that the incident TEM wave is partly reflected. Classically, the macroscopic behavior is described by simple transmission line theory, taken out of any textbook on the subject. However, at the single photon level the situation is much different and not covered by regular transmission line theory. If we visualize an influx of N photons per second, a certain number N_{trans} is *fully* absorbed in the load, and a number N_{refl} *fully* reflected back to the source, all of them retaining their photon energy $\hbar\omega$ in the process. This general picture is maintained regardless of the particular load impedance Z_L. The additional boundary conditions imposed by these energy integrity requirements can not be satisfied by the regular classical TEM modes referred to in (5.37). Each photon engages supplementary support of propagating doublet modes in addition to a set of local evanescent *doublet* modes excited at the output end of the line, providing exactly the required fields needed for these tasks.

The theory of photon routing and absorption in the stripline embodiment is just a special case of the general theory presented in the first part of the chapter. A complete description of photon routing to a *single* spot X_q, Y_p on the resistor sheet requires a fairly complex three-dimensional mathematical procedure involving Transverse Magnetic (TM) as well as Transverse Electric (TE) evanescent doublet modes; however, the mathematical complexity of a full three-dimensional procedure tends to obscure the physics behind the process. For this reason it was decided to model the photon fields near the resistive load as two-dimensional distributions with no variations perpendicular to the metallic plates. This procedure, which is merely a mathematical convenience, serves as a demonstration of an equivalent *two-dimensional* wave to particle transition. In no ways does this simplification detract from an accurate analysis of the fundamentals of the process, which serves as an illuminating example of photon transition from a wavelike to a *linelike* entity. In a true three-dimensional simulation of photon transmission and absorption the photon would also be squeezed in the perpendicular direction, dumping its energy at a point X_q, Y_p to a single charge carrier rather than spreading it out over the line charge X_q.

5.5.1 Evanescent doublet supplements at the load

The polarization of the main TEM propagating modes in the Y-direction restricts the polarizations of the evanescent doublet fields to the same direction. In what follows the doublet TE field components are referred to by Greek letters κ_z, κ_x, and ϵ_y rather than h_z, h_x, and e_y for regular fields. From standard electromagnetic wave theory all the field components of the TE doublet set are derived from the longitudinal magnetic field $\kappa_z(X, Z)$ which is conveniently specified by the following infinite sum of spatial harmonics:

$$\kappa_z(\bar{X}, Z) = \sum_{n=1}^{\infty} -\frac{B_n}{B_z^{1/2}} \sin(B_n \bar{X}) \alpha_n \, e^{B_z Z} \tag{5.40}$$

where $\bar{X}$ is the X-coordinate measured from the left hand edge of the stripline.

$$\bar{X} = X - \frac{D}{2} \tag{5.41}$$

In this sum the expansion coefficients α_n represent harmonic doublet amplitudes with normalized wave numbers B_n and B_z in the $\bar{X}$ and

Z - directions specified by

$$B_n = \frac{\pi n}{D} \qquad\qquad B_z = (B_n^2 - 1)^{1/2} \tag{5.42}$$

The dimensional variables in (5.40) are in normalized form: $D = kd$, $\bar{X} = k\bar{x}$, and $Z = kz$. Boundary conditions, compatible with the assumption (5.38), stipulate no current across the metallic edges of the interior metallic plates, a condition implying that magnetic fields are *normal* to the edges everywhere.

$$\kappa_z(0, Z) = \kappa_z(D, Z) = 0 \tag{5.43}$$

It is observed that the form of $\kappa_z(\bar{X}, Z)$ specified in (5.40) satisfies these conditions everywhere along the transmission line. The corresponding transverse doublet components ϵ_y and κ_x are readily obtained from the appropriate field equations taken together with (5.40).

$$\begin{aligned} \epsilon_y(\bar{X}, Z) &= \sum_{n=1}^{\infty} j\, B_z^{-1/2} \cos(B_n \bar{X})\, \alpha_n\, e^{B_z Z} \\ \kappa_x(\bar{X}, Z) &= \sum_{n=1}^{\infty} B_z^{1/2} \cos(B_n \bar{X})\, \alpha_n\, e^{B_z Z} \end{aligned} \tag{5.44}$$

Provided the longitudinal propagation factor B_z is real, the TE modes of (5.40) are evanescent, a condition requiring

$$B_n^2 > 1 \qquad \text{or} \quad n > \frac{d}{\lambda/2} \tag{5.45}$$

Presumably the linewidth d is less or equal to one half free space wavelength, so that the range of harmonic evanescent TE components is specified by

$$n = 1, 2, 3 ... \infty \qquad\qquad n \neq 0 \tag{5.46}$$

The doublet $n = 0$ component must be added as a separate propagating TEM component

$$\begin{aligned} \epsilon_{y0}(Z) &= \alpha_0^f\, e^{-jZ} + \alpha_0^b\, e^{jZ} \\ \kappa_{x0}(Z) &= \alpha_0^f\, e^{-jZ} - \alpha_0^b\, e^{jZ} \end{aligned} \tag{5.47}$$

The normalized wave impedances of the TEM and the TE modes, defined as usual by the ratio $(\epsilon_y/\kappa_x)_n$, are readily evaluated to

$$\begin{aligned} z^{TEM} &= \frac{Z^{TEM}}{Z_0} = 1 \\ z^{TE} &= j(B_n^2 - 1)^{-1/2} = jB_z^{-1} \end{aligned} \tag{5.48}$$

As expected, the wave impedances of the evanescent components are pure imaginary, reflecting the fact that these components carry no power in the *longitudinal* direction. Their function is to provide *lateral* power flow in the $\bar{X}$-direction, thereby deflecting the approaching photon to specific points on the resistor sheet. The lateral power flow comes about through the product of the *Z-directed* magnetic doublets $\kappa_z(\bar{X}, Z)$ and the *y-directed* electric field $e_0^f(Z)$ of the regular *TEM* photon (5.37) propagating down the line. The details of this process will be clearer from the discussions in the subsequent sections. At this point let us just demonstrate that the lateral $\bar{X}$ - directed power flow is nonzero. From *Poynting*'s theorem:

$$S_x(\bar{X}, Z) = \mathrm{Re}\,[\kappa_z(\bar{X}, Z)\, e_0^f(Z)^*] \neq 0 \tag{5.49}$$

The doublet component $\epsilon_{y0}(Z)$ in (5.47) is in time quadrature to $\kappa_z(\bar{X}, Z)$, contributing nothing to the lateral flow.

The evanescent modes at the resistor sheet are standing waves in the transverse $\bar{X}$-direction, decaying exponentially away from the sheet according to the factor $\exp(B_z Z)$. For the higher harmonics the factor B_z is approximately equal to the lateral wave number B_n. Thus, the longitudinal 'skin depth' field penetration from the sheet into the free space between the plates is roughly equal to the periodicity in the lateral dimension $\bar{X}$. As the harmonic numbers n increase without limit the corresponding wave harmonics take on the appearance of 'creeping modes' along the surface.

The total Z-directed power per unit area of a composite wave containing the incident plane wave (5.37) and possible doublet supplements (5.47) is specified by the general relation

$$p = \frac{P}{dh} = \frac{1}{2DH} \int\limits_{\bar{X}=0}^{D} \int\limits_{Y=-H/2}^{H/2} \mathrm{Re}\,[e_y(\bar{X}, Z)\, h_x(\bar{X}, Z)^*]\, d\bar{X}\, dY \tag{5.50}$$

yielding

$$p = \frac{1}{2}(e_0^f\, e_0^{f*} - e_0^b\, e_0^{b*} + \alpha_0^f\, \alpha_0^{f*} - \alpha_0^b\, \alpha_0^{b*}) \tag{5.51}$$

The evanescent components (5.40)-(5.44) contribute nothing to the Z-directed power. It was also proved, in the theory of photon doublets in Chapter 3, that the *TEM* doublets carry no power

$$\alpha_0^f\, \alpha_0^{f*} = \alpha_0^b\, \alpha_0^{b*} = 0 \tag{5.52}$$

so that (5.51) reduces to

$$p = \frac{1}{2}(e_0^f e_0^{f*} - e_0^b e_0^{b*}) \tag{5.53}$$

The question whether the incident single photon is fully reflected or fully absorbed at the plate is decided from the composite field distribution at the resistor sheet located at $Z = 0$. Let us first consider full reflection of the photon.

5.5.2 Complete photon reflection

If the full energy $\hbar\omega$ of the incident photon is going to be reflected, there certainly can be no absorption at the surface, a condition requiring the total transverse fields to be zero.

$$e_y(\bar{X},0) \equiv 0 \qquad h_x(\bar{X},0) \equiv 0 \qquad \text{for all } \bar{X} \tag{5.54}$$

Accordingly:

$$\begin{aligned} e_0^f + e_0^b + \alpha_{0,refl}^f + \alpha_{0,refl}^b + \varepsilon_y(\bar{X},0)_{refl} &\equiv 0 \\ e_0^f - e_0^b + \alpha_{0,refl}^f - \alpha_{0,refl}^b + \kappa_x(\bar{X},0)_{refl} &\equiv 0 \end{aligned} \tag{5.55}$$

Because the only function of the evanescent doublet components is to provide lateral deflection, which is not taking place in the case of photon reflection, it is expected that the set of equations is satisfied without the need of the doublet components $\varepsilon_y(\bar{X},0)_{refl}$ and $\kappa_x(\bar{X},0)_{refl}$, so that we can put

$$\varepsilon_y(\bar{X},0)_{refl} = \kappa_x(\bar{X},0)_{refl} \equiv 0 \tag{5.56}$$

which results in the following doublet supplements:

$$\alpha_{0,refl}^f = -e_0^f \qquad\qquad \alpha_{0,refl}^b = -e_0^b \tag{5.57}$$

The photon enters as a plane wave and reflects as a plane wave. The conceptual picture evolving from this particular result, based directly on the stipulation of no fields at the resistor sheet surface, is just a variation of the transmission matrix formulation of the same problem discussed in Chapter 2. The results are identical, demonstrating the adequacy of the simple one-dimensional model of Chapter 2 for handling the case of full photon reflection. The reason for this perhaps surprising result is the complete absence of interaction with the resistive load. However, the situation is entirely different in the case of full absorption, discussed next.

5.5.3 Complete photon absorption

As noted in the earlier sections the routing of the incident photon to a specific point $\bar{X}_q$ on the resistive sheet requires engagement of evanescent doublet modes. Full absorption at $\bar{X}_q$ implies a delta function distribution of the lateral magnetic field:

$$h_x(\bar{X},0) = e_0^f - \alpha_0^b + \kappa_x(\bar{X},0) = C_h\delta(\bar{X}-\bar{X}_q) \tag{5.58}$$

In this equation C_h is a constant to be determined and the delta function $\delta(\bar{X}-\bar{X}_q)$ is conveniently taken as a Gaussian defined by the usual formula

$$\delta(\bar{X}-\bar{X}_q) = \frac{1}{\Delta X} e^{-\pi(\frac{\bar{X}-\bar{X}_q}{\Delta X})^2} \tag{5.59}$$

The half width ΔX of the Gaussian defines the resolution of the point $\bar{X}_q$. Regardless of the actual half width the Gaussian satisfies

$$\frac{1}{\Delta X}\int_{-\infty}^{\infty} e^{-\pi(\frac{\bar{X}-\bar{X}_q}{\Delta X})^2}\, d\bar{X} = 1 \tag{5.60}$$

We shall also need the Fourier transform which is also Gaussian.

$$\frac{1}{\Delta X}\int_{-\infty}^{\infty} e^{-\pi(\frac{\bar{X}-\bar{X}_q}{\Delta X})^2}\, e^{jB_m\bar{X}}\, d\bar{X} = e^{jB_m\bar{X}_q}\, e^{-\pi(B_m\Delta X)^2} \tag{5.61}$$

The stipulation of the total lateral field as a Gaussian is an assurance that the field vanishes everywhere except at $\bar{X}_q$. The lateral resolution is the half width ΔX which, as demonstrated shortly, can be made as small as needed. In the limit of ΔX approaching zero the incident photon as a function of $\bar{X}$ has shrunk to a *pointlike* wave disturbance at $\bar{X}_q$, representing the *particle costume* of the original *plane wave* photon. But due to the original restriction of the analysis to two dimensions, it still is *linelike* in the tree-dimensional frame, extending across the transmission line from $y=-H/2$ to $H/2$.

Going back to (5.58) it is noticed that the components e_0^b and α_0^f are missing, because these are not needed in the absorption case. The reflected regular photon part e_0^b vanishes by assumption; its role is replaced by the doublet backward component α_0^b. It remains of course to show that the field specification (5.58) is consistent and that the incident photon energy is fully absorbed at $\bar{X}_q$ with no energy reflected. This requirement implies that the incident photon 'wave packet' must be *matched* to the absorbing mechanism at $\bar{X}_q$.

Let us invert (5.58) through multiplication with $\cos(B_m\bar{X})$ followed by integration over the full range of $\bar{X}$. In the first step of this operation, let the selectable integer m be qual to zero, in which case the described procedure results in the following relation:

$$e_0^f - \alpha_0^b = \frac{1}{D}C_h \tag{5.62}$$

yielding

$$\frac{1}{D}C_h = e_0^f\,[1-\rho(\bar{X}_q)] \tag{5.63}$$

where $\rho(\bar{X}_q)$ is the *reflection coefficient* $\rho(\bar{X}_q)$ at point $\bar{X}_q$.

$$\rho(\bar{X}_q) = \frac{z(\bar{X}_q)-1}{z(\bar{X}_q)+1} \tag{5.64}$$

In this equation $z(\bar{X}_q)$ is the normalized impedance of the absorbing member at the point $\bar{X}_q$. The required plane wave doublet addendum is specified by

$$\alpha_0^b = \rho(\bar{X}_q)\,e_0^f \tag{5.65}$$

In classical electromagnetism the reflection coefficient specifies the required reflected component e_0^b for providing match between the net impedance of the composite forward and backward wave and the load impedance. In the neoclassical theory of single photons the matching component is not e_0^b which is zero, but the photon doublet α_0^b. This nontrivial difference between classical and neoclassical electromagnetism is a key element in satisfying boundary conditions for single photons. The reflected doublet α_0^b carries no energy, in accordance with (5.52), and e_0^b is zero by assumption, so that the total z-directed power density p is specified by (5.53) to

$$p = \frac{1}{2}e_0^f\,e_0^{f*} \tag{5.66}$$

which is the incident photon power density. This perfect adaptability of matching to any load, regardless of the actual impedance, is unheard of in classical electromagnetism. It means that a single photon can always be absorbed by photosensitive molecules or by charge carriers through the intermediary of an appropriate ingredient of photon doublets. But to complete the proof we also have to show that the photon, in addition to dumping its full energy at the plate, is also focused and routed to some arbitrarily small spot $\bar{X}_q$ on the sheet. For this task the evanescent modes are needed. Their contributions are determined following a normal inversion procedure of (5.44). In this procedure the expansion coefficients α_n are first determined by multiplication of the lower equation by $\cos(B_m\bar{X})$ followed by integration over the full range of $\bar{X}$ making use of (5.58) and (5.61). Noting the following orthogonality condition:

$$\frac{1}{D}\int_0^D \cos(B_m\bar{X})\cos(B_n\bar{X})d\bar{X} = \frac{1}{2}\delta(m-n) \tag{5.67}$$

the integration procedure results in

$$\alpha_n = \frac{2}{B_z^{1/2}}\frac{C_h}{D}\cos(B_n\bar{X}_q)e^{-\pi(B_n\Delta X)^2} \tag{5.68}$$

With the expansion coefficients determined, these are plugged back into (5.44), a procedure yielding the following expression for the evanescent doublet set of magnetic components:

$$\kappa_x(\bar{X},Z) = 2\frac{C_h}{D}\sum_{n=1}^{\infty}\cos(B_n\bar{X}_q)\cos(B_n\bar{X})\,e^{-\pi(B_n\Delta X)^2+B_zZ} \tag{5.69}$$

$$\kappa_z(\bar{X},Z) = -2\frac{C_h}{D}\sum_{n=1}^{\infty}\frac{B_n}{B_z}\cos(B_n\bar{X}_q)\,\sin(B_n\bar{X})\,e^{-\pi(B_n\Delta X)^2+B_zZ} \tag{5.70}$$

where the factor C_h/D is given by (5.63). With the doublet supplement specified by these equations the incident photon is routed to the point $\bar{X}_q$. We saw that the impedance $z(\bar{X}_q)$ represents the properties of the absorbing member at the point $\bar{X}_q$. Regardless of its value the doublet magnetic field $\kappa_x(\bar{X},Z)$ in (5.69) adjusts itself through the factor C_h/D so that the equation is always satisfied.

5.5.4 Lateral resolution of the particlelike photon beam

The required number of terms in the expansion (5.69) for a desired resolution ΔX is determined essentially by the Gaussian factor. The Gaussian decays very rapidly with increasing mode number n. Let us tentatively, and somewhat arbitrarily, specify the maximum $N_{\max} = n_{\max}$ by

$$B_{N_{\max}}\Delta X = \pi \tag{5.71}$$

which certainly results in a very low value of the Gaussian, namely

$$e^{-\pi(B_{N_{\max}}\Delta X)^2} = e^{-\pi^3} = 3.4 \times 10^{-14} \tag{5.72}$$

We take (5.71) as defining the nominal maximum $N_{\max}$:

$$N_{\max} = \frac{1}{2}\frac{D}{\Delta X} \tag{5.73}$$

For values of n beyond $N_{\max}$ the terms in the sums of (5.69) and (5.70) are negligibly small and can safely be rejected. The lateral resolution is then

$$\frac{\Delta X}{D} = \frac{1}{2N_{\max}} \tag{5.74}$$

which perhaps is more meaningful if specified relative to the wavelength

$$\frac{\Delta x}{\lambda} = \frac{1}{2N_{\max}}\frac{d}{\lambda/2} \approx \frac{1}{2N_{\max}} \tag{5.75}$$

because the ratio $d/(\lambda/2)$ is less or equal to one for the stripline. It is immediately clear that there is no limit on the resolution because there is no upper limit on $N_{\max}$. Thus, the incident TEM type photon can shrink to an arbitrarily small size Δx at the location $\bar{X}_q$ through interaction with a sufficiently large number of evanescent doublet harmonics located at the resistance sheet. The spot size is expected to be determined by the size of the absorbing members, in the present case the charge carriers in the sheet. In a typical microwave system take the free space wavelength to be $0.01m$ and assume Δx equal to the electron Compton wavelength of $2.42 \times 10^{-12}m$. This extreme example results in

$$N_{\max} \approx 10^{10} \tag{5.76}$$

The corresponding number Q_x of independent spots along x, defined by nonoverlapping, follows from (5.74):

$$Q_x = \frac{d}{\Delta x} = N_{\max} \tag{5.77}$$

which for the example above would amount to a total of 10^{10} independent spots. With resolutions of this small magnitude the modeling of the resistance sheet as a *plane* surface is of course questionable. Although a more sophisticated model might be needed, this point is of no concern for the general interpretation predicting no lower limit for the lateral resolution. The environmental PDE evanescent doublet fields route the incident photon to the particular spot $\bar{X}_q$ that minimizes doublet expenditure, maintaining zero time average over some extended time range. It will be proved shortly that this condition simultaneously provide uniform average absorption over the full surface, under assumption of constant spot impedance $z(\bar{X}_q)$. But first let us look at the energy flux pattern.

5.5.5 Numerical simulations of photon flow pattern

In its transition from the incident plane wave to the particlelike photon at the resistor sheet the energy flux pattern is representative of the undergone transformation. The flux of energy is specified by the two components of the Poynting vector $\vec{S}(\bar{X}, Z)$. As shown earlier by (5.66) the z-component is constant and equal to

$$S_z(\bar{X}, Y, Z) = S_z = \frac{1}{2} e_0^f e_0^{f*} \tag{5.78}$$

The lateral component is given by the equation

$$S_x(\bar{X}, Z) = \mathrm{Re}\left[\kappa_z(\bar{X}, Z)\, e_0^f(Z)^*\right] \tag{5.79}$$

The equations serve as the basis for the computer simulations illustrated in Fig. 5.5, demonstrating the theoretical results of photon routing analyzed in this chapter. The figure displays a set of four computer generated simulations of photon flux lines down the stripline towards arbitrarily selected arrival positions $\bar{X}_q$ at the resistor sheet. In these diagrams the power flux density $\vec{S}$ is inversely proportional to the separation between adjacent lines. In all simulations the computer program initiates the plotting at the very left, far away from the $\bar{X}_q$-position specified in advance as one of the program input parameters. Note how accurately the stream lines converge to the specified position, where the photon is fully absorbed regardless of the load impedance Z_L if only $R_L = \mathrm{Re}(Z_L) > 0$. In this sense the behavior is markedly different from the macroscopic case requiring perfect match if the incident wave is going to be fully absorbed. The flux patterns of *reflected* single photons exhibit no constrictions towards the sheet like the ones in Fig. 5.5, and are not shown. In the case of complete reflection the forward and backward fluxes are simply equal and uniform over the cross section.

The plots in Fig. 5.5 give rise to vivid visualizations of wavelike to particlelike transition of incident photons administered by the environmental *PDE* fields subject to the principle of time average minimization. In the event that a quantum physicist was exposed to the very classical and realistic stripline configuration of Fig. 5.4 and asked to interpret

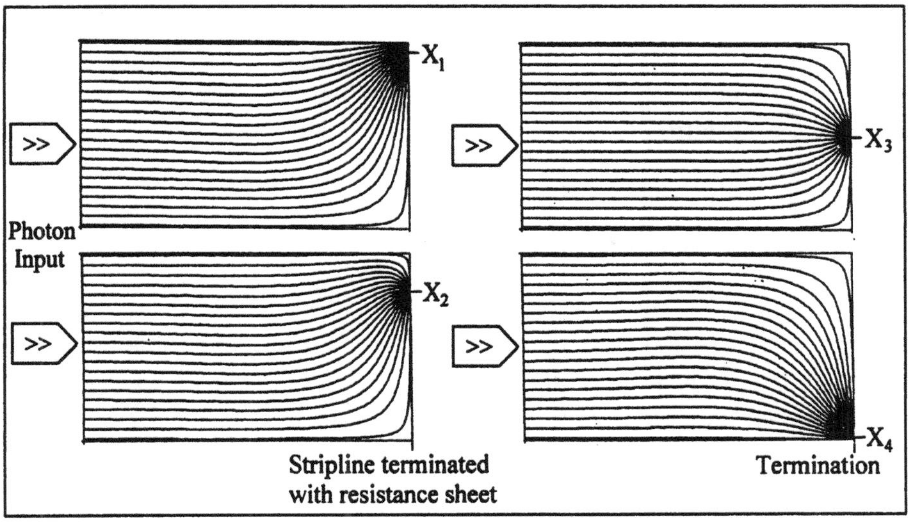

Figure 5.5. Examples of computer generated photon streamlines near the end section of a parallel plate transmission line (stripline) viewed from above. The line is a quarter of a wavelength wide and loaded with a uniform resistance sheet of arbitrary impedance per square. The diagrams show flux lines of single photons routed to four different and equally probable positions $X_1 \ldots X_4$ at the resistance sheet.

the photon behavior in quantum language, he might be compelled to introduce the *duality principle* explaining the photon transition from wave to particle, with the whole process governed by the *correspondence principle.* Perhaps the physicist, if he belongs to the orthodox Copenhagen school, would go even further, claiming that the incident photons propagating down the line really do not exist until they are measured, i.e., registered by the quantum mechanical *collapse principle* into the load impedance at $\bar{X}_q$. Assertions of these kinds would probably be denounced by many quantum physicists as being overly accentuated. Nevertheless, such statements are commonplace in quantum physics although they are more often directed at different experimental situations bearing little resemblance to the stripline configuration.

5.5.6 Verification of zero time average doublet admixture

Before concluding the chapter let us demonstrate that the general prediction of vanishing time average of the doublet engagement in the routing process is satisfied for the strip line. The proof is limited to the simplest case of constant spot impedance $z(X_q)$, which implies *uniform* macroscopic absorption over the sheet, i.e., all microscopic absorbers soak up the same average number of photons. With this specification the time average doublet field in (5.69) is obtained by summing over all the spots $\bar{X}_q$ and all harmonics n. Replacing the sum over $\bar{X}_q$ by an integral we obtain:

$$\langle \kappa_x(\bar{X}, 0) \rangle \sim \sum_{n=1}^{\infty} \int_{-D/2}^{D/2} \cos(B_n \bar{X}_q)\, dX_q = 0 \tag{5.80}$$

showing that the time average of the evanescent doublet supplement vanishes. From (5.57) and (5.65) it is readily observed that also the TEM doublets average out to zero provided the number ratio N_{refl}/N_{trans} corresponds to the macroscopic ratio of reflected and transmitted power.

$$\left\langle \begin{bmatrix} \alpha_0^f \\ \alpha_0^b \end{bmatrix} \right\rangle = 0 \tag{5.81}$$

This equation concludes the proof of vanishing time averages of the photon doublet supplements. This criterion is synonymous with the statement that the average photon numbers absorbed and reflected correspond to the macroscopic absorption and reflection. The two concepts are intimately entwined by the principle of zero time average PDE fluctuations.

The physical picture emerging from the neoclassical theory applied to this simple configuration bears very little resemblance to the classical description of wave propagation in the stripline. In classical electromagnetics we are used to think of the TEM wave hitting the resistance sheet as a plane electromagnetic wave, part of it being uniformly absorbed at the sheet and part of it uniformly reflected. And on the *average,* or macroscopic, level this is also the prediction of the neoclassical theory presented here. But on the microscopic level the picture is entirely different. Each single photon absorbed in the sheet engages an incredibly large number of photon doublet supplements in the form of evanescent modes located at or near the sheet serving the purpose of squeezing the photon from a plane wave to its particlelike costume adapted for absorption at one of the tiny absorbing members such as charge carriers in the resistance sheet. In this scenario the immediate space above the sheet is seething with evanescent PDE electromagnetic fields whose existence is expected to be manifest only by doublet noise i.e., energy fluctuations around the average of zero energy.

The situation envisaged from this simple example would be exactly the same in optical devices where the resistance sheet might be replaced by photosensitive sheets such as the regular film in your photographic camera, or the digital equivalents in newer cameras such as Charge Coupled Devices (CCD) or $CMOS$ transistor sensors. The picture projected on the film or the semiconductor array by the lens is the average of the incoming photon stream, with each photon blackening only one silver atom or engaging one pixel point in the digital camera. Again the process is orchestrated by the PDE environmental fluctuations governed by the principle of zero time average energy. The average spot distribution corresponds to the picture you see on your prints. Without the principle of zero time average PDE fluctuations there would be no organization of the process, no coherent picture, and photography would be dead.

Chapter 6

PHOTONS AND DOUBLETS IN GENERAL NETWORKS

'Original thought
is a straightforward process.
It's easy enough
when you know what to do.
You simply combine
in appropriate doses
the blatantly false
and the blatantly true.'

Grook
by Piet Hein

6.1 CLASSICAL CIRCUIT MODELING

The analysis of networks by the scattering formalism is a standard technique in ordinary electromagnetic theory. In the present chapter much of the same formalism is employed as the basis for diversion into the largely unknown territory of single photon behavior in general networks and circuits. In the treatment we shall relax the generality observed in Chapter 3 by specializing the virtual circuit terminals to well defined and physically separated transmission lines or waveguides. The many circuit configurations in both quantum and macroscopic experiments dealt with in the remainder of the book all belong to this category. The generality of the interior circuit itself, as defined in Chapter 3, is fully retained.

Figure 6.1 shows a schematic of a general circuit represented by an arbitrary enclosure with any number n of independent terminals connected to the inner general circuit. It was demonstrated in Chapter 3 that the electromagnetic properties of the circuit is fully specified by its $n \times n$ scattering matrix $\mathbf{S}(\omega)$. The scattering relation satisfying the classical field equations is given by the familiar formula

$$\mathbf{b}(\omega)_{out} = \mathbf{S}(\omega)\,\mathbf{a}(\omega)_{in} \tag{6.1}$$

which relates the in- and outgoing wave components $\mathbf{a}(\omega)_{in}$ and $\mathbf{b}(\omega)_{out}$ expressed as n-dimensional column vectors. The relation specifies the scattering of input wave components $\mathbf{a}(\omega)_{in}$ into output wave components $\mathbf{b}(\omega)_{out}$. It is basically a macroscopic representation of electromagnetic theory and circuit theory, but fails completely when applied to *single* photons. The simple reason for this failure is the photons' energy integrity, preventing fragmentation into components scattered between terminals. Therefore, single photons do not satisfy the scattering relation (6.1), instead being forced to somehow negotiate their route to one single terminal of their choice. The organization of this routing and distribution of photons to the various terminals by the PDE environmental doublet fields is the theme of this chapter.

Even in its classical form, the basic scatter formula (6.1) is of general stature and needs modification in order to be applicable to the configuration of Fig. 6.1. In particular, this applies to the input column $\mathbf{a}(\omega)_{in}$ which is conveniently split into two components $\mathbf{a}_g(\omega)$ and $\mathbf{a}(\omega)$ according to

$$\mathbf{a}(\omega)_{in} = \mathbf{a}_g(\omega) + \mathbf{a}(\omega) \tag{6.2}$$

In this modified input formulation $\mathbf{a}_g(\omega)$ is a column vector representing *constant amplitude* generators located at some of the terminals, in Fig. 6.1 indicated by arrows. The designation of $\mathbf{a}_g(\omega)$ as constant amplitude generators implies complete independence of the load configuration. They are specified by the details of excitation, with a selected set of terminals engaged from a common coherent source, such as a macroscopic signal, or as discussed later, from a single photon. The constant generator amplitude $\mathbf{a}_g(\omega)$ is related to the generator voltage $\mathbf{V}_g(\omega)$ and the generator current $\mathbf{I}_g(\omega)$ through

$$\mathbf{V}_g(\omega) = \mathbf{Y}_0^{-1/2}\,\mathbf{a}_g(\omega) \qquad\qquad \mathbf{I}_g(\omega) = \mathbf{Y}_0^{1/2}\,\mathbf{a}_g(\omega) \tag{6.3}$$

where $\mathbf{Y}_0$ is the real characteristic admittance matrix of the input terminals. The total generator power P_g is given by

$$P_g = \tilde{\mathbf{V}}_g^*(\omega)\,\mathbf{Y}_0\,\mathbf{V}_g(\omega) = \tilde{\mathbf{I}}_g^*(\omega)\mathbf{Y}_0^{-1}\,\mathbf{I}_g(\omega) = \tilde{\mathbf{a}}_g^*(\omega)\mathbf{a}_g(\omega) \tag{6.4}$$

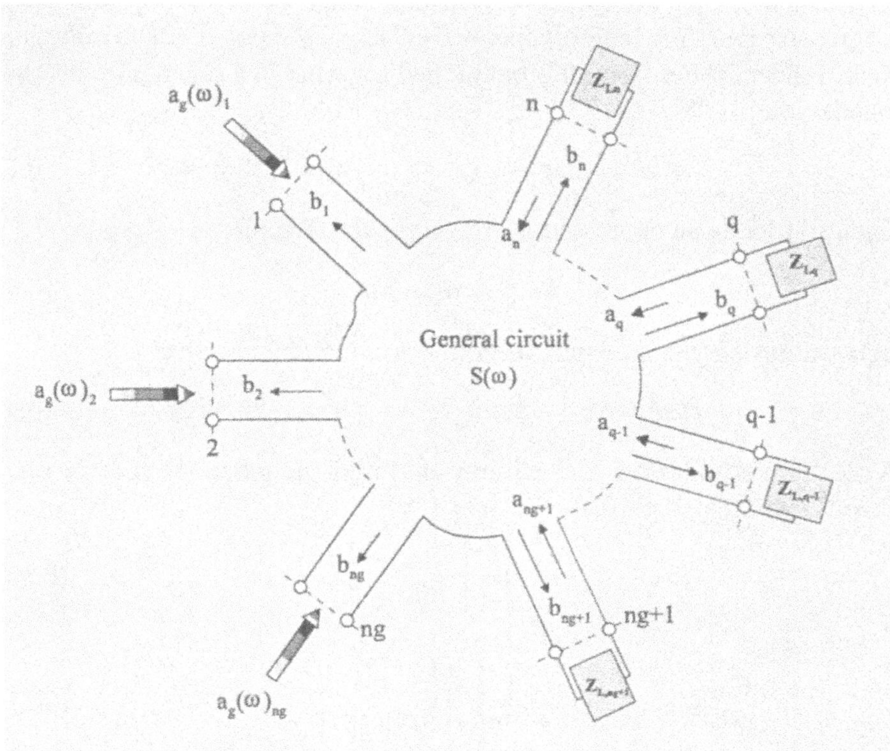

Figure 6.1. **Schematics of a general n-terminal lossless, non-uniform, and reciprocal network analyzed in the text by the *scattering* formulation. The circuit is driven by one single coherent input signal through one or several terminals by constant power generators $a_g(\omega)_1$, $a_g(\omega)_2 ... a_g(\omega)_{ng}$, in the figure represented by shaded arrows.**

Assuming the circuit excited by a coherent input signal $\bar{\mathbf{a}}_g(\omega)$ through the first $n_g < n$ terminals, the column $\mathbf{a}_g(\omega)$ is specified by

$$\mathbf{a}_g(\omega) = \begin{bmatrix} \bar{\mathbf{a}}_g(\omega) \\ \mathbf{0} \end{bmatrix} \tag{6.5}$$

where the truncated generator column $\bar{\mathbf{a}}_g(\omega)$ is of dimension n_g.

The second vector $\mathbf{a}(\omega)$ in (6.2) represents the part of the total input vector $\mathbf{a}(\omega)_{in}$ which is related to the output vector $\mathbf{b}(\omega)_{out}$ through reflections at the terminal loads. If terminal q is loaded with the impedance Z_{Lq} the following familiar relation holds:

$$\frac{a(\omega)_q}{b(\omega)_q} = \rho(\omega)_q = \frac{Z_{Lq} - Z_q}{Z_{Lq} + Z_q} \qquad q = 1, 2, ... n \tag{6.6}$$

where $Z_q = 1/Y_q$ is the characteristic impedance of terminal q, and $\rho(\omega)_q$ is the corresponding macroscopic *reflection coefficient.* Let the reflection coefficients of the n terminals be lumped together in a diagonal reflection matrix $\boldsymbol{\rho}(\omega)$.

$$\boldsymbol{\rho}(\omega) = \text{diag}[\rho(\omega)_q] \qquad q = 1, 2 \ldots n \tag{6.7}$$

Then (6.6) can be expressed by the compact formula

$$\mathbf{a}(\omega) = \boldsymbol{\rho}(\omega)\, \mathbf{b}(\omega) \tag{6.8}$$

By assumption the n_g *input* terminals are matched so that

$$\rho(\omega)_q = 0 \qquad a\,(\omega)_q = 0 \qquad q = 1, 2 \ldots n_g \tag{6.9}$$

With these stipulations the column $\mathbf{a}(\omega)$ and the reflection matrix $\boldsymbol{\rho}(\omega)$ have the truncated forms

$$\mathbf{a}(\omega) = \begin{bmatrix} \mathbf{0} \\ \mathbf{a}(\omega)_L \end{bmatrix} \tag{6.10}$$

$$\boldsymbol{\rho}(\omega) = \begin{bmatrix} \mathbf{0} & \mathbf{0} \\ \mathbf{0} & \boldsymbol{\rho}(\omega)_L \end{bmatrix} \tag{6.11}$$

The column vector $\mathbf{a}(\omega)_L$ is of dimension $n{-}n_g$, and the matrix $\boldsymbol{\rho}(\omega)_L$ of dimension $(n{-}n_g) \times (n{-}n_g)$. As a consequence of these stipulations the generator column $\mathbf{a}_g(\omega)$ and the column $\mathbf{a}(\omega)$ are orthogonal.

$$\tilde{\mathbf{a}}_g(\omega)^* \mathbf{a}(\omega) = 0 \tag{6.12}$$

which turns out to be a prerequisite for the following analytical approach. With the definitions (6.2) to (6.11) the basic scatter relation (6.1) is converted to

$$\mathbf{b}(\omega) = [\tilde{\mathbf{S}}(\omega)^* - \boldsymbol{\rho}\,(\omega)\,]^{-1}\, \mathbf{a}_g(\omega) \tag{6.13}$$

where we have used the unitary condition $\mathbf{S}(\omega)^{-1} = \tilde{\mathbf{S}}(\omega)^*$ satisfied by the scattering matrix. The equation specifies the output column $\mathbf{b}(\omega)$ as functions of the input $\mathbf{a}_g(\omega)$ and the load configuration expressed by $\boldsymbol{\rho}(\omega)$. However, the equation is still not in the desired form because part of the signal represented by the vector $\mathbf{b}(\omega)$ is reflected back into the vector $\mathbf{a}(\omega)$ at the terminal loads. We need a scatter relation between the dissipated power in the loads and the input generator power $\mathbf{a}_g(\omega)$. The power dissipated in the load of terminal q is given by

$$P(\omega)_q = b(\omega)_q^*\, b(\omega)_q - a\,(\omega)_q^*\; a\,(\omega)_q \tag{6.14}$$

It is convenient to define a *power* column $\mathbf{c}(\omega)$ with elements $c(\omega)_q$ specified by

$$c(\omega)_q = b(\omega)_q \,[1 - \rho\,(\omega)_q \; \rho\,(\omega)^*_q]^{1/2} \qquad q = 1, 2 ... n \tag{6.15}$$

and a diagonal *transmission matrix* $\tau(\omega)$ by

$$\boldsymbol{\tau}(\omega) = \begin{bmatrix} 1 & 0 \\ 0 & \boldsymbol{\tau}(\omega)_L \end{bmatrix} = \text{diag}[(1 - \rho\,(\omega)_q \; \rho\,(\omega)^*_q)^{1/2}] \tag{6.16}$$

The power column $\mathbf{c}(\omega)$ is then expressed in compact form by the relation

$$\mathbf{c}(\omega) = \boldsymbol{\tau}(\omega)\, \mathbf{b}\,(\omega) \tag{6.17}$$

or, from (6.13)

$$\mathbf{c}(\omega) = \boldsymbol{\tau}(\omega)\, [\tilde{\mathbf{S}}(\omega)^* - \boldsymbol{\rho}\,(\omega)\,]^{-1}\, \mathbf{a}_g(\omega) \tag{6.18}$$

The n-dimensional column $\mathbf{c}(\omega)$ specifies the square root of the macroscopic power delivered to the n terminal loads as a response to the input power column $\mathbf{a}_g(\omega)$. This scatter equation, which is a special formulation of the basic macroscopic scatter relation (6.1), contains several parameters relating to the input, the circuit, and the loads. The circuit properties are specified by the scattering matrix $\mathbf{S}(\omega)$, and the load by the reflection matrix $\boldsymbol{\rho}\,(\omega)$ and the transmission matrix $\boldsymbol{\tau}(\omega)$. It is readily verified that the total power dissipated in all the loads equals the total input power. The proof goes as follows. From the equation

$$\mathbf{b}(\omega) = \mathbf{S}(\omega)\, [\mathbf{a}(\omega) + \mathbf{a}_g(\omega)] \tag{6.19}$$

we obtain

$$\tilde{\mathbf{b}}(\omega)^*\mathbf{b}(\omega) = \tilde{\mathbf{a}}(\omega)^*\mathbf{a}(\omega) + \tilde{\mathbf{a}}_g(\omega)^*\mathbf{a}_g(\omega) + \tilde{\mathbf{a}}_g(\omega)^*\mathbf{a}(\omega) + \tilde{\mathbf{a}}(\omega)^*\mathbf{a}_g(\omega) \tag{6.20}$$

In view of (6.8) and the orthogonality condition (6.12) the expression converts to

$$\tilde{\mathbf{b}}(\omega)^*\, [\mathbf{1} - \boldsymbol{\rho}(\omega)^*\boldsymbol{\rho}(\omega)]\, \mathbf{b}(\omega) = \tilde{\mathbf{a}}_g(\omega)^*\mathbf{a}_g(\omega) \tag{6.21}$$

Upon substitutions from (6.16) and (6.17) the equation reduces to

$$\tilde{\mathbf{c}}(\omega)^*\mathbf{c}(\omega) = \tilde{\mathbf{a}}_g(\omega)^*\mathbf{a}_g(\omega) \tag{6.22}$$

which verifies the statement.

Up to this point the derivations relate strictly to classical macroscopic electromagnetics. As pointed out in the introduction the macroscopic scattering formulae (6.1) and (6.18) fail totally in predicting single photon behavior. The energy integrity of photons prohibits scattering to presumably independent terminal loads. Defining the terminals as independent means that no coupling exists between their load impedances. If N is the number of photons entering the circuit per unit time, and N_q the number absorbed at terminal q, their ratio is specified by the following relation:

$$\frac{N_q}{N} = \frac{|c(\omega)_q|}{|\mathbf{a}_g(\omega)|} \qquad q = 1, 2, ...n \tag{6.23}$$

Any information on single photon behavior beyond (6.23) is unobtainable from the classical macroscopic scatter relation (6.18). With these preliminaries, relating to the macroscopic case, let us discuss the situation faced by a single photon entering the circuit through the input terminals.

6.2 SINGLE PHOTON PREDICAMENT IN GENERAL CIRCUITS

The photon enters the network as a spatially coherent electromagnetic wave channeled through the n_g input terminals. Regardless of the frequency ω, there is no experimental evidence and no question, even in quantum physics, about the photon not satisfying Maxwell's equations. Let us visualize the photon propagating along the input transmission lines, with $\vec{E}, \vec{H}$ field distributions appropriate to the line geometries. Upon entering the inner circuit the photon is faced with a dilemma concerning its options with regard to terminal selection. The photon's integrity with regard to retaining its energy $\hbar\omega$ forbids the splitting between independent terminals. On the other hand, if the photon is going to satisfy the classical Maxwell's equations, which is equivalent to satisfying the scattering relation (6.18), it simply has to subdivide its energy between the terminals. But the photon is unable to act in this manner because the constituents would have correspondingly lower frequencies compatible with the energy subdivisions. Moreover, such splitting requires a nonlinear or time varying circuit which is not part of the network configuration considered here.

If this subdivision problem is not enough, the photon is faced with another predicament after it somehow has negotiated the routing process to one of the output terminals and is propagating down the terminal guide towards the load impedance. We saw in (6.8) that the load was specified by the macroscopic reflection coefficient $\rho(\omega)$ which is never exactly zero, as any circuit designer can bear witness to. Classical field theory requires a fraction $\rho(\omega)$ of the photon to be reflected back and the rest dissipated in the load. But the photon can not accept this as a possible solution. For the photon it is an either or situation. The reader who has studied Chapter 5 will realize that the present situation is completely covered by the general wave to particle transition theory of that chapter. The reason for dwelling at some length with the quandaries experienced by the photon in regular circuits is merely to establish the inadequacy of classical electromagnetics to handle the stated problems. Let us next ask how quantum physics is treating the problem.

6.2.1 Quantum mechanical interpretation

The stated problems are well recognized in quantum physics, although perhaps more often in reference to photons in the optical range. But the problems remain the same. The explanation is linked to the concept of duality or complementarity. The photon is considered a dual entity, sometimes a particle and sometimes a wave satisfying Maxwell's equations like any electromagnetic field. A typical explanation offered by a quantum physicist might be the following: The photon, upon entering the network interior through the input guide, transforms itself from a regular electromagnetic wave to its complementary photon particle. The particle finds itself perhaps bouncing back and forth a few times between the circuit walls until it hits one of the open terminal apertures, at which time it transforms itself to a wave propagating down the guide towards the load. Before hitting the load it transforms back to its particle costume in order to be absorbed by one of the microscopic charge carriers constituting the load impedance. In this explanation there certainly are a lot of costume changes taking place, all in accordance with the principle of complementarity. But there are more quantum principles involved. The quantum physicist must explain the experimental fact that the scattered distribution of a large number of photons between the n terminals corresponds to the *macroscopic* scattering relation (6.1). He gets away with this problem by invoking Bohr's *correspondence* principle, that the ensemble average, or the time average, of a quantum process must correspond to its macroscopic representation.

This perhaps oversimplified picture is essentially the quantum explanation although it is often formulated in more abstract language describing the photon by its state function Ψ specified by solutions of the Schrødinger's wave equation. The state function evolves as a function of time in a unitary manner into a superposed state consisting of all the alternatives open to the photon in the specified system configuration. Expressed in the usual bra-ket formulation the state function has the form

$$|\Psi >= \sum_q b_q |q > \qquad (6.24)$$

According to this picture the photon exists in a superposed state, with $|b_q|^2$ representing probabilities for the photon to exit from terminal q. Now, for any given network the Schrødinger equation is completely unable to determine the probability factors b_q in (6.24) because the equation does not basically account for relevant electric and magnetic boundary conditions in the electromagnetic circuit. As a substitute for this lack of information the quantum *correspondence principle* asserts that the probabilities must be compatible with macroscopic field theory. In the present case this simply means that the coefficients $b_q s$ in (6.24) are the same as the $b's$ in the macroscopic scatter relation (6.1) or perhaps the $c's$ in (6.18). But there are more quantum 'principles' involved having to do with the exit of the photon from terminal q into the load. It is an experimental fact that the photon dumps its full energy at one minute single atomic or charge absorber out of a very large number of absorbers constituting the overall load, say a simple flat resistive plate terminating the waveguide terminal. The quantum physicist would have to expand the number of states of the photon to account for all the possible tiny absorbers on the plate, describing the process itself as a reduction of this enlarged state function Ψ to the single state described by the particular spot selected. Quantum physics refers to the process as state function 'collapse' to its macroscopic level, another principle that is a major controversial problem in quantum physics. The state function collapse is intimately related to 'the measurement problem' attempting to bridge or mediate the deterministic development of the state function taking place prior to the collapse with the empirical observations at the macroscopic level. It is an adjunct to the correspondence principle, and as a concept it merely illustrates the ineptness of quantum physics to provide a satisfactory description of the world around us. For a more comprehensive discussion of these topics in the context of the neoclassical theory the reader is referred to Chapters 5 and 14.

6.2.2 Classical electromagnetics—at a loss

In fairness to all, the understanding of photon behavior is not much better at the classical electromagnetics level. Standard electromagnetic theory is unable to explain what happens to single photons in regular circuits. It certainly is an accurate statement that the art of microwave and low frequency network designs has advanced to a high level of perfection with regards to their *macroscopic* electromagnetic performance, as witnessed by the jungle of electronic gadgets and highly advanced systems surrounding us in everyday life. But if you ask typical system designers about the photons in their circuits, pointing out to them the problem of single photon behavior, they are at a loss. Questions such as these are entirely foreign in regular circuit theory, as they are in regular classical electromagnetic theory. Problems of this nature are not addressed, simply because they are not needed in circuit designs based exclusively on the macroscopic field equations and their many approximate formulations. For this reason they represent unimportant nuisances that are entirely irrelevant and therefore just as well forgotten. The circuit designers might respond to your question with phrases such as: "Photons, isn't that something having to do with optical frequencies—we never saw anything like photons in our circuits". The truth of the matter is of course that photons are the basic entities of electromagnetic phenomena at *all* frequencies, even down to *DC*, i.e., direct current. There are no theoretical or experimental evidence predicting a lower limit of photon frequency. Therefore, macroscopic circuit theory is facing exactly the same quandary as quantum physics in its ineptness of explaining photon behavior in regular circuits. From the neoclassical viewpoint the fundamental reason for their failure is that both are basically causal ensemble theories and as such unable to handle single events properly.

6.3 NEOCLASSICAL THEORY OF SINGLE PHOTONS IN NETWORKS

The concept of photon doublets was introduced in earlier chapters as a consequence of extending classical electromagnetics to bidirectional time. In the following sections we shall demonstrate explicitly how the photon doublets provide the required complements to regular circuit theory. In the general theory of Chapter 2 and 3 we saw that the doublets have the peculiar property that they carry no energy, but are still associated with electromagnetic fields of frequency ω and $-\omega$, each frequency component satisfying Maxwell's equations. We also ventured the proposition that the doublet fields encompass all space as a photon doublet energy (*PDE*) distribution, in equilibrium with all specific circuit configurations that we can think of, in particular the general circuit configuration of Fig. 6.1. The relevant parts of the general neoclassical theory needed for the present discussion are the scattering formulations for positive and negative frequencies.

For positive frequencies:

$$\mathbf{b}(\omega) = \mathbf{S}(\omega)\,\mathbf{a}(\omega) \tag{6.25}$$

For negative frequencies:

$$\mathbf{b}(-\omega) = \mathbf{S}(-\omega)\,\mathbf{a}(-\omega) \tag{6.26}$$

As shown in Chapter 3 the scattering matrix $\mathbf{S}(-\omega) = \mathbf{S}(\omega)^{-1}$. Therefore, upon inversion the equation takes the form

$$\mathbf{a}(-\omega) = \mathbf{S}(\omega)\,\mathbf{b}(-\omega) \tag{6.27}$$

In comparison to the basic scatter relation (6.25) the formal interchange of input and output variables expressed by (6.27) is a direct consequence of the negative energy carried by the negative frequency components. The input $\mathbf{a}(-\omega)$ means negative energy fed *into* the circuit which is equivalent to positive energy fed *out* of the circuit. The proper *doublet* scatter relations are formed by adding and subtracting (6.25) and (6.27).

$$\mathbf{b}(\omega) + \mathbf{a}(-\omega) = \mathbf{S}(\omega)\,[\mathbf{a}(\omega) + \mathbf{b}(-\omega)] \tag{6.28}$$

$$\mathbf{b}(\omega) - \mathbf{a}(-\omega) = \mathbf{S}(\omega)\,[\mathbf{a}(\omega) - \mathbf{b}(-\omega)] \tag{6.29}$$

The energy relations are further illuminated by forming the equivalent of the expression $\tilde{\mathbf{b}}(\omega)^*\mathbf{b}(\omega) - \tilde{\mathbf{a}}(\omega)^*\mathbf{a}(\omega)$ of the regular scatter case making use of the unitary properties of the scattering matrix.

$$\tilde{\mathbf{b}}(\omega)^*\mathbf{b}(\omega) - \tilde{\mathbf{b}}(-\omega)^*\mathbf{b}(-\omega) = \tilde{\mathbf{a}}(\omega)^*\mathbf{a}(\omega) - \tilde{\mathbf{a}}(-\omega)^*\mathbf{a}(-\omega) \tag{6.30}$$

The left side is the total power *out* of the circuit, the right side the total power *into* the circuit. The doublet scatter relations (6.28) and (6.29) hold for any relative excitations of positive and negative frequency components. However, the expected excitations from the environmental *PDE* fields are of the pure doublet types, earlier referred to as doublet resonances, and these are specified by zero inputs and outputs. The *current doublet* excitation (zero electric field) corresponds to putting the input and output in (6.28) equal to zero

$$\mathbf{b}(\omega) + \mathbf{a}(-\omega) = \mathbf{a}(\omega) + \mathbf{b}(-\omega) = \mathbf{0} \tag{6.31}$$

The *voltage doublet* excitation (zero magnetic field) corresponds to vanishing input and output in (6.29).

$$\mathbf{b}(\omega) - \mathbf{a}(-\omega) = \mathbf{a}(\omega) - \mathbf{b}(-\omega) = \mathbf{0} \tag{6.32}$$

For both kind of resonances the total input and output power vanish but the *fields* are nonzero. In Chapter 3 attention was called to the interpretation of (6.31) and (6.32) as resonances in bidirectional time, resonant phenomena that in no ways depend on the particulars of the scattering matrix $\mathbf{S}(\omega)$. This unique feature has the expected and far reaching consequences that the doublet resonances can be excited in *any* circuit at *any* frequency, serving as supplementary field entities in the overall specifications of boundary conditions for photon behavior. This fundamental role of the doublet resonances applies for microwave and low frequency as well as optical circuits. In particular, it is applicable to familiar experimental quantum configurations, treated by the same formalism in some of the subsequent chapters.

The sketch in Fig. 6.2 serves to illustrate the excitation of an arbitrary two-port circuit by doublet resonances satisfying (6.31) or (6.32). The signal paths inside the circuit of the positive and negative frequency components of the doublet are indicated separately. The resonance can be visualized as a round transit inside the circuit. The positive component $a(\omega)_1$ propagates from terminal 1 causing the output $b(\omega)_2 = S(\omega)a(\omega)_1$ at terminal 2. This output is reflected into $a(-\omega)_2$ which propagates back, causing the output $b(-\omega)_1 = S(-\omega)a(-\omega)_2 = S(\omega)^{-1}\,a(-\omega)_2$ at terminal 1, where it reflects into $a(\omega)_1$ closing the resonant loop. The circuit can be excited in these doublet resonances at any frequency supported by the circuit configuration.

The physical picture evolving from this exposition describes any network as seething with electromagnetic fields associated with the doublet resonances in equilibrium with the environmental *PDE* fields of surrounding space. The corresponding nonzero fields exist at all frequencies supported by the network in accordance with its scattering matrix $\mathbf{S}(\omega)$, have zero time average energy and are therefore not observable by direct measurement. The resonance conditions (6.31) and (6.32) ensure complete independence from terminal load conditions. The doublet resonance forward and backward constituents in all terminal guides are always equal, irrespective of the load. What this means is that the usual reflection condition expressed by (6.8) does not apply to the doublets. The doublets are insensitive to load conditions; they can be created or annihilated spontaneously at any spatial location as the need arises. It is interesting to observe that the phenomena appear to bear a certain conceptual resemblance to certain processes in quantum theory, described by creation and annihilation operators.

In the sections to follow we shall demonstrate that the doublets consistently provide the lacking mechanism discussed in the introductory part for routing the incoming stream of photons to the various terminals and, as an additional service, let the photon be fully absorbed at the load with no energy being reflected, regardless of load impedance.

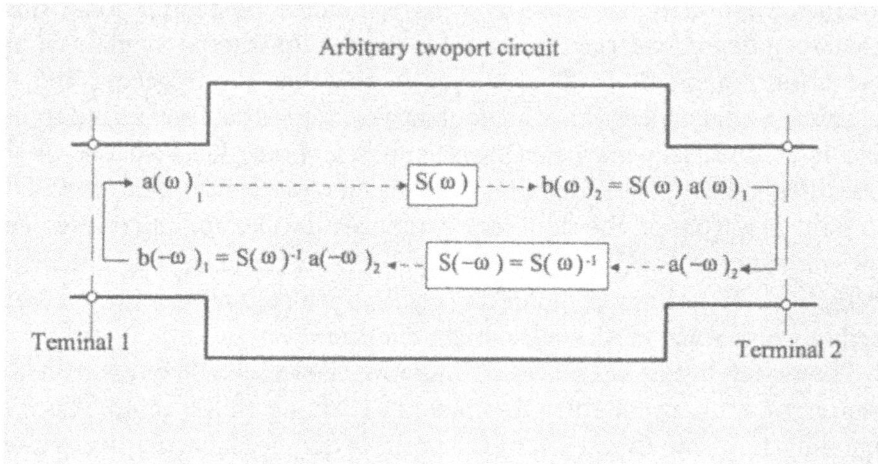

Figure 6.2. Illustration of a photon doublet resonance in an arbitrary twoport circuit. The net doublet inputs at both terminals satisfy the resonance condition (6.31) or (6.32). The positive and negative frequency photons are subject to *total* reflections at the terminals regardless of frequency because the resonance occurs in bidirectional time.

6.3.1 Routing of photons in the network

The time average fraction of photons routed to each terminal was derived from the macroscopic scattering matrix, according to (6.18) and (6.23). But these equations provide no explanation for the stated subdivision beyond the act of a mere number juggling based on correspondence with the macroscopic scattering relation. In this respect classical electromagnetics sides with the *correspondence* principle in quantum mechanics in its inability to provide a detailed physical explanation of the process.

The explanation offered by the neoclassical theory involves photon doublets supplementing the regular input generators $\mathbf{a}_g(\omega)$ with a set of doublet input generators as shown in Fig. 6.3. The generators are of a type providing composite inputs $\boldsymbol{\alpha}_g(\omega, -\omega)$ in the form

$$\boldsymbol{\alpha}_g(\omega, -\omega) = \begin{bmatrix} \alpha_{g1} \\ \alpha_{g2} \\ \vdots \\ \alpha_{gn} \end{bmatrix} = \boldsymbol{\alpha}(\omega) + \boldsymbol{\beta}(-\omega) \tag{6.33}$$

Let a single photon entering the network through one or several terminals be represented by the generator column $\mathbf{a}_g(\omega)$. In view of the fact that the composite doublets satisfy the scatter relations (6.28) and

(6.29) the regular photon input column $\mathbf{a}_g(\omega)$ and the composite but unknown doublet column $\boldsymbol{\alpha}_g(\omega)$ can be added directly. Instead of (6.18) the corresponding composite response $\mathbf{c}(\omega,-\omega)$ is given by

$$\mathbf{c}(\omega,-\omega)=\boldsymbol{\tau}(\omega)\,[\tilde{\mathbf{S}}(\omega)^{*}-\boldsymbol{\rho}(\omega)\,]^{-1}\,[\mathbf{a}_g(\omega)+\boldsymbol{\alpha}_g(\omega,-\omega)] \tag{6.34}$$

The basic procedure for resolving the subject matter is to determine the specific doublet $\boldsymbol{\alpha}_g(\omega,-\omega)]$ needed for routing a photon to a given terminal. Inversion of the equation results in the desired formula:

$$\boldsymbol{\alpha}_g(\omega,-\omega)=[\tilde{\mathbf{S}}(\omega)^{*}-\boldsymbol{\rho}(\omega)\,]\boldsymbol{\tau}(\omega)^{-1}\,\mathbf{c}(\omega,-\omega)\;-\mathbf{a}_g(\omega) \tag{6.35}$$

If the incoming photon is to be routed to terminal q, with no power to the remaining n–1 terminals, the form of the power column $\mathbf{c}(\omega,-\omega)$ must be

$$\mathbf{c}(\omega,-\omega)^{q}=\begin{bmatrix}0\\ \vdots\\ c\,(\omega,-\omega)^{q}\\ \vdots\\ 0\end{bmatrix} \tag{6.36}$$

The required doublet generator supplement $\boldsymbol{\alpha}_g(\omega,-\omega)^q$ achieving the routing of the photon to terminal q is obtained by insertion of this column into (6.35).

$$\boldsymbol{\alpha}_g(\omega,-\omega)^{q}=[\tilde{\mathbf{S}}(\omega)^{*}-\boldsymbol{\rho}(\omega)\,]\boldsymbol{\tau}(\omega)^{-1}\begin{bmatrix}0\\ \vdots\\ c\,(\omega,-\omega)^{q}\\ \vdots\\ 0\end{bmatrix}-\mathbf{a}_g(\omega) \tag{6.37}$$

Of the total of N photons per second entering the network let N_q be routed to terminal q, each of them requiring the same doublet generator admixture $\boldsymbol{\alpha}_g(\omega,-\omega)^q$. The cumulative doublet excitation is the sum of the individual excitations.

$$\boldsymbol{\alpha}_g(\omega,-\omega)^{q}_{tot}=N_q\boldsymbol{\alpha}_g(\omega,-\omega)^{q} \tag{6.38}$$

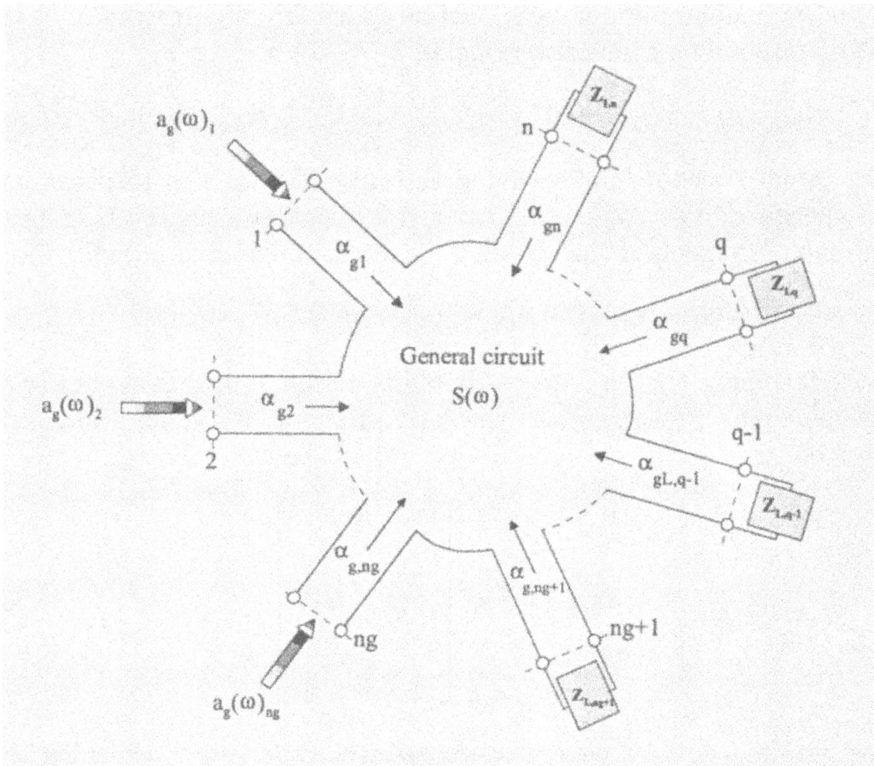

Figure 6.3. **Sketch showing the set of photon doublet generators $\alpha_{g1}...\alpha_{gn}$ supplementing the regular photon input generators $a_g(\omega)_1...a_g(\omega)_{ng}$ at the network terminals.**

yielding

$$\boldsymbol{\alpha}_g(\omega,-\omega)^q_{tot} = [\tilde{\mathbf{S}}(\omega)^* - \boldsymbol{\rho}(\omega)]\boldsymbol{\tau}(\omega)^{-1} \begin{bmatrix} 0 \\ \vdots \\ N_q\, c(\omega,-\omega)^q \\ \vdots \\ 0 \end{bmatrix} - N_q\, \mathbf{a}_g(\omega) \tag{6.39}$$

Through addition of the corresponding doublet admixtures for all the remaining N–N_q photons routed to the other terminals, the following equation is obtained:

$$\sum_{q=1}^{n} N_q \boldsymbol{\alpha}_g(\omega,-\omega)^q = [\tilde{\mathbf{S}}(\omega)^* - \boldsymbol{\rho}(\omega)]\boldsymbol{\tau}(\omega)^{-1} \begin{bmatrix} N_1\, c(\omega,-\omega)^1 \\ \vdots \\ N_q\, c(\omega,-\omega)^q \\ \vdots \\ N_n\, c(\omega,-\omega)^n \end{bmatrix} - \sum_{q=1}^{n} N_q\, \mathbf{a}_g(\omega) \tag{6.40}$$

In interpreting this formula let us assume, for a moment, that the photon numbers N_q differ somewhat from the expected macroscopic averages, writing

$$N_q = N_{q,av} + \Delta N_q \qquad q = 1, 2, \ldots n \tag{6.41}$$

Because the total photon number N is retained we must take

$$\sum_{q=1}^{n} \Delta N_q = 0 \tag{6.42}$$

Upon substitution into (6.40) and dividing by N the equation transforms to

$$\sum_{q=1}^{n} \frac{N_q}{N} \boldsymbol{\alpha}_g(\omega,-\omega)^q = [\tilde{\mathbf{S}}(\omega)^* - \boldsymbol{\rho}(\omega)]\, \boldsymbol{\tau}(\omega)^{-1} \begin{bmatrix} \frac{\Delta N_1}{N} c(\omega,-\omega)^1 \\ \vdots \\ \frac{\Delta N_q}{N}\, c(\omega,-\omega)^q \\ \vdots \\ \frac{\Delta N_n}{N} c(\omega,-\omega)^n \end{bmatrix} \tag{6.43}$$

Compatibility with macroscopic relations require all the ΔN_q to be zero. Hence, the following conclusion must be drawn:

$$\boldsymbol{\alpha}_g(\omega,-\omega)_{tot} = \sum_{q=1}^{n} \frac{N_q}{N} \boldsymbol{\alpha}_g(\omega,-\omega)^q = \mathbf{0} \tag{6.44}$$

Thus, the time average of the doublet generators vanishes, a result that is in full accord with the formal structure of the neoclassical theory expressed by (1.1)-(1.2) of Chapter 1.

The routing of the photons to specific terminals, as described in this section, is just the first step in the overall routing process. In the subsequent step the photon routed to a specific terminal must undergo the 'wave to particle' transition described in Chapter 5 in order to deliver its full energy to a singular point on the plate forming the terminal load. The enormous number of minuscule absorbing entities form a subgroup of independent terminals, one group for each of the n regular terminals. For details on this additional routing process the reader is referred to Chapter 5.

The conceptual picture emerging from the described process is a scenario with photon doublets supplementing the regular input photons so that each photon is routed to one single terminal only, allowing its full energy to be dumped at the terminal load. The photon doublets are fluctuations in the normally neutral environmental PDE fields, which orchestrate the routing process in a manner that maintains zero time average doublet energy. This modus operandi implies the existence of a *memory* process providing information to the next photon in line about the cumulative past history of the preceding ones with regard to terminal selections. The guiding principle is to route the incident photon to the particular terminal that minimizes the overall doublet energy expenditure.

The neoclassical theory on photon routing presented here links Bohr's quantum mechanical *correspondence principle* directly to a physical phenomenon, namely the minimization of the time average of the environmental PDE fluctuations. Macroscopic phenomena involving photons "happen" the way they do because no average environmental energy is permissible. The stationary state of macroscopic phenomena is guaranteed through the direct relationship with a principle of energy minimization, which inherently is insensitive to minor perturbations around the minimum.

In this neoclassical theory it is apparent that the vanishing of the time average photon doublet energy appears as a fundamental general principle in electromagnetics at the photon level. Implicit in this process is the occurrence of fluctuations of photon doublet energy around the zero energy equilibrium. The fluctuations appear at the terminals as photon induced noise superimposed on the signal, and is an unavoidable accomplice generated in the routing process. Terminal noise excitation is a well recognized phenomenon in quantum physics, where it is attributed to spontaneous fluctuations of the zero point energy (ZPE) of the vacuum space. The consensus amongst quantum physicists is that the fluctuations have no classical explanation following from Maxwell's equations. The neoclassical theory is at variance with this thesis. Not only are the fluctuations derived directly from Maxwell's equations, admittedly with the extension to bidirectional time and assuming photon energy integrity, but the theory also relates the fluctuations directly to photon routing, a process having no immediate explanation in orthodox quantum physics.

6.3.2 Concept of independent network terminals

The presentations in this chapter repeatedly stressed the importance of *independent* terminals or loads, a concept intimately tied in with photon energy integrity. In a linear system a photon is prohibited from splitting its energy between several independent loads with each part contributing to the total photon energy $\hbar\omega$. If such a process were allowed to happen, a particular terminal with its load would receive some fraction of the photon energy $\hbar\omega$, but the frequency would still be ω. Clearly, in a linear system this is an impossibility. If photons are going to split into constituents of lower frequencies proportional to their energies, a nonlinear or a time varying system is a prerequisite. Even if the theory of such processes are beyond the scope of this book, they should not be discounted as entirely unimportant for the major topics that we are dealing with. With reference to the discussion on spectral energy distribution in connection with (3.69)-(3.72) in Chapter 3, it can be speculated that various nonlinear and time varying processes are the main players in the process of establishing a minimum of the environmental doublet energy fluctuations, in Chapter 3 characterized by spectral *rms* energy being inversely proportional to frequency.

The definition and explanation of what is meant by independent terminals seem clear enough. However, in practical circuit configurations the concept is not always that simple. As an illustration of some of the difficulties which might be encountered, Fig. 6.4 shows two circuit configurations both behaving in accordance with the present theory. In circuit (a) the single photon entering at terminal 1 is routed to one of the independent terminals 1, 2, or 3. Routing to terminal 1 simply means reflection. Each incident photon goes to one terminal only, propagates down the terminal guide, and dumps its energy at its load. During this process the regular electromagnetic fields are zero in the remaining terminal guides.

In Fig. 6.4b the same circuit is embedded with an extra network, increasing the number of terminals from three to an arbitrary selection of six, which could be extended to any number. For this composite circuit the present routing theory is of course still valid. But now an incident photon at terminal 1, routed to one of the terminals 2 to 6 has to propagate through the *old* terminals 2" and 3". In circuit (*a*) the photon goes to *either* terminal 2 *or* terminal 3, but never through both. In circuit (*b*) the same photon behaves quite differently. The old terminals 2" and 3" inherited from circuit (*a*) are now part of the *internal* structure of the

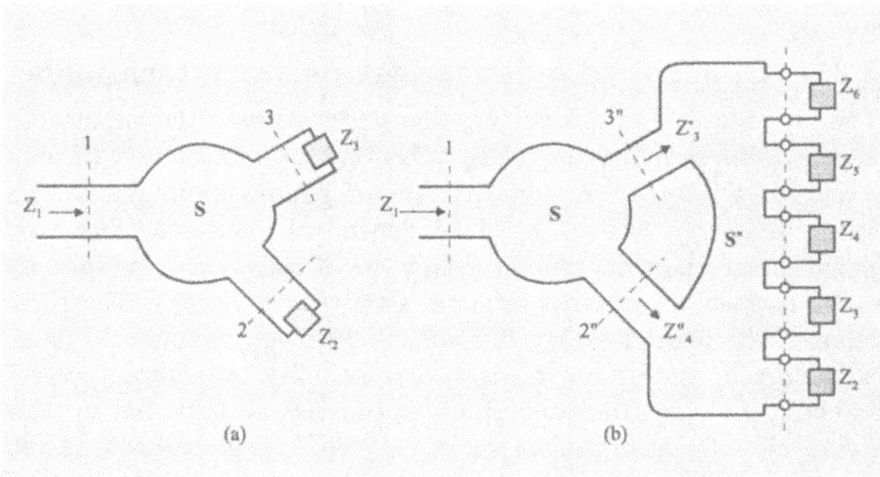

Figure 6.4. **(a) Simple three-terminal network excited by single photons. The incident photons at terminal 1 are routed to the independent terminals 2 or 3, or reflected back to terminal 1. (b) Embedded new network with 6 independent terminals. The photons are routed to terminals 1, 2...6. The old terminals 2" and 3" are not independent even if they are physically separated.**

new circuit. The two terminals are coupled together by the right-hand add-on circuit, and do not any longer represent independent terminals, in spite of the fact that the left part of circuit (*b*) looks *identical* to circuit (*a*). In this new configuration the photon has no qualms splitting its energy between the two old terminals 2" and 3" in its flight to one of the presumably independent right hand terminals 2 to 6.

This example of photon behavior in a regular microwave or low frequency network, stressing the concept of independent terminals, has relevance far beyond this simple configuration. It touches a fundamental schism between the quantum theory and the neoclassical theory. If the configuration in Fig. 6.4 were a "Gedanken" experiment or a real optical quantum experiment on photons, optical counters placed at all the terminals would presumably register the sequential arrivals of photons at the respective terminals. Additional coincidence counters would show that each photon dumps its energy at one single terminal, not splitting it between terminals. Observation over an extended period of time would verify the macroscopic averages, a fact that quantum physics ascribes to the correspondence principle, which is simply what it says here, a statement of fact with no real physics behind it. Quantum physics interprets the failure of photons to split their energy between terminals as an indication of the particle nature of photons. And this interpretation fits well with the simple configuration of Fig. 6.4a. But the particle interpretation runs into serious trouble in the embedded circuit of Fig. 6.4b, first of all the "which way" problem. Does the photon go through terminal 2" or terminal 3" in its flight towards the right? And if this puzzle is not enough, the photon really has to go through *both* terminals in order to produce the observed average photon distribution between the right hand terminals. Quantum physics ends up with a paradoxical situation that can not be resolved in objective terms within the theoretical framework of quantum mechanics.

The observant reader might recognize these same questions from a famous and much celebrated configuration on optical photons, namely the *double slit* experiment. And no wonder he does, because the circuit in Fig. 6.4b *is* the double slit configuration, albeit in slightly different costume. The two guides 2" and 3" represent the two slits in the slitplate. The terminal loads $Z_2 ... Z_6$, which could be extended to any number, represent the presumably independent microcrystals on the screen located behind the slitplate, where the photons dump their energy, each photon at a single crystal. The macroscopic response column $\mathbf{c}(\omega)$ in (6.18) specifies the equivalent macroscopic *diffraction pattern*, i.e., the over time build up of blackened spots on the screen.

If the reader is not familiar with the double slit experiment for photons, which is discussed at length in Chapter 7, he may perhaps recognize the posed questions from other experiments in quantum physics built around *beam splitters* and the Max-Zehnder interferometer discussed in Chapter 12. At this point the equivalence with the general network configuration of Fig. 6.4b is merely pointed out. For more comprehensive analyses the reader is referred to the respective chapters. Because both these optical circuits—the double slit configuration and the Max-Zehnder device—are special cases of the general circuits of the present chapter, the general results from the neoclassical theory are applicable to these special devices. In particular, we shall find that the paradoxical behaviors attributed to single photon experiments in both these devices are absent.

6.3.3 Microscopic causality gone

Let us return to the two networks in Fig. 6.4 and ask a very relevant question that was touched upon earlier. How can a single photon entering terminal 1 make up its mind with regard to its subsequent behavior relative to the two terminals 2 and 3, which for a photon entering from the left look exactly the same in both networks? More specifically, how can the photon know that the terminals are independently loaded in

Fig. 6.4a and therefore insist on going through *either* terminal 2 *or* terminal 3, whereas the photon happily divides its energy between the two terminals 2" and 3" in Fig. 6.4b? Questions like these are commonplace in quantum physics which is unable to provide physically acceptable answers but typically ascribes the observations to some kind of weird behavior. Almost any quantum mechanical experiment on single entity events run into interpretational difficulties which regularly are ascribed to the paradoxical nature of the microcosm. "We simply have to accept that the microcosm is paradoxical. That is how the world is" might be a typical statement by quantum physicists and philosophers concerned with interpretational problems. In Chapter 1 the conjecture was put forth that single quantum events can not be properly interpreted in objective terms by any causal theory such as quantum mechanics and classical electromagnetics. The interpretations offered by the neoclassical theory are founded on the hypothesis of a bitemporal microcosm in which the new composite photon entities—the photon doublets—contain positive as well as negative time components or, equivalently, positive and negative frequencies and energies. The doublets certainly violate classical electromagnetism because they violate causality, but they do not violate Maxwell's field equations which are symmetric in time. The neoclassical theory accepts that causality is violated in the microscopic world at the photon level.

For the uninhibited observer there appears to be ample experimental evidence of causality violation in the microworld, and numerous indirect evidence through the capability of photon doublets to explain seemingly paradoxical observations in quantum experiments and in regular circuit theory, as exemplified in this and forthcoming chapters. Specifically, the organization of photon routing to independent terminals is described adequately by environmental field supplements consisting of photon doublets with positive as well as negative time components. It is clear that microscopic causality is violated in these processes. However, the success of doublets explaining phenomena such as photon wave to particle transition, photon routing, and many more to be demonstrated in the forthcoming chapters provides strong support for the neoclassical bitemporal theory.

The scenario evolving in this causality violated microworld is quite different from the world of standard quantum physics. In this new kind of microworld, at the photon level, the overall field pattern is described as solutions of Maxwell's equations in bidirectional time, subject to a complete set of boundary conditions in space and bidirectional time. The concepts of cause and effect loose their usual meanings. In such a world it is meaningless to ask questions like: "how can the photon know which way to take, etc.", nor does it make sense to design experiments of the 'delayed choice type', quite common in quantum physics, with the hope of fooling the photon. We must look upon such experiments with new and fresh eyes discarding the causality restriction inherited from the macroworld. This new look, in fact, is the very core of the material presented in the following chapters of this book.

Chapter 7

THE DOUBLE SLIT EXPERIMENT FOR PHOTONS

Two roads diverged in a yellow wood
and sorry I could not travel both
and be one traveller, long I stood
And looked down one as far as I could
To where it bent in the undergrowth..

Two roads diverged in a wood, and I–
I took the one less travelled by,
And that has made all the difference.

Robert Frost
in
'The Road Not Taken'

7.1 INTRODUCTION—THE PARADOXES

The double slit experiment plays a central and famous role in quantum mechanics. It portraits a behavior of quantum entities pointing to a dualistic interpretation of nature. The quantum entities are photons and electrons, both showing the same paradoxical conduct. Figure 7.1 shows the experimental set up for photons. A coherent optical beam impinging on a plate with two narrow slits is partly reflected and partly transmitted through the slits. The transmitted part forms a typical diffraction pattern on a screen located behind the slit plate, in perfect agreement with regular optical diffraction theory and classical electromagnetism.

The problem arises when the intensity of the incident radiation is gradually reduced until only one single photon at a time arrives at the slit plate. Such a condition can be realized experimentally and is not a theoretical construct. Under these circumstances it is observed that the well separated individual photons dump their energy at separate localized spots on the photosensitive screen. This observation, quoted in any textbook on quantum physics, certainly projects the notion of a particlelike behavior of the photon. On the other hand, as more and more photons are arriving at the screen, the accumulated set of spots builds up to the typical diffraction pattern shown in Fig. 7.1. And this experimental result, being in full agreement with optical theory based on wave propagation, certainly points to the photon as a wave. These contradictory statements—the photon behaving as both a particle and a wave—are the source of the celebrated dilemma in quantum physics, and a major contribution to the launching of the duality principle. This principle interprets the photon as a dual particle, behaving sometimes as a wave and sometimes as a particle, depending on circumstances. The argument goes something like this: the photon has to posses wave nature in order to pass through both slits, which is necessary in order to produce the observed diffraction pattern. On the other hand, the experiment shows that each photon arrives at the screen as a localized pointlike entity producing single and well separated black spots on the screen, thereby indicating a particlelike behavior. If the photon is a particle, it must have gone through one of the two slits, in clear contradiction to the first statement. This is the celebrated 'which way' paradox. Does the photon, now as a particle, go through the first or the second slit? And this problem is accentuated with the additional paradox that the particle really has to go through both slits in order to produce the diffraction pattern. This paradox is a major dilemma in quantum physics, and lots of ink has flowed in futile attempts to resolve the paradox.

7.1.1 Neoclassical interpretation

The experimental fact that the diffraction pattern at the screen is built up of tiny single blackened spots indicating arrival and conversion of photon energy at single spots rather than a continuous distribution should not be the least surprising. It is a consequence of photon energy integrity, an experimental fact repeatedly stressed throughout the book. An incident photon can not spread its energy amongst several independent silver atoms on the screen, which in the parlance of this book represent *independent* terminals of the circuit formed by the slit plate and the screen. The photon simply has no other choice than to dump its energy at individual silver atoms. Hence, the photon, whether wave or particle, is bound to appear at the screen in the same costume, as tiny spots indicating the positions of individual photon arrivals.

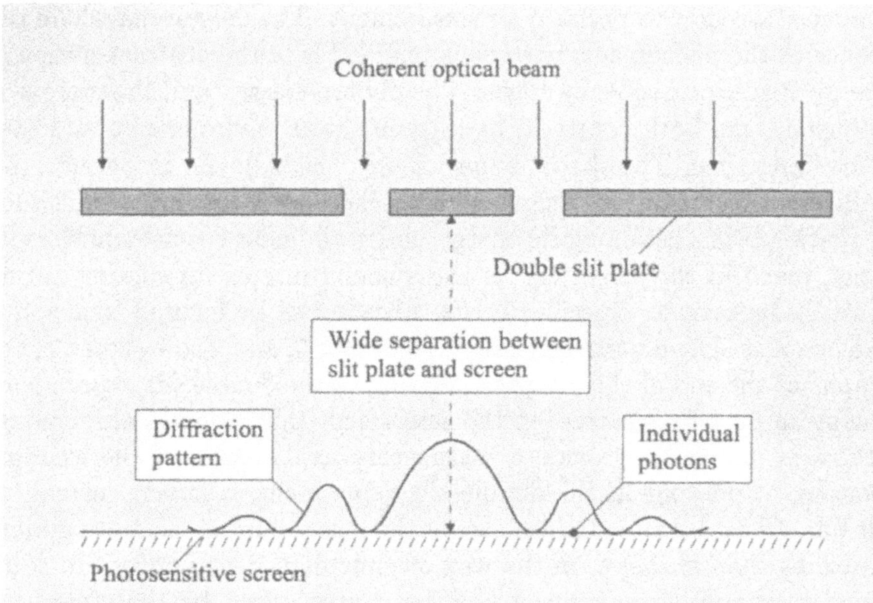

Figure 7.1. **Basic two slit configuration for photon experiments, with diffraction pattern formed at the photosensitive screen.**

Perfect spotlike focusing of individual wave type photons has, for whatever reason, been considered foreign to the wave nature. But in a general sense this notion is incorrect because Maxwell's equations predict no finite diffraction limit, only practical limits due to various constraining parameters. The neoclassical theory of Chapter 5 on wavelike to particlelike photon transition certainly adds credence to this point. Thus, there is absolutely no sound logical reason to link a particlelike behavior at the screen to the implication of a particlelike photon behavior during its flight through the slits towards the screen. In fact, there does not appear to exist any experimental observations supporting a particlelike photon in situations other than during excitations (in radiation from atoms) or during conversion to other energy forms. The reason for the absence of measurements on photons during flight is quite fundamental and tied in with photon energy integrity. The spatial distribution of the photon can not be measured for the simple reason that transfer of information requires transfer of at least a tiny amount of energy. The energy must be extracted from the photon, which in the process is modified to a constituent with slightly lower frequency corresponding to the small loss of energy. Thus, any observable photon attribute, like diffraction pattern, or pointlike behavior at the screen, is doomed to be upset by the mere attempt to perform a measurement. The only arena where reference to the photon as a particle is justified is in the abstract notion of energy space, or frequency space. The photon energy, and therefore also frequency, are both constant, at least in linear environments and over some finite time. The spatial behavior, i.e., the behavior in k-space, can be likened to that of an amoeba, with the photon having a multitude of spatial forms, retaining its energy under all linear spatial transformations, much in the same way as the amoeba retains its volume during spatial adaptations. Specifically, the photon can be focused to a pointlike beam as shown quite generally in Chapter 5, and also by some of the graphs at the end of this chapter, relating to the double slit experiment. Quantum physics is correct in the assessment that classical electromagnetism is unable to provide a satisfactory explanation of the extreme focusing of photons in the double slit experiment, resorting instead to the duality explanation, which carries the heavy burden of creating more problems than it solves, in the way of unexplained paradoxes. In summing up the situation, it can best be characterized by the statement that neither classical electromagnetics nor quantum theory are able to provide a satisfactory explanation of the experimental observations of photon behavior in the double slit experiment. This unsatisfactory situation looks strikingly different from the viewpoint of the neoclassical theory, in which the admixture of photon doublets serves to explain the experimental observations. In the following exposition the unexplainable photon behavior is fully resolved with no paradoxes remaining.

The slit plate configuration is a special kind of electromagnetic circuit that necessarily must fit into the general circuit theory developed in Chapters 3, 5, and 6. In this theory the electromagnetic properties including single photon behavior are described by the macroscopic scattering matrix S of the circuit. The procedure of determining the scattering matrix of the slit plate configuration is rather involved, but the problem is amendable to mathematical analysis as shown separately in the appendix to this chapter. Once the scattering properties expressed by the scattering matrix are established, the general circuit theory of the earlier chapters is directly applicable. For that matter we could follow the theory of Chapter 6, modeling the double slit configuration as a multiple terminal network, with terminals represented by the individual photosensitive photon recipients on the screen. Although the general results of the double slit experiment would emerge from such a procedure, the detailed description would certainly be lacking. Moreover, the particular status of the double slit experiment in quantum theory, with its many unresolved paradoxes, warrants the more detailed treatment presented in this chapter.

The study discusses how the photons are transmitted and reflected at the slit plate under two different environments—the classical macroscopic model—and the neoclassical single photon model. The main objective is to demonstrate the differences between the two models, in particular how the neoclassical model fully supports the experimental observations of single photon behavior in the double slit configuration.

7.2 CLASSICAL MACROSCOPIC MODEL

Figure 7.2 shows a drawing of the double slit configuration. All dimensions are normalized with respect to wavelength, with slit spacing $L_0 = k\ell_0$, slit width $L = k\ell$, and slit plate thickness $\Delta Z = k\Delta z$, where $k = \omega/c$ is the wavenumber in free space. The incident illumination is taken to be a coherent optical beam of light referred to a virtual plane some large but arbitrary distance $Z = Z_1 = kz_1$ above the slit plane. Fourier series resolution of the double slit plate shown in Appendix A of this chapter requires a minimum fundamental width $2(L_0 + L)$. The nominal beam width D is taken much larger than the minimum width, so that we are guaranteed that perturbations in beam density etc. due to the two slits are well contained within the beam width D, which then represents the window over which the analysis is made. The slit plate and the screen, considered as an electromagnetic circuit, are analyzed in the appendix, employing a regular scattering description of Transverse Magnetic (TM) wave components. Although the scattering model is classical in the sense that it is based on conventional electromagnetic theory, it is rather unconventional in its particular formulation in terms of the scattering matrix of the slit plate considered as an electromagnetic circuit. The formulation serves as a conceptual link to the general circuit theory developed in Chapters 3, 5, and 6. In this respect the macroscopic model differs from optical diffraction theory, but the results are essentially the same except that the approach provides supplementary information specific to a circuit description such as detailed electromagnetic field distributions, exact conditions for transmission and reflection at the slit plate, and energy flow visualization.

7.2.1 Slit plate scattering of space harmonics

The scattering relation for the overall slit plate configuration is expressed by relation (A.80) in the appendix, for convenience repeated

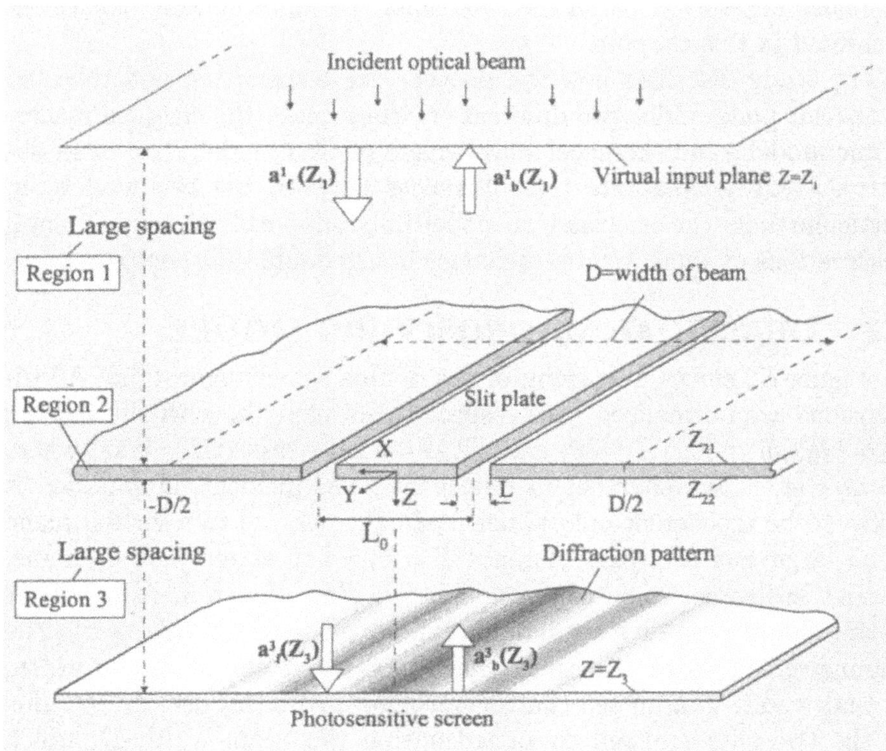

Figure 7.2. **Schematic view of the double slit configuration illuminated from above with a coherent beam of light. All dimensions are normalized as exemplified by $D = kd$ and $Z = kz$, where $k = \omega/c$ is the free space wave number. The model is laterally one-dimensional, with no parameter variations with regard to the Y-coordinate.**

here.

$$\begin{bmatrix} \mathbf{a}_b^1(Z_1) \\ \mathbf{a}_f^3(Z_3) \end{bmatrix} = \mathbf{S} \begin{bmatrix} \mathbf{a}_f^1(Z_1) \\ \mathbf{a}_b^3(Z_3) \end{bmatrix} \tag{7.1}$$

where $\mathbf{S}$ is the slit plate scattering matrix. In the column variables superscripts 1 and 3 refer to the two free space regions, the subscripts f and b to forward and backward harmonic wave components. The formulation differs somewhat from regular scatter formulations in that the variables are spatial *harmonic* vectors rather than vectors referring to spatial *terminal* variables. Thus, the equation describes the scattering between the spatial harmonics referred to the virtual input plane Z_1 and the screen plane Z_3, rather than scattering between terminals. In the scattering formulation (7.1) the column vectors representing 'inputs' are $\mathbf{a}_f^1 = \mathbf{a}_f^1(Z_1)$ and $\mathbf{a}_b^3 = \mathbf{a}_b^3(Z_3)$, whereas $\mathbf{a}_b^1 = \mathbf{a}_b^1(Z_1)$ and $\mathbf{a}_f^3 = \mathbf{a}_f^3(Z_3)$

are the 'output' column vectors, all shown by white arrows in Fig. 7.2. Based on the direction of energy flow of the forward and backward wave components this representation is consistent with the scatter description. Following the general scatter theory of Chapter 6 the input variables are decomposed into two components, according to

$$\begin{bmatrix} \mathbf{a}_f^1 \\ \mathbf{a}_b^3 \end{bmatrix} = \boldsymbol{\rho} \begin{bmatrix} \mathbf{a}_b^1 \\ \mathbf{a}_f^3 \end{bmatrix} + \begin{bmatrix} \mathbf{a}_{f,in}^1 \\ \mathbf{0} \end{bmatrix} \tag{7.2}$$

The last column in the equation is the *structure-independent* input vector $\mathbf{a}_{f,in}^1$ representing the incident beam. Because there is no input at the screen the lower element $\mathbf{a}_{f,in}^3 = \mathbf{0}$. The second column in the equation is the *load-dependent* component proportional to the *reflection matrix* $\boldsymbol{\rho}$, which we shall return to in a moment. Combination of the two equations results in the following modified scatter relation:

$$\begin{bmatrix} \mathbf{a}_b^1 \\ \mathbf{a}_f^3 \end{bmatrix} = (\mathbf{1} - \mathbf{S}\,\boldsymbol{\rho})^{-1}\,\mathbf{S} \begin{bmatrix} \mathbf{a}_{f,in}^1 \\ \mathbf{0} \end{bmatrix} \tag{7.3}$$

The equation is in a convenient form expressing the *transmitted* part $\mathbf{a}_f^3$ and the *reflected* part $\mathbf{a}_b^1$ in terms of whatever input $\mathbf{a}_{f,in}^1$ is driving the double slit configuration. Through regular inversion the equation is expressed in the alternate way

$$(\tilde{\mathbf{S}}^* - \boldsymbol{\rho}) \begin{bmatrix} \mathbf{a}_b^1 \\ \mathbf{a}_f^3 \end{bmatrix} - \begin{bmatrix} \mathbf{a}_{f,in}^1 \\ \mathbf{0} \end{bmatrix} = \mathbf{0} \tag{7.4}$$

where use is made of the unitary property $\mathbf{S}^{-1} = \tilde{\mathbf{S}}^*$ of the scattering matrix. In the second part of the chapter the particular formulation (7.4) is needed in the comparison between macroscopic and single photon analyses. Quite generally, the submatrices $\boldsymbol{\rho}^1$ and $\boldsymbol{\rho}^3$ in the diagonal reflection matrix $\boldsymbol{\rho}$ specify the backscatter from whatever loads are terminating the circuit.

$$\boldsymbol{\rho} = \begin{bmatrix} \boldsymbol{\rho}^1 & 0 \\ 0 & \boldsymbol{\rho}^3 \end{bmatrix} \tag{7.5}$$

In the present case there is no backscatter from free space at the virtual input plane Z_1, so that $\boldsymbol{\rho}^1 = \mathbf{0}$. In the case of single photons it is imperative to retain the general expression (7.3) with a nonzero reflection matrix $\boldsymbol{\rho}^3$ at the screen, a statement that the reader at this point is asked to accept at face value. Accordingly, the reflection matrix has the general form

$$\boldsymbol{\rho} = \begin{bmatrix} 0 & 0 \\ 0 & \boldsymbol{\rho}^3 \end{bmatrix} \tag{7.6}$$

On the other hand, in the macroscopic case, inclusion of reflections from the screen is not essential, merely tending to confuse the basic issue. Therefore, in this case let us put $\rho^3 = 0$ so that the full reflection matrix vanishes.

$$\rho = 0 \tag{7.7}$$

meaning that the transmitted part merely propagates into free space beyond the slit or, if you like, is perfectly matched to the screen plate with no reflections taking place.

7.2.2 Macroscopic transmission and reflection

With the assumption (7.7) the scatter relation (7.3) simplifies to

$$\begin{bmatrix} \mathbf{a}_b^1 \\ \mathbf{a}_f^3 \end{bmatrix} = \mathbf{S} \begin{bmatrix} \mathbf{a}_{f,in}^1 \\ 0 \end{bmatrix} = \begin{bmatrix} \mathbf{S}_{11} & \mathbf{S}_{12} \\ \mathbf{S}_{21} & \mathbf{S}_{22} \end{bmatrix} \begin{bmatrix} \mathbf{a}_{f,in}^1 \\ 0 \end{bmatrix} \tag{7.8}$$

where the scatter submatrices are specified in the appendix to this chapter. The equation is decomposed into the following constituents:

Input component:

$$\mathbf{a}_f^1 = \mathbf{a}_{f,in}^1 \tag{7.9}$$

Reflected component:

$$\mathbf{a}_b^1 = \mathbf{S}_{11}\, \mathbf{a}_{f,in}^1 \tag{7.10}$$

Transmitted component:

$$\mathbf{a}_f^3 = \mathbf{S}_{21}\, \mathbf{a}_{f,in}^1 \tag{7.11}$$

In these equations all scattered harmonic components are expressed as functions of the incident illumination $\mathbf{a}_{f,in}^1$, with $\mathbf{S}_{11}$ and $\mathbf{S}_{21}$ being the submatrices of the scattering matrix $\mathbf{S}$. The equations tell us exactly how much of the incident beam is reflected and how much is transmitted through the slits.

The reflected power:

$$P_r = \tilde{\mathbf{a}}_b^{1*}\, \mathbf{a}_b^1 = \tilde{\mathbf{a}}_{f,in}^{1*}\, \tilde{\mathbf{S}}_{11}^*\, \mathbf{S}_{11}\, \mathbf{a}_{f,in}^1 \tag{7.12}$$

The transmitted power:

$$P_t = \tilde{\mathbf{a}}_f^{3*}\, \mathbf{a}_f^3 = \tilde{\mathbf{a}}_{f,in}^{1*}\, \tilde{\mathbf{S}}_{21}^*\, \mathbf{S}_{21}\, \mathbf{a}_{f,in}^1 \tag{7.13}$$

The sum of reflected and transmitted power::

$$P_r + P_t = \tilde{\mathbf{a}}_{f,in}^{1*}\, (\tilde{\mathbf{S}}_{11}^*\, \mathbf{S}_{11} + \tilde{\mathbf{S}}_{21}^*\, \mathbf{S}_{21})\, \mathbf{a}_{f,in}^1 = \tilde{\mathbf{a}}_{f,in}^{1*}\, \mathbf{a}_{f,in}^1 = P_{input} \tag{7.14}$$

which follows from the unitary condition (A.75) of the scattering matrix, together with (A.79) in the appendix.

There are certain conceptual advantages in specifying the reflected and transmitted parts a little different than in (7.10) and (7.11). In this alternative description the input column $\mathbf{a}^1_{f,in}$ is subdivided into two mutually orthogonal columns.

$$\mathbf{a}^1_{f,in} = \mathbf{a}^1_{f,in,refl} + \mathbf{a}^1_{f,in,trans} \tag{7.15}$$

The two constituents are determined from the eigenvector set Ψ of the submatrix $\mathbf{S}^p_{11}$ of the slit plate scattering matrix $\mathbf{S}^p$. According to (A.70)-(A.75) the matrix $\mathbf{S}^p_{11}$ is real, symmetric, and singular so that the eigenvalue relation is

$$\mathbf{S}^p_{11}\,[\Psi_0, \Psi_1] = [\Psi_0, \Psi_1] \begin{bmatrix} \mathbf{0} & \mathbf{0} \\ \mathbf{0} & \mathbf{-1} \end{bmatrix} \tag{7.16}$$

where Ψ_0 and Ψ_1 are the eigenvectors of the *nullspace* and the *range*, respectively. As shown in the appendix all eigenvalues in the range are equal to minus one. The nullspace of $\mathbf{S}^p_{11}$ is equal to the range of $\mathbf{S}^p_{12}$ and vice versa. Therefore

$$\mathbf{S}^p_{12}\,[\Psi_1, \Psi_0] = [\Psi_1, \Psi_0] \begin{bmatrix} \mathbf{0} & \mathbf{0} \\ \mathbf{0} & \mathbf{-1} \end{bmatrix} \tag{7.17}$$

The rank and nullity are simply related to the relative slit plate aperture L/D by (A.40).

$$\begin{aligned} \text{rank} &= N_1 = N\,(1 - \frac{2L}{D}) \\ \text{nullity} &= N_0 = N - rank = N\,\frac{2L}{D} \end{aligned} \tag{7.18}$$

In these equations $N = 2n_{\max}+1$ is the total number of spatial harmonics, where

$$n_{\max} = \frac{d}{\lambda} \tag{7.19}$$

is the width of the window measured in free space wavelengths. If the eigenvalue relations (7.16) and (7.17) are decomposed into their constituents

$$\begin{aligned} \mathbf{S}^p_{11}\,\Psi_0 &= \mathbf{0} \qquad & \mathbf{S}^p_{11}\,\Psi_1 &= -\Psi_1 \\ \mathbf{S}^p_{12}\,\Psi_1 &= \mathbf{0} \qquad & \mathbf{S}^p_{12}\,\Psi_0 &= -\Psi_0 \end{aligned} \tag{7.20}$$

it follows that

$$\mathbf{S}_{11}^{p} = \mathbf{\Psi}_1 \tilde{\mathbf{\Psi}}_1 \qquad \tilde{\mathbf{\Psi}}_1 \mathbf{\Psi}_0 = \mathbf{0} \qquad \mathbf{\Psi}_1 \tilde{\mathbf{\Psi}}_1 + \mathbf{\Psi}_0 \tilde{\mathbf{\Psi}}_0 = \mathbf{1} \tag{7.21}$$

With these preliminaries, we proceed by expanding (7.15) in terms of the *complete* set Ψ of eigenvectors. It follows then, from (7.10) and (7.21) that

$$\begin{aligned} \mathbf{a}_{f,in,trans}^{1} &= \mathbf{\Psi}_0 \tilde{\mathbf{\Psi}}_0 \, \mathbf{a}_{f,in}^{1} \\ \mathbf{a}_{f,in,refl}^{1} &= \mathbf{\Psi}_1 \tilde{\mathbf{\Psi}}_1 \, \mathbf{a}_{f,in}^{1} \end{aligned} \tag{7.22}$$

Because Ψ_1 and Ψ_0 are orthogonal, the two input constituents are also orthogonal regardless of the form or distribution of the incident illumination $\mathbf{a}_{f,in}^{1}$.

$$\tilde{\mathbf{a}}_{f,in,trans}^{1*} \, \mathbf{a}_{f,in,refl}^{1} = 0 \tag{7.23}$$

Equation (7.22) tells us the same as (7.10) and (7.11) in a slightly different form, carrying a bit more information with regard to the role of the incident vector $\mathbf{a}_f^1(Z_1)_{in}$ of the illuminating beam. The information can be summarized as follows: All incident vectors belonging to the nullspace Ψ_0 are completely transmitted to the screen with nothing reflected at the slit plate. Conversely, incident vectors belonging to the range Ψ_1 are completely reflected at the slit plate with no part of it transmitted to the screen.

Up to this point the analysis belongs in the domain of classical macroscopic electromagnetics, but the scattering approach formulated in the appendix is by no means a standard method. Although several new results and interpretations have been demonstrated, the real advantage of the approach shows up in the treatment of single photons following the present macroscopic analysis.

7.2.3 Macrophotons—a macroscopic concept

The splitting of the incident illumination into two mutually orthogonal sets, each with clear physical significance, promotes introduction of the concept of *macrophotons.* With reference to (7.22) the *transmitted macrophoton* is associated with the space harmonic vector $\mathbf{a}_{f,in,trans}^{1}$. The *reflected macrophoton,* associated with the vector $\mathbf{a}_{f,in,refl}^{1}$, is orthogonal to the transmitted macrophoton. Regardless of the particular harmonic distribution $\mathbf{a}_{f,in}^{1}$ of the incident illumination, the procedure specified in (7.22) splits the incident beam into two orthogonal components, one of which is completely transmitted to the screen and one completely reflected.

The expansions in (7.22) supplemented by (A.11) and (A.21) let us determine the detailed power flux intensity of transmitted and reflected 'macrophotons'. In region 1, above the slit plate, the actual field components derived from the full set of space harmonics are given by

$$h_y(X,Z) = \tilde{\mathbf{w}}(X)\,\boldsymbol{\eta}^{-1/2}\,[e^{-j\boldsymbol{\eta}(Z-Z_1)}\,\mathbf{a}_f^1 - e^{j\boldsymbol{\eta}(Z-Z_1)}\,\mathbf{a}_b^1]$$

$$e_x(X,Z) = \tilde{\mathbf{w}}(X)\,\boldsymbol{\eta}^{1/2}\,[e^{-j\boldsymbol{\eta}(Z-Z_1)}\,\mathbf{a}_f^1 + e^{j\boldsymbol{\eta}(Z-Z_1)}\,\mathbf{a}_b^1] \qquad (7.24)$$

$$e_z(X,Z) = -\tilde{\mathbf{w}}(X)\,\mathbf{B}\,\boldsymbol{\eta}^{-1/2}\,[e^{-j\boldsymbol{\eta}(Z-Z_1)}\,\mathbf{a}_f^1 - e^{j\boldsymbol{\eta}(Z-Z_1)}\,\mathbf{a}_b^1]$$

In these equations $\mathbf{w}(\mathbf{X})$ is a column vector defined in the appendix by

$$\mathbf{w}(X) = \text{col}[e^{-jB_nX}] \qquad (7.25)$$

The diagonal wave impedance matrix $\boldsymbol{\eta}$ is defined by

$$\boldsymbol{\eta} = (\mathbf{1} - \mathbf{B}^2)^{1/2} \qquad (7.26)$$

where $\mathbf{B}$ is the normalized diagonal wave number matrix expressed in terms of the harmonic number n.

$$\mathbf{B} = \text{diag}[B_n] = \text{diag}[2\pi n/D] \qquad (7.27)$$

With region 3 variables Z_3, $\mathbf{a}_f^3$, and $\mathbf{a}_b^3$ replacing Z_1, $\mathbf{a}_f^1$, and $\mathbf{a}_b^1$ the field equations *below* the slit plate have the same form as (7.24).

The power density at any position X, Z is specified by Poynting's vector $\vec{\mathbf{S}}(X,Z)$:

$$\vec{\mathbf{S}}(X,Z) = \frac{1}{2}\,\text{Re}\,[\vec{e}(X,Z)^* \times \vec{h}(X,Z)] \qquad (7.28)$$

whose components are

$$S_z(X,Z) = \tfrac{1}{2}\,\text{Re}\,[e_x(X,Z)^* h_y(X,Z)]$$
$$\qquad (7.29)$$
$$S_x(X,Z) = \tfrac{1}{2}\,\text{Re}\,[e_z(X,Z)^* h_y(X,Z)]$$

By substituting the field equations (7.24) into the two last equations and, furthermore, applying (7.22) the power density components are readily evaluated. If we take the transmitted macrophoton as an example, the power density components in region 1 are specified by the quadratic sums

$$S_z(X,Z) = \tfrac{1}{2}\tilde{\mathbf{a}}_{f,in}^{1*}\,\boldsymbol{\Psi}_0\tilde{\boldsymbol{\Psi}}_0\,\mathbf{F}(X,Z)\boldsymbol{\Psi}_0\tilde{\boldsymbol{\Psi}}_0\,\mathbf{a}_{f,in}^1$$
$$\qquad (7.30)$$
$$S_x(X,Z) = -\tfrac{1}{2}\tilde{\mathbf{a}}_{f,in}^{1*}\,\boldsymbol{\Psi}_0\tilde{\boldsymbol{\Psi}}_0\,\mathbf{G}(X,Z)\boldsymbol{\Psi}_0\tilde{\boldsymbol{\Psi}}_0\,\mathbf{a}_{f,in}^1$$

The expressions are real so that the real signs in (7.29) are superfluous. The functional dependence of power flux components on the coordinates X, Z are contained in the matrices $\mathbf{F}(X,Z)$ and $\mathbf{G}(X,Z)$ given by

$$\begin{aligned}\mathbf{F}(X,Z) &= e^{j\eta(Z-Z_1)}\,\eta^{1/2}\,\mathbf{w}^*(X)\,\tilde{\mathbf{w}}(X)\,\eta^{-1/2}\,e^{-j\eta(Z-Z_1)}\\ \mathbf{G}(X,Z) &= e^{j\eta(Z-Z_1)}\,\eta^{-1/2}\,\mathbf{w}^*(X)\,\tilde{\mathbf{w}}(X)\,\eta^{1/2}\,e^{-j\eta(Z-Z_1)}\end{aligned} \tag{7.31}$$

Similar expressions apply to the transmitted macrophoton in region 3, and to the reflected macrophoton in region 1, with obvious substitutions of appropriate harmonic amplitudes.

7.2.4 Numerical simulations of macrophoton power flux

From computer simulations, based on the equations above, the flux density patterns of the macrophotons are illustrated in several graphs assuming a simple plane wave input $\mathbf{a}^1_{f,in}$. Figure 7.3 shows the flux lines of the transmitted macrophoton above and below the slit plate with relevant slit plate parameters listed in the caption. Note that the vertical plot scale is greatly compressed. The flux density is inversely proportional to the spacing between adjacent flux lines. In the computational simulation the plotting of stream lines progresses from the top towards the slit plate and beyond, all the way to the screen. For specified slit plate parameters the procedure begins by calculation of the eigenvector sets Ψ_0 and Ψ_1 from the equations in the appendix, followed by calculation of the longitudinal flux $S_z(X, Z_1)$ at the input plane $Z = Z_1$ from (7.30) and (7.31). In the next step the input plane is subdivided into segments of equal power serving as the initial starting X-coordinates for the flux lines. The power flux components $S_x(X,Z)$ and $S_z(X,Z)$ evaluated from the same equations define the local direction of a particular stream line along its trajectory, according to

$$\frac{dX}{dZ} = \frac{S_x(X,Z)}{S_z(X,Z)} \tag{7.32}$$

It is remarkable how all the flux lines converge to the two slits, none of them hitting the solid parts of the plate. It is equally remarkable that the continuation lines behind the slit plate all originate from the slits although this was not imposed as a an initial precondition for locations of lines below the slits. The intricate striations in the flux lines were found in all computer runs covering wide ranges of relevant parameters. In a true vertical scale display the striations would of course be correspondingly less pronounced.

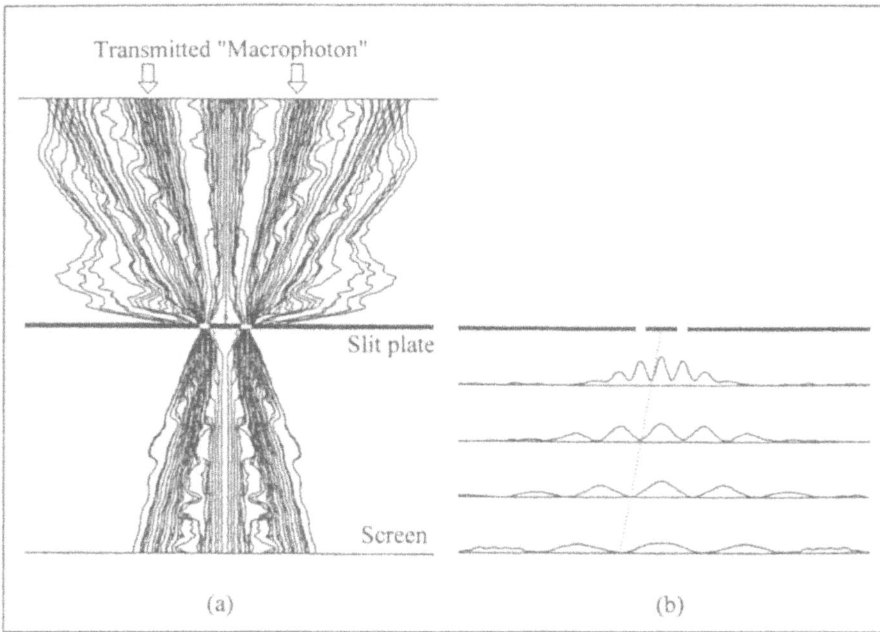

Figure 7.3. **Numerical simulation of transmitted "macrophoton" through a double slit plate with slit specifications: $L_0/D = 0.1$, $L/D = 0.025$, and vertical scale compression approximately equal to 50. (a) Computer generated macroscopic flux lines above and below the double slit plate. (b) Diffraction pattern at four equally spaced planes, obtained from the same data applying low pass filtering.**

The computer generated diffraction pattern in Fig. 7.3b represents the low pass filtered Z-component $S_z(X, Z)$ of the Poynting flux plotted versus X at four equally spaced planes below the slit plate. Their general shapes correspond exactly to diffraction patterns predicted from conventional optical diffraction theory with the dotted line representing the first cancellation points predicted from the optical theory.

The flux pattern of the reflected macrophoton is obtained by a similar procedure. It follows from (7.22) that we simply have to replace the nullspace vector set Ψ_0 in (7.30) by the vector set Ψ_1 of the range. Figure 7.4 shows the corresponding flux lines for the reflected 'macrophoton'. The plot in Fig. 7.4a represents the *forward* flux plotted from top towards the slit plate. Again, notice the perfect termination of all lines on the solid part of the slit plate, with none hitting the slits. The *backward* reflected flux plotted from the slit plate towards the top, illustrated in Fig. 7.4b, shows a pattern similar to the *forward* flux.

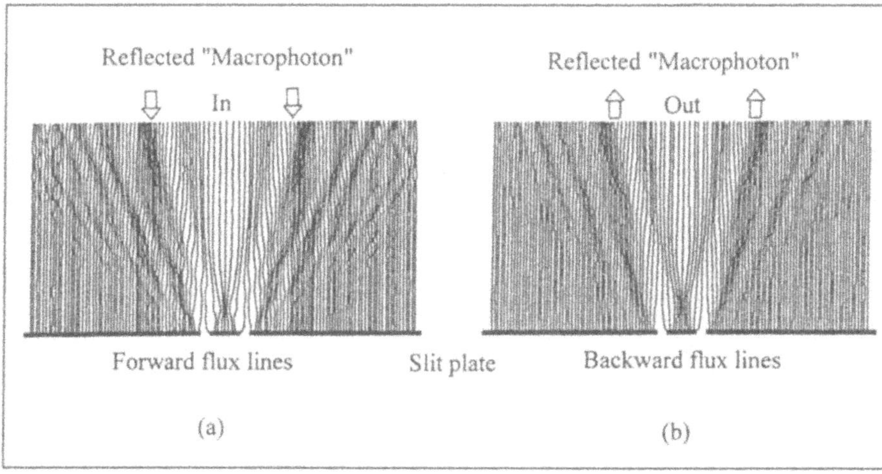

Figure 7.4. Numerical simulation of reflected "macrophoton" from the double slit plate. The configuration data are the same as those of Fig. 7.3. **(a)** Computer generated forward flux lines of macrophoton. **(b)** Backward, i.e., reflected flux lines of macrophoton.

The flow patterns of Figs. 7.3 and 7.4 give rise to vivid mental images of the power flow patterns associated with the two macrophotons, in particular that of the transmitted macrophoton in Fig. 7.3. In assessing these plots it should be kept in mind that the notion of 'macrophotons' is a strictly macroscopic concept. Each of the macrophoton plots represent the macroscopic averages of the total number of transmitted and reflected photons, respectively. There is nothing wrong with this notion as long as we talk about macroscopic behavior, based on the described macroscopic scattering process of the configuration. The same macroscopic theory also predicts the following relation between the time average number of transmitted and reflected photons as

$$\lim_{N\to\infty}\left(\frac{N_{trans}}{N_{refl}}\right)=\left(\frac{\tilde{\mathbf{a}}^{1*}_{f,in}\,\Psi_0\tilde{\Psi}_0\,\mathbf{a}^1_{f,in}}{\tilde{\mathbf{a}}^{1*}_{f,in}\,\Psi_1\tilde{\Psi}_1\,\mathbf{a}^1_{f,in}}\right)^{1/2} \tag{7.33}$$

which follows from (7.22). If N is finite, the integer relation is satisfied only to some degree of accuracy.

What happens then if the optical illumination $\mathbf{a}^1_{f,in}$ is reduced to the point where perhaps only one photon arrives at the slit plate at a time, with the single photon containing a mixture of both components $\mathbf{a}^1_{f,in,trans}$ and $\mathbf{a}^1_{f,in,refl}$? If we elect to apply the macroscopic theory to this situation, (7.22) tells us that part of the photon should be transmitted and part of it reflected. But for the photon this outcome is prohibitive because it violates its energy integrity. The photon can not subdivide its energy by dissipating a fraction at the screen and reflecting the rest back from the slit plate. Therefore, the scatter relation in its macroscopic form can not hold for single photons. The reader will recall that the problem raised here about photon energy integrity was dealt with in general terms in the earlier chapters. The slit plate configuration, with the experimental observations discussed in the introduction to this chapter, serves as a typical and illuminating example of the general theory. In the following the observations will be explained by interaction of the photons with the environmental doublet fluctuations (PDE) through a process in which appropriate admixtures of photon doublets serve to route and focus the photons to localized single spots on the screen or, alternatively, completely reflect them back at the slit plate. Furthermore, it will be demonstrated by analytical means that the distribution of photon arrivals at the screen, averaged over a large number of photons, builds up to the *macroscopic* diffraction pattern shown in Fig. 7.3.

In the demonstrations to follow, referring to (1.1) in Chapter 1, the theoretical platform is being changed from $(em)_{classical}$ to $(em)_{neoclassical}$. Realizing that the slit plate configuration is a special case of the general circuit configuration treated in Chapter 6, results from the general theory could be applied right off. However, the considerable attention devoted to the double slit configuration in quantum physics, along with the unresolved paradoxes, warrants a more detailed analysis based on the platform of the neoclassical theory.

7.3 NEOCLASSICAL THEORY OF SINGLE PHOTON ROUTING

The lateral routing of photons to single spots on the screen is a two-dimensional process that requires a set of electromagnetic modes more general than the Transverse Magnetic (TM) set which specified the basic modes in the macroscopic treatment. This particular set was perfectly adequate for the macroscopic case, but needs to be supplemented by a Transverse Electric (TE) set for the single photon case. Appropriate superpositions of the two sets will provide routing in the X as well as the Y- direction. Theoretically, the extension is straight forward, but the basic principles of routing tend to be obscured by the mere mathematical complexity of the added TE set, without providing much in the way of added physical understanding. For this reason the analysis is being confined to variations in the X-direction only. This restriction does not materially limit the validity of the principal results. Extension to two lateral dimensions does not introduce new concepts, only fairly trivial modifications.

We saw that the *macroscopic* behavior of photons in the slit plate configuration was determined from the scattering relation (7.3), with right hand input drive $\mathbf{a}^1_{f,in}$ corresponding to the illuminating optical beam at the upper virtual terminal $Z = Z_1$. In order to account for supplemental doublet excitation in the single photon case, the equation has to be modified to

$$\begin{bmatrix} \mathbf{a}^1_b \\ \mathbf{a}^3_f \end{bmatrix} = (\mathbf{1} - \mathbf{S}\,\rho)^{-1}\mathbf{S}\left\{\begin{bmatrix} \mathbf{a}^1_{f,in} \\ \mathbf{0} \end{bmatrix} + \begin{bmatrix} \boldsymbol{\alpha}^1_f \\ \boldsymbol{\alpha}^3_b \end{bmatrix}\right\} \qquad (7.34)$$

The input column at the extreme right is a generalized form of the input column of (7.3), with supplemental photon doublet components $\boldsymbol{\alpha}^1_f$ and $\boldsymbol{\alpha}^3_b$ at the virtual input plane Z_1 and the screen position Z^3, respectively. The component $\boldsymbol{\alpha}^3_b$ at the screen is in apparent contradiction to the fact that there is no *actual* photon input at this plane. But the photon doublets organizing the transmission or reflection of the photons are quite independent of the actual photon input $\mathbf{a}^1_{f,in}$, and we shall indeed find that both $\boldsymbol{\alpha}^1_f$ and $\boldsymbol{\alpha}^3_b$ are nonzero. The explicit form of the doublet supplement vector is determined by inversion of (7.34). Noting that $\mathbf{S}^{-1} = \tilde{\mathbf{S}}^*$ the result is

$$\begin{bmatrix} \boldsymbol{\alpha}^1_f \\ \boldsymbol{\alpha}^3_b \end{bmatrix} = \begin{bmatrix} \tilde{\mathbf{S}}^*_{11} & \tilde{\mathbf{S}}^*_{12} \\ \tilde{\mathbf{S}}^*_{21} & \tilde{\mathbf{S}}^*_{22} - \rho^3 \end{bmatrix}\begin{bmatrix} \mathbf{a}^1_b \\ \mathbf{a}^3_f \end{bmatrix} - \begin{bmatrix} \mathbf{a}^1_{f,in} \\ \mathbf{0} \end{bmatrix} \qquad (7.35)$$

where the scatter submatrices are specified by (A.79) in the appendix. The equation is applicable for transmitted as well as reflected single photons, with appropriate forms of the components $\mathbf{a}^1_b$ and $\mathbf{a}^3_f$ for either case stipulated. If we know the characteristics of these components for full transmission and full reflection, then the corresponding doublet admixture follows from the equation. These particular questions are discussed in the following sections, where we take full advantage of the scatter formulations and results stated in the appendix to this chapter.

7.3.1 General criteria for single photon transmission

By definition, the transmitted photon has no backscattered components at the virtual input plane Z_1, so that we must put

$$\mathbf{a}^1_b = \mathbf{0} \qquad (7.36)$$

With this condition (7.35) is readily solved with regard to the doublet components. The solutions are expressed in the convenient form

$$\begin{aligned} \boldsymbol{\alpha}_f^1 + \mathbf{a}_{f,in}^1 &= \tilde{\mathbf{S}}_{12}^* \mathbf{a}_f^3 \\ \boldsymbol{\alpha}_b^3 + \mathbf{a}_b^3 &= \tilde{\mathbf{S}}_{22}^* \mathbf{a}_f^3 \end{aligned} \tag{7.37}$$

Referring to (A.79) in the appendix, let us first note that the submatrices $\tilde{\mathbf{S}}_{12}^*$ and $\tilde{\mathbf{S}}_{22}^*$ of the overall configuration are singular matrices with the same nullspace and range as the simpler real matrices $\mathbf{S}_{11}^p = \mathbf{S}_{22}^p$ and $\mathbf{S}_{12}^p = \mathbf{S}_{21}^p$ of the slit plate itself. We found that the range of $\mathbf{S}_{11}^p$ is equal to the nullspace of $\mathbf{S}_{12}^p$ and vice versa. Let us expand the forward harmonic column $\mathbf{a}_f^3$ in terms of the presumably complete eigenvector set of $\mathbf{S}_{11}^p$.

$$\mathbf{a}_f^3 = \Psi_0 \mathbf{A} + \Psi_1 \mathbf{B} \tag{7.38}$$

where $\mathbf{A}$ and $\mathbf{B}$ are expansion columns, the dimension of $\mathbf{A}$ being equal to the nullity and that of $\mathbf{B}$ equal to the rank of $\mathbf{S}_{11}^p$. In view of the eigenrelations (7.20), the two relations (7.37) reduce to

$$\begin{aligned} \boldsymbol{\alpha}_f^1 + \mathbf{a}_{f,in}^1 &= -\Psi_0 \mathbf{A} \\ \boldsymbol{\alpha}_b^3 &= -\Psi_1 \mathbf{B} - \mathbf{a}_b^3 \end{aligned} \tag{7.39}$$

The first equation in (7.39) tells us that the incident photon $\mathbf{a}_{f,in}^1$ together with the doublet admixture $\boldsymbol{\alpha}_f^1$ must belong to the nullspace of $\mathbf{S}_{11}^p$. In the unlikely case that the primary incident photon already happens to belong in the nullspace, there would be no need for any doublet supplement $\boldsymbol{\alpha}_f^1$, and the photon would happily travel to the screen with no doublet admixture required. In the normal situation the input $\mathbf{a}_{f,in}^1$ requires a doublet supplement $\boldsymbol{\alpha}_f^1$ compensating exactly the part $\Psi_1 \mathbf{B}$ belonging to the *range* of $\mathbf{S}_{11}^p$. With this stipulation satisfied, the overall input $\boldsymbol{\alpha}_f^1 + \mathbf{a}_{f,in}^1$ is confined to the nullspace, as required by (7.39). The additional condition of photon energy conservation requires the following equation to hold:

$$P = \tilde{\mathbf{a}}_{f,in}^{1*} \mathbf{a}_{f,in}^1 = \tilde{\boldsymbol{\alpha}}_f^{1*} \boldsymbol{\alpha}_f^1 + \tilde{\boldsymbol{\alpha}}_f^{1*} \mathbf{a}_{f,in}^1 + \tilde{\mathbf{a}}_{f,in}^{1*} \boldsymbol{\alpha}_f^1 + \tilde{\mathbf{a}}_{f,in}^{1*} \mathbf{a}_{f,in}^1 \tag{7.40}$$

The photon doublet energy is always zero so that the first term $\tilde{\boldsymbol{\alpha}}_f^{1*} \boldsymbol{\alpha}_f^1 = \mathbf{0}$. Therefore, the equation is satisfied provided the doublet vector $\boldsymbol{\alpha}_f^1$ is orthogonal to the incident photon vector $\mathbf{a}_{f,in}^1$.

$$\tilde{\boldsymbol{\alpha}}_f^{1*} \mathbf{a}_{f,in}^1 = \mathbf{0} \tag{7.41}$$

The orthogonality condition assures that the photon can be supplemented with the doublet compensation from the first equation of (7.39) and retain its energy in the process. The second equation of (7.39) demonstrates the compensation affected by the doublet $\boldsymbol{\alpha}_b^3$ of *any* photon reflection $\mathbf{a}_b^3$ which might occur at the screen, as well as that particular part $\Psi_1\mathbf{B}$ of $\mathbf{a}_f^3$ which belong to the *range* of $\mathbf{S}_{11}^p$. Apparently, the overall compensation of reflected wave components in the double slit configuration is conceptually similar to the simpler case of photon absorption and reflection at a plane metallic surface treated in Chapter 2 (see Fig. 2.5).

The conceptual picture emerging from this discussion is a scenario in which the arbitrary incident photon $\mathbf{a}_{f,in}^1$ at the virtual input plane Z_1 is moulded by the doublet admixture $\boldsymbol{\alpha}_f^1$ into a form appropriate to the slit plate geometry, with full transmission through the slits. At the screen itself the doublet supplement $\boldsymbol{\alpha}_b^3$ functions to mould the photon further by compensating any reflections $\mathbf{a}_b^3$ that might occur, as well as the undesirable part $\Psi_1\mathbf{B}$ of the photon harmonic column $\mathbf{a}_f^3$. As a result of these processes the photon is guaranteed full transmission and absorption at the screen. The nullspace set Ψ_0 is the active set of harmonics of the moulded input photon. The inactive set Ψ_1, belonging to the range, is not taking part in transmission, but this set is expected to be active in total *reflection* of photons. In this sense the nullspace and the range are complementary sets, associated with transmission and reflection, respectively. Before discussing the actual routing process to single spots on the screen let us look at the criteria for full photon reflection.

7.3.2 General criteria for single photon reflection

With the photon completely reflected from the slit plate, there can be no fields in region 3 below the slit plate; therefore, $\mathbf{a}_f^3$ and $\mathbf{a}_b^3$ are both zero, and only the first part of (7.35) applies

$$\boldsymbol{\alpha}_f^1 + \mathbf{a}_{f,in}^1 = \tilde{\mathbf{S}}_{11}^* \mathbf{a}_b^1 \tag{7.42}$$

With the backward column $\mathbf{a}_b^1$ expressed in terms of the full eigenvector set,the appropriate part is seen to belong in the range of $\mathbf{S}_{11}^p$.

$$\mathbf{a}_b^1 = \Psi_{11}\mathbf{C} \tag{7.43}$$

where $\mathbf{C}$ is some expansion column of dimension equal to the rank of $\mathbf{S}_{11}^p$. In view of (7.20) the following relation applies:

$$\mathbf{a}_b^1 = -(\boldsymbol{\alpha}_f^1 + \mathbf{a}_{f,in}^1) \tag{7.44}$$

showing that the reflected component $\mathbf{a}_b^1$ is equal to the composite input column $\boldsymbol{\alpha}_f^1 + \mathbf{a}_{f,in}^1$, both belonging to the *range* of $\mathbf{S}_{11}^p$.

Following the same line of arguments as in the case of transmission, the equation shows that the doublet admixture $\boldsymbol{\alpha}_f^1$ serves to nullify the nullspace part of the incident photon $\mathbf{a}_{f,in}^1$. In the unlikely case that the incident photon already *happens* to belong to the range of $\mathbf{S}_{11}^p$, the photon would be reflected from the slit plate with no doublet supplement required. But this is not expected to be the normal situation. As an additional condition the doublet supplement $\boldsymbol{\alpha}_f^1$ must be orthogonal to $\mathbf{a}_{f,in}^1$, a stipulation similar to (7.41), assuring that photon energy is conserved in transmission as well as reflection.

With the general criteria for single photon transmission and reflection established, the following sections show the details of how transmitted and reflected photons are routed to their respective destinations.

7.3.3 Transmission of photons to local screen spots

Referring to region 3, between the slit plate and the screen, the photon is specified by its harmonic set $\mathbf{a}_f^3$ which we found to belong to the nullspace of $\mathbf{S}_{11}^p$. The task at hand is to determine the specific distribution of the harmonic $\mathbf{a}_f^3$ which causes routing of the photon to a particular spot X_q on the screen. As a preliminary to the process of determining the particular $\mathbf{a}_f^3$ performing this task we have to be a little more specific about defining the resolution of a point X_q on the screen. Sufficient resolution of the basic routing process is achieved by the set of harmonics $\mathbf{a}_f^3$ alone without the need of supplementary evanescent modes at the screen. The finite spatial bandwidth of the optical beam is equal to $N = 2n_{\max}+1$ so that the spatial resolution is specified by

$$\Delta X = k\,\Delta x = \frac{D}{N} = \frac{2\pi n_{\max}}{N} \simeq \pi \tag{7.45}$$

The smallest resolution Δx is then

$$\Delta x = \frac{\pi}{k} = \frac{\lambda}{2} \tag{7.46}$$

Thus, the inherent uncertainty in the routing process is half a wavelength. This limit is a consequence of the assumption that only propagating modes specified by the harmonic column $\mathbf{a}_f^3$ take part in the routing process. If nonpropagating doublet evanescent modes located at the screen surface were taken into consideration, there would be no finite resolution limit, as indeed demonstrated in Chapters 3 and 5. The reason for not including evanescent modes in the present analysis is a simple one. It adds mathematical complexity without a corresponding gain of basic understanding.

Thus, we can assume that the photon is routed to a specified point X_q with accuracy $\lambda/2$. The statement implies that the transverse photon fields $h_y(X)$ and $e_x(X)$ associated with the harmonics $\mathbf{a}_f^3$ of the photon must vanish everywhere except at $X = X_q$. This condition prompts us to express the fields as delta functions. Referring to (A.11) and (A.21) in the appendix to this chapter, the fields are specified by:

$$\begin{aligned} h_y(X) &= \tilde{\mathbf{w}}(X)\,\boldsymbol{\eta}^{-1/2}\,(\mathbf{1} - \boldsymbol{\rho}^3)\,\mathbf{a}_f^3 = C_h\,\delta(X - X_q) \\ e_x(X) &= \tilde{\mathbf{w}}(X)\,\boldsymbol{\eta}^{1/2}\,(\mathbf{1} + \boldsymbol{\rho}^3)\,\mathbf{a}_f^3 = C_e\,\delta(X - X_q) \end{aligned} \tag{7.47}$$

where C_h and C_e are some unknown constants to be determined later. In these equations the backward component $\mathbf{a}_b^3$ has been expressed in terms of $\mathbf{a}_f^3$ by the relation

$$\mathbf{a}_b^3 = \boldsymbol{\rho}^3 \mathbf{a}_f^3 \tag{7.48}$$

where $\boldsymbol{\rho}^3$ is the reflection matrix. Due to the finite bandwidth, the delta function $\delta(X - X_q)$ is approximate to within $\lambda/2$. The normalized impedance z_q of the photon beam at X_q is specified by the ratio of transverse electric and magnetic fields.

$$z_q = \frac{e_x(X_q)}{h_y(X_q)} = \frac{C_e}{C_h} = z(X_q) \tag{7.49}$$

As shown by the equation the photon's impedance z_q must be matched to the equivalent impedance $z(X_q)$ of the power absorbing mechanism at $X = X_q$. The variety of absorbing mechanisms that can be visualized makes it difficult to quote any specific impedance. If the screen were conducting, the photon would dump its energy to single conducting electrons at the screen surface. With a photosensitive screen the photon would interact with microcrystals or single atoms causing blackening due to some fairly complex mechanism that perhaps would be a three level absorbing process. It will be demonstrated shortly that the actual screen impedance $z(X_q)$ at point X_q is irrelevant in the sense that the photon is always matched through the occurrence of an appropriate reflection $\mathbf{a}_b^3$ and a corresponding admixture of photon doublets in accordance with the general relation (7.39). Through this doublet compensation full absorption of the photon energy at the point X_q is guaranteed.

By premultiplication of the two equations (7.47) with the column $\mathbf{w}(X)^*$, followed by integration over the width D, using the orthogonality relation (A.15), they are transformed to

$$\begin{aligned} \mathbf{a}_f^3 &= \tfrac{C_h}{D}(\mathbf{1}-\boldsymbol{\rho}_q^3)^{-1}\,\boldsymbol{\eta}^{1/2}\mathbf{w}(X_q)^* \\ \mathbf{a}_f^3 &= \tfrac{C_h}{D}z(X_q)(\mathbf{1}+\boldsymbol{\rho}_q^3)^{-1}\,\boldsymbol{\eta}^{-1/2}\mathbf{w}(X_q)^* \end{aligned} \tag{7.50}$$

The reflection matrix $\boldsymbol{\rho}_q^3$ is obtained by equating the two expressions.

$$\boldsymbol{\rho}_q^3 = [z(X_q)\mathbf{1}-\boldsymbol{\eta}]\,[z(X_q)\mathbf{1}+\boldsymbol{\eta}]^{-1} \tag{7.51}$$

where $\boldsymbol{\eta}$ is the impedance matrix of the spatial harmonics, defined in (A.22). Readers familiar with transmission line theory will recognize (7.51) as a generalized reflection coefficient of a loaded transmission line whose reflection coefficient ρ is given by

$$\rho = \frac{z_L - z_0}{z_L + z_0} \tag{7.52}$$

Here, z_L is the load impedance and z_0 the characteristic line impedance. The equivalence with (7.51) is only formal because the diagonal reflection matrix $\boldsymbol{\rho}^3$ refers to the reflected space *harmonic* components and not to reflection of the *spatial* components at the spot X_q. Hence, (7.51) tells us that matching of photon impedance at $X = X_q$ requires a certain amount of space harmonic reflections $\boldsymbol{\rho}_q^3$ depending on the spot impedance $z(X_q)$ only. Regardless of the nature of the screen, with uniform or random variation of screen impedance $z(X)$ the photon routed to any spot is always matched through the intermediary of an appropriate amount of reflection. Even if it is a necessary ingredient in the absorption process, the reflection is everywhere nullified by a compensating doublet admixture, according to the general principle (7.39). From the form of (7.50) it is expected that the harmonic column $\mathbf{a}_f^3$ contains components from the nullspace Ψ_0 as well as the range Ψ_1. As shown in the earlier discussion, the nullspace component Ψ_0 is the active part for transmission to the screen. The part contained in the set Ψ_1 is nullified by the doublet admixture, in accordance with (7.39).

Returning to (7.50) it remains to determine the constant C_h/D which is obtained from power conservation principles. Because no power is reflected at the input plane $Z = Z_1$, and any reflection $\mathbf{a}_b^3$ at the screen is nullified by an appropriate doublet admixture, power conservation requires:

$$\tilde{\mathbf{a}}_f^{3*}\,\mathbf{a}_f^3 = \tilde{\mathbf{a}}_{f,in}^{1*}\,\mathbf{a}_{f,in}^1 \tag{7.53}$$

Together with (7.50) and (7.51) the equation determines the constant C_h/D for a specified spot impedance X_q. With this stipulation the complete set of conditions routing the incident photon $\mathbf{a}_{f,in}^1$ to point X_q on the screen has been established.

7.3.4 Numerical simulations of transmitted photons

Figure 7.5 shows computer generated flux lines for single photons routed to specific points on the screen. The trajectories represent the Poynting vector flux of a single photon with subdivisions corresponding to equal power between adjacent flux lines. The program calculates the various variables from the equations derived in the foregoing sections for a selected destination point X_q with assumed impedance $z(X_q)$.

The plotting of flux lines starts at the top, proceeding downward until the photon hits the selected spot X_q as a concentrated deltalike flux entity. Above the slit plate, in region 1, the photon starts out at the virtual input plane Z_1 with the overall input vector $\mathbf{a}^1_{f,in,q} = \mathbf{a}^1_{f,in} + \boldsymbol{\alpha}^1_f$ selected from the nullspace of $\mathbf{S}^p_{11}$ in accordance with (7.39). Under all conditions of simulation the photon trajectories invariably coalesced to single streams before arriving at the slits in the slit plate, some times hitting both slits and some times just one slit, with no apparent systematic pattern. The fact that this behavior was consistent under all operating conditions, regardless of chosen parameters, rules out possible artifacts.

In region 3, below the slit plate, one might think that the confined 'photon string' or 'strings' entering region 3 through the slits should perhaps continue as single strings down to point X_q on the screen. But this is not what happens. The physical situation in region 3 appears to be considerately more complex than in region 1, where the photon propagated to the slit plate as a set of *forward wave* harmonics with no backward waves, albeit with a fairly intricate harmonic distribution. We saw in (7.51) that photon impedance match at the screen point X_q required a certain harmonic reflection specified by the reflection matrix ρ^3. Although the reflected backward component $\mathbf{a}^3_b$ is compensated by the doublet admixture, it plays a role in the matching of field components immediately above and below the slit plate. This effect is showing up in the wide flux pattern immediately below the slit plate. It should be noted that the trajectories in region 3 below the slit plate represent the forward wave components $\mathbf{a}^3_{f,q}$ only. Also in this lower plot the stream lines coalesce into striations which finally merge to one single string at X_q. It is remarkable how accurately the trajectories hit the initially specified final destination point.

The graphs in Fig. 7.6 show *both* forward *and* backward wave constituents for a photon routed to the particular point $X_q = 0.2$ on the screen. No immediate explanation is offered for the peculiar flux 'lumps' appearing in the backward flux component in Fig. 7.6b. The lumps

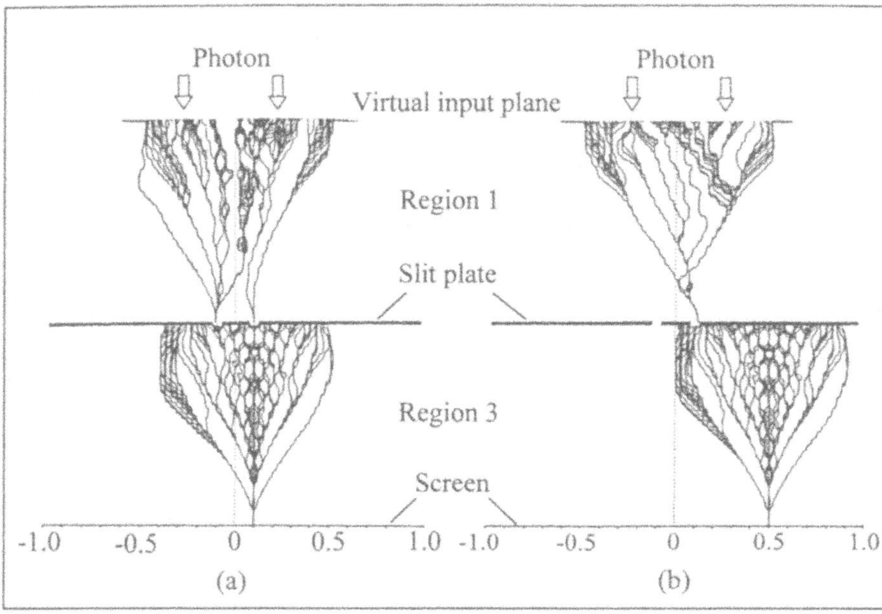

Figure 7.5. **Examples of *single* photons routed to localized spots on the screen. The graphs represent flux lines, with longitudinal power inversely proportional to spacing between lines. The vertical scale is highly compressed by a factor of approximately 60. (a) Horizontal coordinate X_q= 0.1: The input photon goes through both slits. (b) X_q= 0.5: The photon goes through one slit only.**

showed up quite consistently under all operating conditions, regardless of the chosen parameters.

7.3.5 Reflection of photons at the slit plate

The analysis of the full reflection case is considerably simpler than the case of transmission. It follows from (7.43) that the composite incident photon $\mathbf{a}^1_{f,in} + \boldsymbol{\alpha}^1_f$ must belong to the *range* of $\mathbf{S}_{11}$. For the transmitted photons we recall that the input belonged to the nullspace of $\mathbf{S}_{11}$. Thus, the overall input separates into two orthogonal sets, one for transmitted and one for reflected photons, both administered by the environmental photon doublet admixture $\boldsymbol{\alpha}^1_f$. If the *pth* photon is being reflected, the composite input satisfies

$$(\boldsymbol{\alpha}^1_f + \mathbf{a}^1_{f,in})_p = \Psi_1 \mathbf{C}_p \tag{7.54}$$

where $\mathbf{C}_p$ is some expansion column of dimension equal to the rank of $\mathbf{S}^p_{11}$. The equation tells us that the general input of reflected photons can be any linear combination of the eigenvectors Ψ_1 in the range of

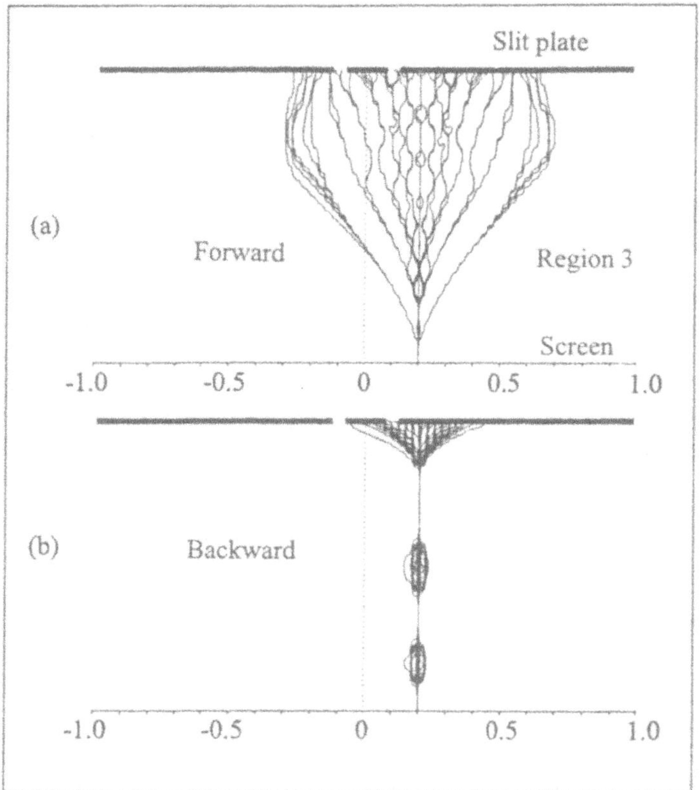

Figure 7.6. **Forward and backward flux pattern of single photon routing to the spot X_q= 0.2 on the screen. The illustration refers to the pattern in region 3 below the slit plate, demonstrating that both wave components are required for proper match of the photon at the screen. (a) Flux lines of forward components. (b) Flux lines of backward components.**

$\mathbf{S}_{11}$, with relative contributions specified by the expansion column $\mathbf{C}_p$. The equation implies that

$$\mathbf{a}_b^1 = -(\boldsymbol{\alpha}_f^1 + \mathbf{a}_{f,in}^1)_p \tag{7.55}$$

showing that the forward and backward components are equal with opposite sign, a result which of course is expected.

7.3.6 Numerical simulations of reflected single photons

Figure 7.7 shows examples of computer generated graphs of flux lines for single photons that are totally reflected from the slit plate. Each of the graphs represents an arbitrary selection $\mathbf{C}_p$ out of the eigenvector set Ψ_1 in (7.54). This must not be construed as meaning that the particular examples shown here are the actual combinations selected by the incident photon. The selection is determined from the principle of *PDE* minimization, to be discussed in a moment. The graphs are merely meant to demonstrate the general character of photon reflections at the slit plate. The solid curves represent the forward wave flux trajectories, the dotted curves the reflected flux lines. Again, observe the pronounced tendencies of the photon to form striations as it approaches the slit plate, very much in the same way as demonstrated in Figs. 7.5 and 7.6 for the transmitted photons. But in this case of full reflection all forward flux lines end up on the solid parts of the slit plate, with none hitting the slits, as indeed expected. A large number of eigenvector selections of the set Ψ_1 were investigated, all displaying this same common feature although none of the detailed flux patterns were identical.

7.3.7 Minimization of environmental energy fluctuations

What physical principle decides whether the incident photon elects to be transmitted to some point X_q on the screen or be reflected from the slit plate? And why are not all the incoming photons transmitted to the screen, or all of them reflected? In a low density coherent stream of well separated incoming photons all the photons present themselves to the configuration with the same polarization, the same energy, and are subject to exactly the same physical conditions in their interaction with the double slit configuration. From a classical point of view any prediction of a photon's fate would apply to *all* the photons since they are identical and subject to the same experimental conditions. This paradox, surfacing in any quantum experiment designed to determine the fate of single photons, is a permanent embarrassment in quantum theory which is unable to answer this question in an objective way. The basic problem is the absence of a specific randomizing process serving to divide the photons into a transmitted group and a reflected group satisfying the macroscopically observed numbers. Instead, quantum theory claims a certain *inherent* probability for a photon being transmitted and a photon being reflected, specified only by the details of the slit plate structure, without pointing to any randomizing process. The *inherent* probabilities of the various outcomes ascribed by quantum theory are expressed by the square of the photon's wave function Ψ which supposedly exists in a superposed state with components specified from the quantum *correspondence principle,* asserting that the overall outcome corresponds to the observed macroscopic pattern. But this kind of quantum probability is much different from classical probabilities. The classical random

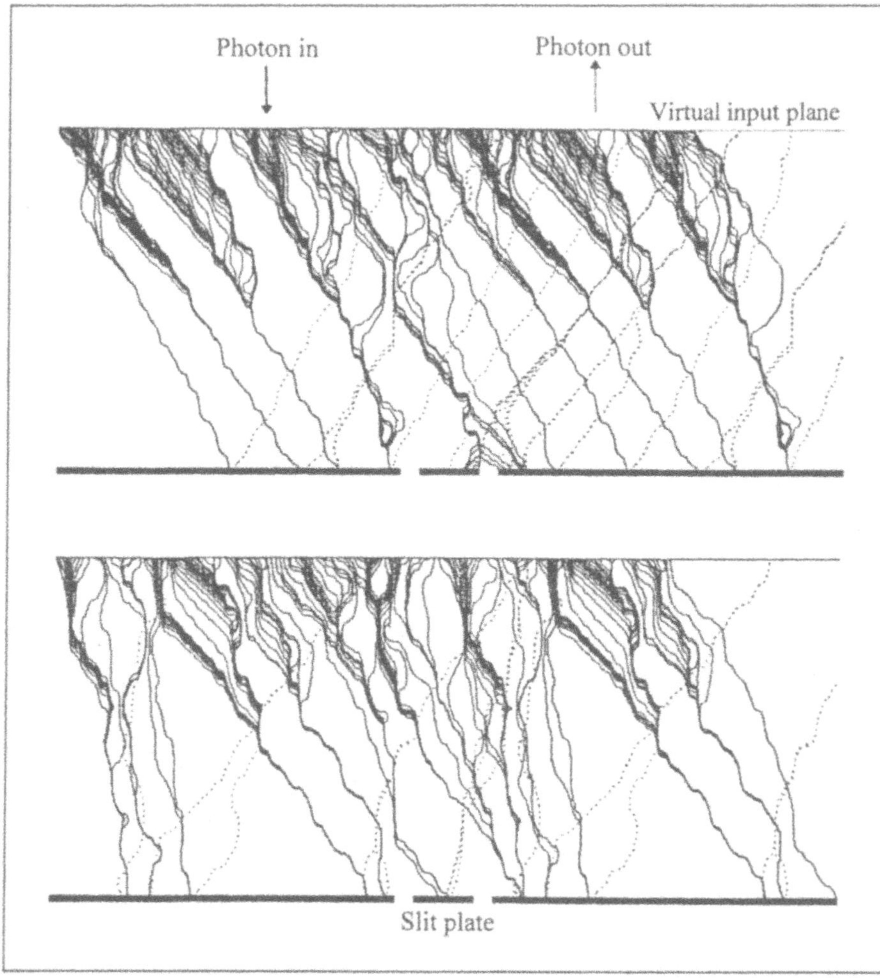

Figure 7.7. **Graphs showing two examples *of total reflections* of single photons from the double slit plate. Each graph represents a particular but arbitrary eigenvector selection $\mathbf{C}_p$ of the set Ψ_1 in (7.54). The solid lines are the *forward* flux lines, the dotted lines the reflected *backward* flux lines. The vertical scale is highly compressed by a factor of approximately 75.**

process has an absolute *deterministic* outcome. Quantum randomness is, in principle, indeterminate. In any classical probability game the background randomizing process is a necessary ingredient, providing minute differences in the initial conditions. As a familiar classical example the roulette game has well established probabilities for each outcome. The outcome of a particular play can not be predicted due to the randomizing process represented by initial conditions provided by the croupier spinning the wheel and launching the ball in a random fashion. But if we visualize a hypothetical roulette system with a mechanical 'croupier' launching subsequent plays with exactly the same input conditions, the ball would necessarily come to rest on the same number each time, provided of course that the wheel and table system are perfectly stable and do not contain additional randomizing factors. The randomizing process has been eliminated. The outcome is the same as long as the input is the same. And this is essentially the relevant situation of well separated *identical* photons impinging on the *same* double slit configuration.

In the neoclassical theory the answer to the question of photon separation into two groups does not rely on any inherent quantum probability and superposed quantum states. Without a proper balance of transmitted and reflected photons the associated admixture of doublet excitation would, over some extended time, add up to unacceptable levels of its time average. The basic requirement of zero time average leads to a balance of transmitted and reflected photons corresponding to the specifications by classical macroscopic electromagnetics. Not only is this requirement in accordance with the formal proposition (1.1) in Chapter 1, which of course is expected if the theory is right, but the detailed analysis presented in this section gives exactly the same conclusion.

The required doublet supplements were specified by (7.35). In the case of single photon *transmission* to point X_q on the screen, the backward part $\mathbf{a}_b^1$ was equal to zero, so that the equation reduces to

$$\begin{bmatrix} \alpha_{f,q}^1 \\ \alpha_{b,q}^3 \end{bmatrix}_{trans} = \left[\tilde{\mathbf{S}}^* - \rho\right] \begin{bmatrix} 0 \\ \mathbf{a}_{f,q}^3 \end{bmatrix} - \begin{bmatrix} \mathbf{a}_{f,in,q}^1 \\ 0 \end{bmatrix} \tag{7.56}$$

In the case of full photon reflection, the forward part $\mathbf{a}_f^3$ was zero, so that the generated doublet admixture for *reflection* of photon number p is given by

$$\begin{bmatrix} \alpha_{b,p}^1 \\ \alpha_{b,p}^3 \end{bmatrix}_{refl} = \left[\tilde{\mathbf{S}}^* - \rho\right] \begin{bmatrix} \mathbf{a}_{b,p}^1 \\ 0 \end{bmatrix} - \begin{bmatrix} \mathbf{a}_{f,in,p}^1 \\ 0 \end{bmatrix} \tag{7.57}$$

Of the overall number N of photons per second suppose that a total of N_q transmitted photons are identically treated in the sense that they incur the same doublet additive (7.56), specified by the same superposition $\mathbf{A}_q$ of Ψ_0 vectors. Also assume that a total of N_p photons are reflected, each with doublet supplement specified by (7.57). Let us add the doublet supplements from all the transmitted photons and all the reflected photons, by summing over all $q's$ and $p's$ and adding the two sets.

$$\alpha_{total} = \sum_q \frac{N_q}{N} \begin{bmatrix} \alpha_{f,q}^{1} \\ \alpha_{b,q}^{3} \end{bmatrix}_{trans} + \sum_p \frac{N_p}{N} \begin{bmatrix} \alpha_{b,p}^{1} \\ \alpha_{b,p}^{3} \end{bmatrix}_{refl}$$

$$= (\tilde{\mathbf{S}}^{*} - \rho) \begin{bmatrix} \sum_p \frac{N_p}{N} \mathbf{a}_{b,p}^{1} \\ \sum_q \frac{N_q}{N} \mathbf{a}_{f,q}^{3} \end{bmatrix} - \sum_q \frac{N_q}{N} \begin{bmatrix} \mathbf{a}_{f,in,q}^{1} \\ 0 \end{bmatrix} - \sum_p \frac{N_p}{N} \begin{bmatrix} \mathbf{a}_{f,in,p}^{1} \\ 0 \end{bmatrix} \tag{7.58}$$

where the total number N is conserved.

$$\sum_q \frac{N_q}{N} + \sum_p \frac{N_p}{N} = 1 \tag{7.59}$$

In the macroscopic case treated earlier the corresponding equation was listed in (7.4), for convenience repeated here.

$$(\tilde{\mathbf{S}}^{*} - \rho) \begin{bmatrix} \mathbf{a}_{b}^{1} \\ \mathbf{a}_{f}^{3} \end{bmatrix} - \begin{bmatrix} \mathbf{a}_{f,in}^{1} \\ 0 \end{bmatrix} = 0 \tag{7.60}$$

The time average of (7.58), taken over a large number N of photons must equal the macroscopic limit (7.60). Direct comparison of the two yields the following results:

$$\lim_{N\to\infty} \left\{ \sum_q \frac{N_q}{N} \mathbf{a}_{f,in,q}^{1} + \sum_p \frac{N_p}{N} \mathbf{a}_{f,in,p}^{1} \right\} = \frac{1}{N} \mathbf{a}_{f,in,macro}^{1} \tag{7.61}$$

$$\lim_{N\to\infty} \sum_p \frac{N_p}{N} \mathbf{a}_{b,p}^{1} = \frac{1}{N} \mathbf{a}_{b,macro}^{1} \tag{7.62}$$

$$\lim_{N\to\infty} \sum_q \frac{N_q}{N} \mathbf{a}_{f,q}^{3} = \frac{1}{N} \mathbf{a}_{f,macro}^{3} \tag{7.63}$$

$$\lim_{N\to\infty} = \left\{ \sum_q \frac{N_q}{N} \begin{bmatrix} \alpha_{f,q}^{1} \\ \alpha_{b,q}^{3} \end{bmatrix}_{trans} + \sum_p \frac{N_p}{N} \begin{bmatrix} \alpha_{b,p}^{1} \\ \alpha_{b,p}^{3} \end{bmatrix}_{refl} \right\} = 0 \tag{7.64}$$

What do the equations tell us? The first point is that the build up of black spots on the screen is distributed exactly as the experimentally observed macroscopic distribution, i.e., the diffraction pattern. This fact follows from (7.63) which perhaps can be clarified further by defining a spot density distribution function $n(X)_{screen}$ and replacing the sum by a continuous integral.

$$\int_{-D/2}^{D/2} n(X)_{screen}\, \mathbf{a}_f^3(X) dX = \frac{1}{N} \mathbf{a}^3_{f,macro} \tag{7.65}$$

And finally, (7.64) tells us that the time average of the generated doublet supplements vanishes, in agreement with he general principle of the neoclassical theory. All these equations are contingent on each other, meaning that the vanishing of time average implies a spot density function equal to the macroscopic observation. In a sense it can be stated that Bohr's correspondence principle of quantum physics is proved, or at least verified, for the double slit configuration. The mental picture projected from these significant results is a scenario in which the routing of the incident photon is orchestrated by the environmental photon doublets in strict adherence to the principle of minimum time average doublet energy. A particular photon is either fully transmitted or fully reflected, with the time average ratio N_{trans}/N_{refl} specified by (7.33). The decision to transmit or reflect hinges on the best alternative for minimizing overall doublet expenditure. If the optimum is transmission, the photon is routed to the particular spot on the screen which is minimizing the expenditure. On the other hand, if the best option is reflection, the process selects the particular linear combinations $\mathbf{C}_p$ in (7.54) that minimizes the accumulated doublet time average. The number of independent eigenvectors Ψ_1 is equal to the rank of $\mathbf{S}_{11}$ which is also the number of independent choices of reflected photons. The number of independent transmissions equals the nullity of $\mathbf{S}_{11}$ so that the ratio between transmitted and reflected photons is given by

$$\lim_{N\to\infty} \left(\frac{N_{trans}}{N_{refl}} \right) = \frac{\text{nullity}}{\text{rank}} \tag{7.66}$$

The entire process is orchestrated by the environmental photon doublets, permitting no excess of either positive or negative energy components. The principle of net zero time average energy is the all encompassing rule.

It has been tacitly assumed that all the incident photons are exactly the same, each one having the *same* input vector $\mathbf{a}^1_{f,in}$, which means a *coherent* macroscopic optical beam. It is a well documented experimental fact that only in this case do we get a diffraction pattern behind the slit plate. The absence of diffraction pattern for an incoherent beam must somehow be reflected in the present single photon theory. And this is indeed the case. With a sequence of *arbitrary* input vectors $\mathbf{a}^1_{f,in,q,p}$ representing the incoherent macroscopic input, the routing to specific points on the screen is affected exactly as before, but with the appropriate doublet admixture now being a function of the actual input vector. If the set of input vectors $\mathbf{a}^1_{f,in,q,p}$ are incoherent, so is the admixture, meaning that the optimum routing is randomized, with all points X_q on the screen equally probable. Thus, with incoherent input the theory predicts no diffraction pattern, just a blurred out screen illumination, in agreement with experimental facts.

7.3.8 Analysis of spot buildup of diffraction pattern

The macroscopic diffraction pattern on the screen is defined as the spot distribution N_q/N which is evaluated from the lower subvector of (7.58), noting that the left hand doublet expenditure vanishes for large N. Thus, the spot distribution function N_q/N is specified by the following implicit equation:

$$(\tilde{\mathbf{S}}^*_{22} - \boldsymbol{\rho}^3) \sum_q \frac{N_q}{N} \mathbf{a}^3_{f,q} = \mathbf{0} \tag{7.67}$$

Of the parameters appearing in the equation the scattering submatrix $\tilde{\mathbf{S}}^*_{22}$ is defined in (A.79) in the appendix to this chapter. The other parameters $\boldsymbol{\rho}^3$ and $\mathbf{a}^3_{f,q}$ depend on the photon arrival point X_q. The reflection matrix $\boldsymbol{\rho}^3$ is specified by (7.51) as a function of the spot impedance $z(X_q)$ which in the present analysis is assumed to be constant, but $z(X_q)$ could well be a function of X_q. The forward wave harmonic vector $\mathbf{a}^3_{f,q}$ is given by (7.50). Equation (7.67) specifies implicitly the spot distribution function $\mathbf{n}(X)$ which is defined as the N-dimensional vector

$$\mathbf{n}(X) = \frac{1}{N} \begin{bmatrix} N_1 \\ N_2 \\ \vdots \\ N_q \\ \vdots \\ N_{2n_{\max}+1} \end{bmatrix} \tag{7.68}$$

It is convenient to normalize $\mathbf{n}(X)$ to length of unity

$$\tilde{\mathbf{n}}(X)\mathbf{n}(X) = 1 \tag{7.69}$$

Let us define the complex vector $\hat{\mathbf{f}}_q$ by

$$\hat{\mathbf{f}}_q = \mathbf{f}_q e^{-j\phi_q} = (\tilde{\mathbf{S}}^*_{22} - \rho^3)\, \mathbf{a}^3_{f,q}\, e^{-j\phi_q} \qquad q = 1, 2, ...2n_{\max}+1 \tag{7.70}$$

where $\exp(-j\phi_q)$ is a phase factor to be determined later, and where

$$\mathbf{f}_q = (\tilde{\mathbf{S}}^*_{22} - \rho^3)\, \mathbf{a}^3_{f,q} \tag{7.71}$$

is a known vector. If the complex quadratic matrix $\mathbf{F}$ is defined by

$$\mathbf{F} = [\mathbf{f}_1, \mathbf{f}_2, ...\mathbf{f}_q...\mathbf{f}_{2n_{\max}+1}] \tag{7.72}$$

then (7.67) is expressed by a compact homogeneous equation in the unknown distribution column $\mathbf{n}(X)$:

$$\mathbf{F}\, \mathbf{e}^{j\Phi}\, \mathrm{n}(X) = 0 \tag{7.73}$$

The diagonal matrix $\mathbf{e}^{j\Phi}$ appearing in the equation is given by

$$\mathbf{e}^{j\Phi} = \mathrm{diag}(e^{j\phi_q}) \tag{7.74}$$

The matrix $\mathbf{F}$ is known, but the matrix $\exp(j\Phi)$ is at this stage unknown. The task at hand is to determine the spot distribution vector $\mathbf{n}(X)$ from the homogeneous equation (7.73). Whereas nontrivial solutions require a singular matrix $\mathbf{F}$, the equation is expected to be exactly satisfied only in the limiting case of very large N. Therefore, in the numerical procedure, with finite N, the matrix $\mathbf{F}$ can not be expected to be singular, only *close* to singular. Instead of the exact (7.73) we must put

$$\mathbf{F}\, \mathbf{e}^{j\Phi}\, \mathrm{n}(X) = \boldsymbol{\delta} \tag{7.75}$$

where $\boldsymbol{\delta}$ is some small N-dimensional complex vector. The numerical problem is to determine $\mathbf{n}(X)$ so that the length of $\boldsymbol{\delta}$ is minimized. From (7.75):

$$\tilde{\boldsymbol{\delta}}^* \boldsymbol{\delta} = \tilde{\mathrm{n}}(X)\, \mathbf{e}^{-j\Phi} \tilde{\mathbf{F}}^* \mathbf{F}\, \mathbf{e}^{j\Phi}\, \mathrm{n}(X) \tag{7.76}$$

where the Hermitian matrix $\mathbf{Q} = \tilde{\mathbf{F}}^*\mathbf{F}$ is positive definite. The minimization problem is resolved through the aid of the eigenvalue relation for the matrix $\mathbf{Q}$.

$$\mathbf{QR} = \mathbf{RD} \tag{7.77}$$

where the N columns of the $N \times N$ matrix $\mathbf{R}$ represent the orthogonal set of eigenvectors, presumably normalized so that $\tilde{\mathbf{R}}^*\mathbf{R} = \mathbf{1}$, and where $\mathbf{D}$ is the diagonal eigenvalue matrix, which is real because $\mathbf{Q}$ is Hermitian. With these preliminaries established, let us expand the spot distribution column $\mathbf{n}(X)$ in terms of the complete set of eigenvectors $\mathbf{R}$, putting

$$\mathbf{n}(X) = \mathbf{e}^{-j\Phi}\mathbf{R}\boldsymbol{\gamma} \tag{7.78}$$

where $\boldsymbol{\gamma}$ is some complex expansion vector to be determined. In view of the normalization condition (7.69) the following holds:

$$\tilde{\boldsymbol{\gamma}}^*\tilde{\mathbf{R}}^*\mathbf{e}^{j\Phi}\mathbf{e}^{-j\Phi}\mathbf{R}\boldsymbol{\gamma} = \tilde{\boldsymbol{\gamma}}^*\boldsymbol{\gamma} = 1 \tag{7.79}$$

showing that the expansion vector $\boldsymbol{\gamma}$ is also of length unity, a fact that will be used later. Substitution of (7.77) and (7.78) into (7.76) results in the equation

$$\tilde{\boldsymbol{\delta}}^*\boldsymbol{\delta} = |\boldsymbol{\delta}|^2 = \tilde{\boldsymbol{\gamma}}^*\mathbf{D}\boldsymbol{\gamma} = \sum_{i=1}^{2n_{\max}+1} D_i\,|\gamma|^2 \tag{7.80}$$

Noting that $\tilde{\boldsymbol{\gamma}}^*\boldsymbol{\gamma} = \mathbf{1}$ and that all eigenvalues D_i are positive, it follows that $|\boldsymbol{\delta}|^2$ is minimized by selecting the particular term in the sum corresponding to the smallest eigenvalue $D_{\min}$. If this particular i-value is i_0, then the minimum is obtained from

$$\begin{aligned} |\gamma_i| &= 1 \qquad i = i_0 \\ |\gamma_i| &= 0 \qquad i \neq i_0 \end{aligned} \tag{7.81}$$

The minimum is given by

$$(\tilde{\boldsymbol{\delta}}^*\boldsymbol{\delta})_{\min} = D_{i_0} \tag{7.82}$$

showing that the minimum is specified directly by the smallest eigenvalue of the known positive definite Hermitian matrix $\mathbf{Q} = \tilde{\mathbf{F}}^*\mathbf{F}$. The corresponding spot distribution function is obtained from (7.78).

$$\mathbf{n}(X) = \mathbf{e}^{-i\Phi}\mathbf{r}_{i_0} \tag{7.83}$$

where $\mathbf{r}_{i_0}$ is the eigenvector corresponding to the smallest eigenvalue D_{i_0}. Noting that the eigenvector is complex and that the spot density vector is real, the requirement imposed by these two stipulations specifies the unknown phase matrix $exp(j\Phi)$. At a particular spot X_q the equation specifies the following density:

$$n(X_q) = e^{-j\Phi_q}r_q \qquad q = 1, 2, \ldots 2n_{\max}+1 \tag{7.84}$$

where r_q is the qth element in the vector $\mathbf{r}_{i_0}$. The requirement of real $n(X_q)$ specifies Φ_q as follows:

$$\Phi_q = \arctan\left(\frac{\mathrm{Im}(r_q)}{\mathrm{Re}(r_q)}\right) \tag{7.85}$$

yielding

$$n(X_q) = |r_q| \qquad q = 1, 2, ...2n_{\max}+1 \tag{7.86}$$

Hence, the spot distribution vector is given by the simple expression

$$\mathbf{n}(X) = \mathrm{modulus}(\mathbf{r}_{i_0}) \tag{7.87}$$

where $\mathbf{r}_{i_0}$ was determined as the eigenvector corresponding to the smallest eigenvalue. The equation predicts the time average, or large scale density, of spots blackened by the single photons arriving at the screen. In the actual physical buildup of spots, as time progresses, sequential selection of photon routing is expected to be governed by the principle of *PDE* minimization, whereas the mathematical minimization procedure leading to (7.87) was based on the *final* condition of zero time average doublet admixture. The two must be compatible and therefore give rise to the same final spot distribution.

7.4 NUMERICAL PROCEDURE OF SPOT BUILD UP PATTERN

The numerical procedure for determination of the spot density function $\mathbf{n}(X)$ is a straight forward application of the theory developed in the preceding sections. With a selected system dimension $N = 141$ the eigenvalue relation (7.77) was solved for all the eigenvectors and eigenvalues. The eigenvectors separate into symmetric and antisymmetric groups, of which only the symmetric group is relevant for the uniform plane wave type input assumed in the present case. The eigenvalues in this group spanned the range from 0.004 to 340. In accordance with the minimization principle (7.82) the eigenvector $\mathbf{r}_{i_0}$ of the smallest eigenvalue $D_{i_0} = 0.004$ was selected as representative for the density vector $\mathbf{n}(X)$.

The results of this procedure are shown in Fig. 7.8. The number of spots set down at one particular location X_q is proportional to the element n_q of the vector $\mathbf{n}(X)$ in (7.87). The total number of points is 10000. As pointed out earlier by (7.46) the spatial resolution is equal to one half wavelength which in the figure will be $1/141$. The program distributes the n_q spots at X_q randomly over the spatial resolution around X_q. In a similar fashion the program distributes the spots randomly over the *length* of the slits, simulating the assumed constant average density in the Y-direction. For the purpose of comparison with the macroscopic diffraction pattern shown earlier in Fig. 7.3 the slit plate configurations were selected identical. Moreover, the two dot patterns (*b*) and (*c*) of Fig. 7.8 refer to the same screen positions as the two upper diffraction patterns in the macroscopic displays of Fig. 7.3. It is remarkable how closely they match each other with regard to the bands of maxima and minima. The results represent a significant demonstration how the neoclassical theory predicts exactly what is observed in the double slit experiment—that the transmitted photons arrive at the screen at localized spots dumping their energy at single photo absorbers, and over time build up the overall blackening to a periodic diffraction pattern that exactly matches the pattern obtained from regular macroscopic electromagnetic theory. It should be mentioned at this point that the diffraction patterns of Fig. 7.8 were unique for the *smallest* eigenvalue selection from the full group of eigenvectors. The few next lowest eigenvalues produced patterns not remotely resembling the ones in Fig. 7.8. The lowest eigenvalue selection turned out to be unique, and it can only be speculated that the eigenvalue approaches zero as the system dimension becomes infinite, a situation that would be fully compatible with *exact* zero time average of the doublet admixture, as expressed in (7.64). Admittedly, the theory is confined to one dimension laterally, a decision made at the beginning to avoid loosing track of the basic photon routing principles caused by the mere mathematical complexity of the added dimensional variation. However, a full two-dimensional analysis is not expected to add anything of importance that is conceptually different from the one-dimensional case.

This brings us to the end of the application of the neoclassical theory to the double slit experiment with single photons. There is an eon between this description and the quantum mechanical picture projecting the photon as a dual entity, sometimes a particle and sometimes a wave. In the neoclassical theory the embarrassing paradoxes such as the 'which way' problem and others, discussed briefly in the introduction, are completely absent. In this theory the photon is an honest and goodness electromagnetic entity focused and routed to localized spots on the screen through interaction with the environmental *PDE* fields. The overall distribution of transmitted, reflected, and deposited photons is orchestrated by the photon doublets under the auspices of the minimization principle of doublet fluctuations. This all encompassing selection mechanism—a basic general principle in the neoclassical theory—is completely foreign to standard quantum physics.

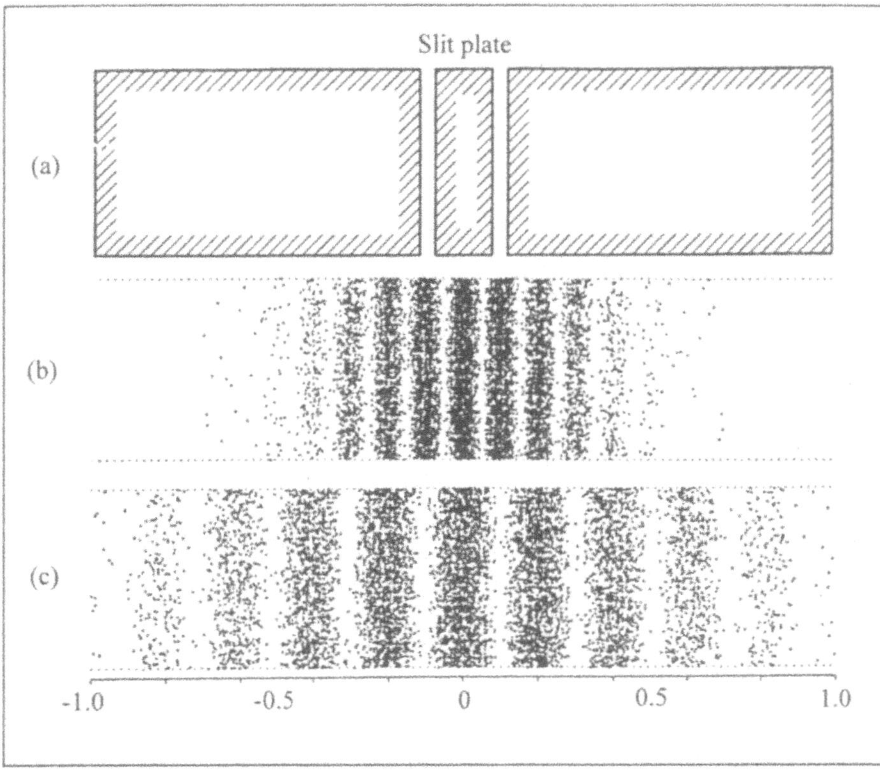

Figure 7.8. **Computer simulation of the build up of diffraction pattern by 10000 photons dumping their energy $\hbar\omega$ at localized spots on the screen. (a) Slit plate configuration. (b) Diffraction pattern at the screen location corresponding to the upper pattern of Fig. 7.3b. (c) Diffraction pattern next to the upper pattern of Fig. 7.3b.**

In the remaining chapters of the book the neoclassical theory of photons is applied to a number of other configurations and experimental situations that are central in discussions on interpretations of quantum mechanics. We shall invariably find that a consistent application of the theory resolves quantum paradoxes linked with these experiments.

APPENDIX A TO CHAPTER 7

A.1 FIELDS IN THE DOUBLE SLIT PLATE SYSTEM

The scattering matrix of the double slit plate is determined from a complete field matching procedure above and below the slit plate. With reference to Fig. 7.2 in Chapter 7 the electromagnetic fields are of the Transverse Magnetic (TM) type, assumed to be polarized in the x-direction, with no variation with respect to the transverse coordinate y.

$$\frac{\partial}{\partial y} = 0 \qquad H_x = H_z = E_y = 0$$

$$\vec{E} = E_x(x,z,t)\,\vec{a}_x + E_z(x,z,t)\,\vec{a}_z \tag{A.1}$$

$$\vec{H} = H_y(x,z,t)\,\vec{a}_y$$

$$\vec{\imath}(x,z,t) = \sigma(x,z)\vec{E}(x,z,t)$$

In the last equation the parameter $\sigma(x,z)$ is the metallic conductivity, being zero everywhere except in the slit plate itself, where it is equal to the constant σ. The equation is the scatter dominated approximation, valid for frequencies ω well below the dielectric relaxation frequency $\omega_{diel} = \sigma/\epsilon$, a condition well satisfied for metallic slit plates at optical frequencies. If the slit plate is dielectric rather than metallic, the dielectric susceptibility χ replaces the conductivity σ. In all essential respects the forthcoming analysis remains valid for both cases except for the detailed field distributions in the slit plate interior which is not an essential point in the forthcoming analysis.

As explained in Chapter 7 the specifications in (A.1) is a minimum set for the problem at hand. The restriction to one single lateral dimension x is not prohibitive for demonstration of the basic principles of photon transmission and reflection at the slit plate. A full three-dimensional analysis might actually confuse the issue without adding much in the sense of understanding.

A.1.1 A special adaptation of the field equations

With reference to Fig. 7.2 the following normalized lengths are used throughout:

$$X = kx \qquad Z = kz \qquad L_0 = k\ell_0 \qquad L = k\ell \qquad D = kd \tag{A.2}$$

Furthermore, it is convenient to normalize the frequency ω by the free space wavenumber $k = \omega/c$, and the magnetic field H_y by $h_y = Z_0 H_y$ where $Z_0 = \sqrt{\mu_0/\varepsilon_0}$ is the impedance of free space. Defining the complex amplitudes by lower case letters according to

$$E_x(X,Z,t) = \mathrm{Re}\left[e_x(X,Z)\,e^{j\omega t}\right] \tag{A.3}$$

and similarly for the other variables, Maxwell's equations are expressed by

$$\frac{\partial h_y(X,Z)}{\partial Z} = \left[-jg(X,Z) - \frac{Z_0}{k}\sigma f(X,Z)\right] e_x(X,Z)$$

$$\frac{\partial h_y(X,Z)}{\partial X} = \left[jg(X,Z) + \frac{Z_0}{k}\sigma f(X,Z)\right] e_z(X,Z) \tag{A.4}$$

$$\frac{\partial e_x(X,Z)}{\partial Z} - \frac{\partial e_z(X,Z)}{\partial X} = -jh_y(X,Z)$$

The functions $g(X,Z)$ and its complementary function $f(X,Z)$ appearing in these equations specify the overall slit plate configuration according to

$$g(X,Z) = \begin{cases} 0 & \text{in solid} \\ 1 & \text{elsewhere} \end{cases} \tag{A.5}$$

$$f(X,Z) = 1 - g(X,Z) = \begin{cases} 1 & \text{in solid} \\ 0 & \text{elsewhere} \end{cases} \tag{A.6}$$

By virtue of the special formulation (A.4) the field equations hold everywhere, in the slit plate region 2 as well as in the free space region 1 and region 3. In the free space regions the function $g(X,Z) = 0$ so that the field equations reduce to the regular free space forms. In the slit plate region $g(X,Z)$ is a function of X only, having the form shown in Fig. A.1. The slit plate function $g(X)$ is of *finite* width and, according to

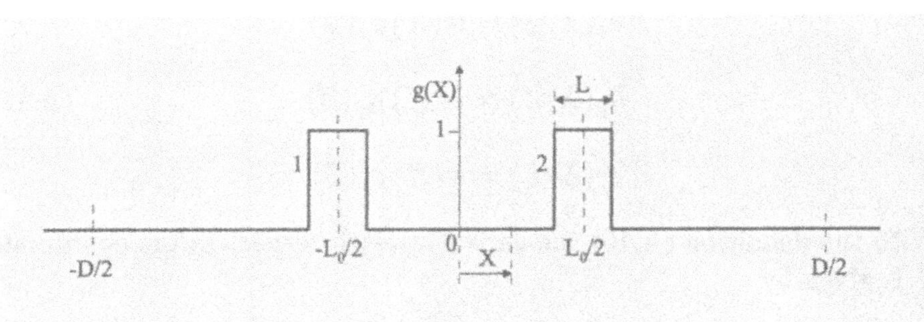

Figure A.1. **Slit plate function $g(X)$ in the slit plate region.**

fundamental Fourier theory, can be expressed as a Fourier series with a selectable fundamental period $D > 2(L_0+L)$. The width D is then the *window* over which the electromagnetic fields are expanded:

$$e_x(X,Z) = \sum_{n=-\infty}^{n=\infty} e_{xn}(Z)\, e^{-j2\pi n \frac{X}{D}} \tag{A.7}$$

and similarly for the other components. All expansions are periodic in X with fundamental period D so that the fields are repeated periodically along X. By selecting D relatively large relative to the slit spacing L_0 the local field perturbations due to the slits are essentially contained within the width D. The periodic repetitions are then of no concern. The requirement of propagating rather than evanescent wave solutions in the free space regions puts an upper limit on the harmonic number n

$$-n_{\max} \leq n \leq n_{\max} \tag{A.8}$$

where $n_{\max}$ will ne determined shortly. The spatial field harmonics are conveniently lumped together in column vectors of dimension $2\,n_{\max}+1$:

$$\mathbf{e}_x(Z) = \mathrm{col}\left[e_{xn}(Z)\right] \tag{A.9}$$

and similarly for the components $\mathbf{h}_y(Z)$ and $\mathbf{e}_z(Z)$. The periodic variations of the field components with X are represented by the column vector

$$\begin{aligned} \mathbf{w}(X) &= \mathrm{col}\left[e^{-jB_n X}\right] \\ B_n &= 2\pi n \tfrac{1}{D} \end{aligned} \tag{A.10}$$

The complex field amplitudes are then expressed in the compact form

$$\begin{aligned} e_x(X,Z) &= \tilde{\mathbf{w}}(X)\, \mathbf{e}_x(Z) \\ h_y(X,Z) &= \tilde{\mathbf{w}}(X)\, \mathbf{h}_y(Z) \\ e_z(X,Z) &= \tilde{\mathbf{w}}(X)\, \mathbf{e}_z(Z) \end{aligned} \tag{A.11}$$

With the definition (A.10) the derivative with respect to the coordinate X is given by

$$\frac{\partial}{\partial X} = -j\, \tilde{\mathbf{w}}(X)\, \mathbf{B} \tag{A.12}$$

where $\partial/\partial X$ operates on $\tilde{\mathbf{w}}(X)$, and $\mathbf{B}$ is the diagonal matrix

$$\mathbf{B} = \mathrm{diag}\left[B_n\right] \tag{A.13}$$

With these preliminaries established, the field equations can be transformed into a convenient form developed by the author for analyzing slow-wave periodic structures for microwave traveling wave tubes [46]. Let us demonstrate the procedure for the first equation in (A.4) which, after substitution from (A.11), transforms to

$$\tilde{\mathbf{w}}(X)\frac{d\,\mathbf{h}_y(Z)}{dZ} = -jg(X,Z)\,\tilde{\mathbf{w}}(X)\,\mathbf{e}_x(Z) - \frac{Z_0}{k}\sigma f(X,Z)\,\tilde{\mathbf{w}}(X)\,\mathbf{e}_x(Z) \tag{A.14}$$

Premultiply by the column $\mathbf{w}(X)^*$ and integrate over the width D. Let us note the following orthonormality relation:

$$\frac{1}{D}\int_{-D/2}^{D/2} \mathbf{w}(X)^*\tilde{\mathbf{w}}(X)dX = \mathbf{1} \tag{A.15}$$

where the right hand side is the unity matrix. Also, let us define the *convolution matrix* $\Phi(Z)$ by

$$\Phi(Z) = \frac{1}{D}\int_{-D/2}^{D/2} [1 - g(X,Z)]\,\mathbf{w}(X)^*\tilde{\mathbf{w}}(X)\,dX \tag{A.16}$$

which in view of (A.15) becomes

$$\Phi(Z) = \mathbf{1} - \frac{1}{D}\int_{-D/2}^{D/2} g(X,Z)\,\mathbf{w}(X)^*\tilde{\mathbf{w}}(X)\,dX \tag{A.17}$$

The convolution matrix $\Phi(Z)$ serves to completely determine the electromagnetic properties of the overall slit plate configuration considered as a circuit. This statement is perhaps more understandable if we proceed to express the field equations (A.4) in terms of $\Phi(Z)$ following the indicated procedure.

$$\begin{aligned}
\frac{d\mathbf{h}_y(Z)}{dZ} &= -j\mathbf{e}_x(Z) + j\,\Phi(Z)\,\mathbf{e}_x(Z) - \frac{Z_0}{k}\sigma\,\Phi(Z)\,\mathbf{e}_x(Z) \\
-j\mathbf{B}\mathbf{h}_y(Z) &= j\mathbf{e}_z(Z) - j\,\Phi(Z)\mathbf{e}_z(Z) + \frac{Z_0}{k}\sigma\,\Phi(Z)\,\mathbf{e}_z(Z) \\
\frac{d\mathbf{e}_x(Z)}{dZ} &+ j\mathbf{B}\,\mathbf{e}_z(Z) = -j\,\mathbf{h}_y(Z)
\end{aligned} \tag{A.18}$$

The fact that the convolution matrix $\Phi(Z)$ is constant in each of the three regions allows us to solve the set of equations separately for each region followed by appropriate matching procedures at the interfaces.

A.1.2 Field solutions in the free space regions

In free space the matrix $\Phi(Z)$ is zero, in accordance with (A.5) and (A.17).

$$\Phi(Z) = \mathbf{0} \tag{A.19}$$

Accordingly, the fields are specified by the following system of differential equations:

$$\frac{d}{dZ}\begin{bmatrix} \mathbf{e}_x(Z) \\ \mathbf{h}_y(Z) \end{bmatrix} = j\begin{bmatrix} \mathbf{0} & \mathbf{1}-\mathbf{B}^2 \\ \mathbf{1} & \mathbf{0} \end{bmatrix}\begin{bmatrix} \mathbf{e}_x(Z) \\ \mathbf{h}_y(Z) \end{bmatrix} \tag{A.20}$$

The general solutions of (A.20) are superpositions of forward and backward traveling components $\mathbf{a}_f$ and $\mathbf{a}_b$.

$$\mathbf{h}_y(Z) = \boldsymbol{\eta}^{-1/2}[e^{-j\boldsymbol{\eta}(\mathbf{Z}-\mathbf{Z}_2)}\mathbf{a}_f - e^{j\boldsymbol{\eta}(\mathbf{Z}-\mathbf{Z}_2)}\mathbf{a}_b]$$

$$\mathbf{e}_x(Z) = \boldsymbol{\eta}^{1/2}[e^{-j\boldsymbol{\eta}(\mathbf{Z}-\mathbf{Z}_2)}\mathbf{a}_f + e^{j\boldsymbol{\eta}(\mathbf{Z}-\mathbf{Z}_2)}\mathbf{a}_b] \tag{A.21}$$

$$\mathbf{e}_z(Z) = -\mathbf{B}\,\mathbf{h}_y(Z)$$

where Z_2 is the slit plate position, to be put equal to Z_{21} above and Z_{22} below the plate. The diagonal matrix $\boldsymbol{\eta}$ appearing in the equations is given by

$$\boldsymbol{\eta} = (\mathbf{1} - \mathbf{B}^2)^{1/2} \tag{A.22}$$

The matrix $\boldsymbol{\eta}$ represents the *impedance matrix* of the spatial harmonics set because

$$\mathbf{e}_x(Z)_{fwd} = \boldsymbol{\eta}\,\mathbf{h}_y(Z)_{fwd} \qquad \mathbf{e}_x(Z)_{bwd} = -\boldsymbol{\eta}\,\mathbf{h}_y(Z)_{bwd} \tag{A.23}$$

In the harmonic range for which $\boldsymbol{\eta}$ is real, i.e., for

$$|B_n| \leq 1 \qquad n = \pm 1, \pm 2, ... \pm n_{\max} \tag{A.24}$$

the components represent *propagating* modes. The condition specifies the acceptable maximum $n_{\max}$ from the equation

$$B_{\max} = 2\pi\frac{n_{\max}}{D} = 2\pi\frac{n_{\max}}{kd} = 1 \tag{A.25}$$

yielding

$$n_{\max} = \frac{d}{\lambda_0} \tag{A.26}$$

where λ_0 is the free space wavelength. Hence, the maximum mode number $n_{\max}$ is equal to the periodic width d measured in wavelengths, and the total number of harmonics is $2n_{\max}+1$.

A.1.3 Slit plate convolution matrix

In the slit plate where $Z_{21} \le Z \le Z_{22}$ the slit plate function $g(X,Z)$ defined in (A.5) is independent of Z so that the convolution matrix Φ is constant and given by

$$\Phi = \mathbf{1} - \frac{1}{D}\int_{slits} g(X)\,\mathbf{w}(X)^*\tilde{\mathbf{w}}(X)\,dX = \mathbf{1} - \frac{1}{D}\int_{slits} \mathbf{w}(X)^*\tilde{\mathbf{w}}(X)\,dX \tag{A.27}$$

The matrix elements Φ_{nm} are readily evaluated. Defining n' by

$$n' = n - m \tag{A.28}$$

we find

$$\begin{aligned} \Phi_{nm} &= -2\frac{L}{D}\frac{\sin(\pi n'\frac{L}{D})}{\pi n'\frac{L}{D}}\cos(\pi n'\tfrac{L_0}{D}), && n' \neq 0 \\ \Phi_{nn} &= 1 - 2\frac{L}{D}, && n' = 0 \end{aligned} \tag{A.29}$$

The Φ-matrix has a number of properties of importance in the solution of the field equations. Some of these are readily apparent from the form of the matrix elements such as:

$$\Phi_{nm} = \Phi_{mn} = \Phi_{n'} \tag{A.30}$$

The matrix is real, symmetric and of the Toeplitz type where all elements along any subdiagonal are the same. This implies that the matrix is specified by a total of $2n_{\max}+1$ elements so that we can write

$$\Phi_{nm} \rightarrow \Phi_n \qquad |n| \le 2n_{\max} \tag{A.31}$$

The subdiagonal elements Φ_n are closely related to the slit function $g(X)$ or rather its complementary function $f(X)$. The column $\mathbf{f}$ specifying the Fourier components of $f(X)$ is given by

$$\mathbf{f} = \frac{1}{D}\int_{-D/2}^{D/2} f(X)\,\mathbf{w}(X)^*\,dX = \mathbf{1} - \frac{1}{D}\int_{slits} \mathbf{w}(X)^*\,dX \tag{A.32}$$

where $\mathbf{1}$ is a unity column. Comparison with (A.27) reveals the following relation:

$$f_n = f_{-n} = \Phi_n = \Phi_{-n} \tag{A.33}$$

demonstrating that the *nth* subdiagonal element of Φ is the *nth* Fourier component of the complementary slit function $f(X) = 1 - g(X)$. This equivalence justifies the mental interpretation of the Φ-matrix as the Fourier space representation of the slit plate, although a more accurate description would be slit plate convolution matrix.

A.1.4 Slit plate orthogonal eigenvectors

The Φ-matrix has a complete set of orthogonal eigenvectors $\mathbf{U}$ with real eigenvalues $\mathbf{D}$ described by the eigensystem relation

$$\Phi \mathbf{U} = \mathbf{U}\mathbf{D} \tag{A.34}$$

Due to the special form of the Φ-matrix the eigensystem splits into two separate classes, one having all eigenvalues equal to zero, and one all eigenvalues equal to unity. This statement is contingent on the matrix being of very large dimension, in fact it holds exactly if

$$\dim \Phi \rightarrow \infty \tag{A.35}$$

and only approximately for finite dimension, as will be demonstrated in a moment. The proof goes as follows: The *ith* eigenvalue relation has the form

$$(\Phi - d_i \mathbf{1})\mathbf{u}_i = \mathbf{0} \tag{A.36}$$

which requires the determinant to vanish

$$\begin{vmatrix} (\Phi_0 - d_i) & \Phi_{-1} & \cdots & \Phi_{-2n_{\max}+1} & \Phi_{-2n_{\max}} \\ \Phi_1 & (\Phi_0 - d_i) & \cdots & \Phi_{-2n_{\max}+2} & \Phi_{-2n_{\max}+1} \\ \vdots & \vdots & \ddots & \vdots & \vdots \\ \Phi_{2n_{\max}-1} & \Phi_{2n_{\max}-2} & \cdots & (\Phi_0 - d_i) & \Phi_{-1} \\ \Phi_{2n_{\max}} & \Phi_{2n_{\max}-1} & \cdots & \Phi_1 & (\Phi_0 - d_i) \end{vmatrix} = 0 \tag{A.37}$$

Adding linear combinations of rows or columns leaves the determinant invariant. Accordingly, let us multiply all rows with $\exp(j2\pi nX/D)$ and all columns by $\exp(j2\pi mX/D)$, n and m being the row and column indices, respectively, and X some argument in D. By this process a particular determinant element n, m is multiplied by $\exp\left[-j2\pi(n-m)\, X/D\right]$ which is the same for all elements along the subdiagonal $n-m$. The proof is now straight forward, if it is recognized that Φ is actually of infinite dimension so that the elements extend beyond those in (A.37). Simply add all the rows in the determinant to the first row, taking care to include supplementary rows above and below those shown in the finite and truncated version (A.37). It follows then that all elements in the first row are equal and given by

$$a = -d_i + \sum_{-2n_{\max}}^{2n_{\max}} \Phi(n) e^{-j2\pi n \frac{X}{D}} = -d_i + f(X) \qquad \text{(A.38)}$$

The latter relation follows from (A.33) relating the matrix elements $\Phi(n)$ to the Fourier components of the slit plate function $f(X)$. The common factor (A.38) can be taken outside the determinant showing that (A.37) is solved by

$$d_i = f(X) = \begin{cases} 0 & \text{in slits} \\ 1 & \text{in solid} \end{cases} \qquad \text{(A.39)}$$

The equation shows that the eigenvalues are either zero or unity depending on the selection of the X-parameter in the multiplication process described above. The $f(X)$ function of (A.38) is the slit plate function evaluated from the finite number of spatial harmonics $N = 2n_{\max}+1$ where $n_{\max} = d/\lambda_0$. The consequence of the finite range is that (A.39) is exact only in the infinite case, requiring an infinite fundamental window d or rather infinite ratio d/λ_0. In principle this condition is possible, but in computational procedures dimensions must be finite. Figure A.2 serves to illustrate this point by showing the eigenvalues in a normalized distribution for a set of finite N, with $N_{\max} = 141$ being the largest number. The slit plate parameters are the same as those of Figs. 7.3-7.8 in Chapter 7.

The analysis proceeds with the assumption of infinite dimension so that (A.39) is satisfied exactly. Computer simulations with finite dimensions are found to converge very rapidly and present no major problems. The eigenvalue specification (A.39) implies that the rank of Φ is equal to its trace.

$$\begin{aligned} \text{rank} &= N_1 = N(1 - 2L/D) \\ \text{nullity} &= N_0 = N - \text{rank} = N\,2L/D \end{aligned} \qquad \text{(A.40)}$$

The corresponding set of eigenvectors $\mathbf{U}_1$ and $\mathbf{U}_0$ span the *range* and the *nullspace*, respectively, satisfying

$$\begin{aligned} \Phi\mathbf{U}_1 &= \mathbf{U}_1 & \qquad \Phi\mathbf{U}_0 &= \mathbf{0} \\ (\mathbf{1} - \Phi)\mathbf{U}_1 &= \mathbf{0} & \qquad (\mathbf{1} - \Phi)\mathbf{U}_0 &= \mathbf{U}_0 \end{aligned} \qquad \text{(A.41)}$$

The two orthogonal sets $\mathbf{U}_0$ and $\mathbf{U}_1$ satisfy a number of relations needed at a later stage. Because Φ is symmetric, the following holds:

$$\tilde{\mathbf{U}}\,\mathbf{U} = \mathbf{U}\,\tilde{\mathbf{U}} = \mathbf{1} \qquad \text{(A.42)}$$

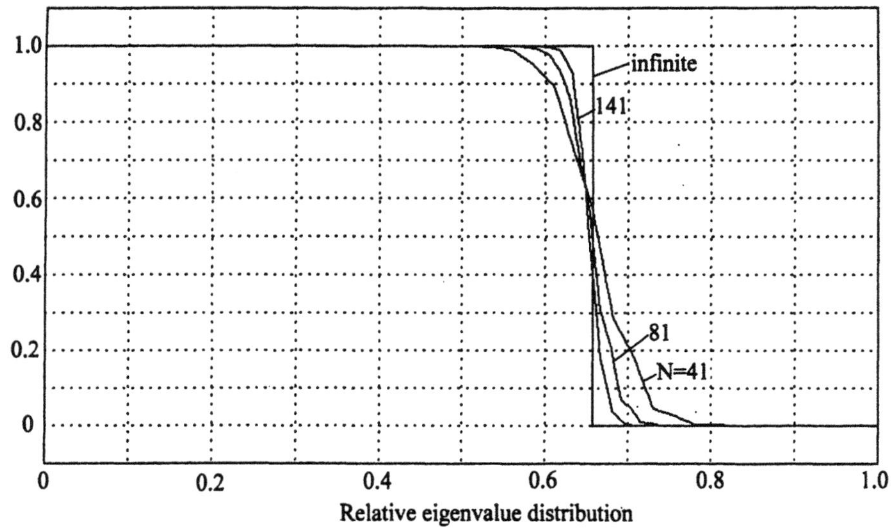

Figure A.2. Computer plots of the eigenvalue distributions plotted vs normalized range of the harmonic number N for $N=41$, 81, 141, and ∞. Slit plate parameters are $L_0/D = 0.1$, $L/D = 0.025$, the same as those of Figs. 7.3-7.8 in Chapter 7.

yielding

$$\begin{aligned} \tilde{\mathbf{U}}_0\mathbf{U}_0 &= \tilde{\mathbf{U}}_1\mathbf{U}_1 = \mathbf{1} \qquad & \tilde{\mathbf{U}}_0\mathbf{U}_1 &= \tilde{\mathbf{U}}_1\mathbf{U}_0 = \mathbf{0} \\ \mathbf{U}_0\tilde{\mathbf{U}}_0 + \mathbf{U}_1\tilde{\mathbf{U}}_1 &= \mathbf{1} \qquad & \mathbf{\Phi} &= \mathbf{U}_1\tilde{\mathbf{U}}_1 = \mathbf{1} - \mathbf{U}_0\tilde{\mathbf{U}}_0 \end{aligned} \tag{A.43}$$

The two sets $\mathbf{U}_0$ and $\mathbf{U}_1$ are instrumental in separating the fields in the slit plate region into two classes with distinctly different properties shown in the following demonstration.

A.1.5 Separation of fields in the slit plate region

The following is a general proof of a significant slit plate property relating to electromagnetic fields belonging to set $\mathbf{U}_0$ and $\mathbf{U}_1$. Let a general electric or magnetic field variable $c(X,Z)$ in the slit plate region be specified by the following superposition of spatial harmonics:

$$c(X,Z) = \tilde{\mathbf{w}}(X)\,\mathbf{c}(Z) \tag{A.44}$$

where X is in the range $-D/2$ to $D/2$ and the expansion coefficients are contained in the column $\mathbf{c}(Z)$, with Z limited to the slit plate thickness Z_{22}–Z_{21}. The column $\mathbf{c}(Z)$ can always be expanded in the *complete* set of eigenvectors $\mathbf{U}_0$ and $\mathbf{U}_1$ as follows:

$$\mathbf{c}(Z) = \mathbf{U}_0\,\hat{\mathbf{c}}_0(Z) + \mathbf{U}_1\hat{\mathbf{c}}_1(Z) \tag{A.45}$$

where $\hat{\mathbf{c}}_0(Z)$ and $\hat{\mathbf{c}}_1(Z)$ are expansion vectors for the two sets. Let us square the arbitrary field component $c(X,Z)$ and integrate the result over the solid part and the slits, respectively.

$$I_{solid} = \frac{1}{D} \int_{solid} |c(X,Z)|^2 \, dX = \tilde{\mathbf{c}}(Z)^* \left[\frac{1}{D} \int_{solid} \mathbf{w}(X)^* \tilde{\mathbf{w}}(X) \, dX \right] \mathbf{c}(Z)$$

$$I_{slits} = \frac{1}{D} \int_{slits} |c(X,Z)|^2 \, dX = \tilde{\mathbf{c}}(Z)^* \left[\frac{1}{D} \int_{slits} \mathbf{w}(X)^* \tilde{\mathbf{w}}(X) \, dX \right] \mathbf{c}(Z) \tag{A.46}$$

Using (A.27) the equations can be expressed by the convolution matrix Φ in the forms

$$\begin{aligned} I_{solid} &= \tilde{\mathbf{c}}(Z)^* \Phi \, \mathbf{c}(Z) \\ I_{slits} &= \tilde{\mathbf{c}}(Z)^* (\mathbf{1} - \Phi) \, \mathbf{c}(Z) \end{aligned} \tag{A.47}$$

Substituting for $\mathbf{c}(Z)$ from (A.45) and making use of the relations (A.43) we find

$$\begin{aligned} I_{solid} &= \tilde{\mathbf{c}}_1(Z)^* \mathbf{U}_1 \, \mathbf{c}_1(Z) \\ I_{slits} &= \tilde{\mathbf{c}}_0(Z)^* \mathbf{U}_0 \, \mathbf{c}_0(Z) \end{aligned} \tag{A.48}$$

Recalling that I_{solid} and I_{slits} represent the square of the general field amplitude $c(X,Z)$ integrated over the solid part and the slits, respectively, we have proved the following:

- *Any field distribution of the* $\mathbf{U}_0$ *class vanishes everywhere in the solid and is nonzero in the slits.*
- *Any field distribution of the* $\mathbf{U}_1$ *class vanishes everywhere in the slits and is nonzero in the solid.*

The two classes represent complementary distributions with clear physical interpretations with regard to photon behavior. Electromagnetic fields of photons propagating *through* the slits must necessarily belong to the nullspace $\mathbf{U}_0$ of $\mathbf{U}$. Fields of photons being *reflected* from the slit plate must belong to the range $\mathbf{U}_1$ of $\mathbf{U}$. Note that the proof has been derived entirely from the slit plate *geometry* and is not contingent on the material properties such as conductivity σ or susceptibility χ. Only the details of field distributions inside the solid depend on material properties, as demonstrated shortly.

The properties of the convolution matrix Φ specified in (A.43) as a set of relations between the two complementary classes $\mathbf{U}_0$ and $\mathbf{U}_1$ are necessary preliminaries for determining the fields in the slit plate region. By substitution of the appropriate relations from (A.43) into the general field equations (A.18) the following equations result:

$$\frac{d\,\mathbf{h}_y(Z)}{dZ} = -j\mathbf{U}_0\tilde{\mathbf{U}}_0\,\mathbf{e}_x(Z) - \frac{Z_0}{k}\sigma\,\mathbf{U}_1\tilde{\mathbf{U}}_1\,\mathbf{e}_x(Z)$$

$$-j\mathbf{B}\mathbf{h}_y(Z) = j\mathbf{U}_0\tilde{\mathbf{U}}_0\,\mathbf{e}_z(Z) - \frac{Z_0}{k}\sigma\mathbf{U}_1\tilde{\mathbf{U}}_1\;\mathbf{e}_z(Z) \qquad \text{(A.49)}$$

$$\frac{d\,\mathbf{e}_x(Z)}{dZ} + j\mathbf{B}\,\mathbf{e}_z(Z) = -j\,\mathbf{h}_y(Z)$$

The next step is to expand, as in (A.45), the general harmonic field vectors $\mathbf{e}_z(Z)$, $\mathbf{h}_y(Z)$, and $\mathbf{e}_z(Z)$ appearing in these equations in terms of the *complete* set $\mathbf{U} = [\mathbf{U}_0, \mathbf{U}_1]$.

$$\mathbf{e}_x(Z) = \mathbf{U}_0\,\hat{\mathbf{e}}_{x0}(Z) + \mathbf{U}_1\,\hat{\mathbf{e}}_{x1}(Z)$$

$$\mathbf{h}_y(Z) = \mathbf{U}_0\,\hat{\mathbf{h}}_{y0}(Z) + \mathbf{U}_1(Z)\,\hat{\mathbf{h}}_{y1}(Z) \qquad \text{(A.50)}$$

$$\mathbf{e}_z(Z) = \mathbf{U}_0\,\hat{\mathbf{e}}_{z0}(Z) + \mathbf{U}_1(Z)\,\hat{\mathbf{e}}_{z1}(Z)$$

It follows, from the general properties expressed by (A.48), that the expansion columns satisfy the following relations.

All fields in the slits:

$$\hat{\mathbf{e}}_{x1}(Z) = \hat{\mathbf{h}}_{y1}(Z) = \hat{\mathbf{e}}_{z1}(Z) = \mathbf{0} \qquad \text{(A.51)}$$

All fields in the solid:

$$\hat{\mathbf{e}}_{x0}(Z) = \hat{\mathbf{h}}_{y0}(Z) = \hat{\mathbf{e}}_{z0}(Z) = \mathbf{0} \qquad \text{(A.52)}$$

The absence of overlapping of the two classes of field components shown above rules out any volume coupling between them. However, the slits and the solid parts have a common boundary represented by the slit walls giving rise to *surface* coupling due to continuity of tangential magnetic fields across the boundary. The coupling strength is expected to be proportional to the slit wall area, i.e., to the plate thickness and is therefore negligible for the thin plate of the double slit configuration. On the other hand, the two $\mathbf{U}_1$ class distributions on either side of the slit plate are probably coupled through the continuity of surface currents at the slits. The assumption of a thin slit plate with negligible volume coupling allows separate processing of the field equations for the two classes, as shown in the following.

A.1.6 Open slit fields of the nullspace set $\mathbf{U}_0$

Substituting from (A.50) into (A.49), noting that the $\mathbf{U}_1$ expansion vectors (A.51) are zero, and then premultiplying the equations with $\tilde{\mathbf{U}}_0$, we obtain the following differential equations for the expansion vectors of the $\mathbf{U}_0$ system:

$$\frac{d}{dZ}\begin{bmatrix} \hat{\mathbf{e}}_{x0}(Z) \\ \hat{\mathbf{h}}_{y0}(Z) \end{bmatrix} + j\begin{bmatrix} \mathbf{0} & \mathbf{1} - (\tilde{\mathbf{U}}_0\mathbf{B}\mathbf{U}_0)^2 \\ \mathbf{1} & \mathbf{0} \end{bmatrix}\begin{bmatrix} \hat{\mathbf{e}}_{x0}(Z) \\ \hat{\mathbf{h}}_{y0}(Z) \end{bmatrix} = \mathbf{0} \qquad \text{(A.53)}$$

The equation, subject to appropriate boundary conditions, specifies the fields everywhere in the slits. By comparison with the free space solutions (A.20) it is immediately apparent that the solutions are of the same form, the only difference being the *nondiagonal* propagation matrix $(\tilde{\mathbf{U}}_0\mathbf{B}\mathbf{U}_0)^2$ replacing the *diagonal* matrix $\mathbf{B}^2$. The physical implication of this difference is readily recognized as a coupling affected by the slit plate between the otherwise orthogonal space harmonics of free space. If the slit plate is thin, it is unessential to solve (A.53) in detail. It suffices to specify the $\mathbf{U}_0$ class fields as being continuous and constant from above to below the slit plate.

A.1.7 Solid slit plate fields of the range set $\mathbf{U}_1$

Putting the $\mathbf{U}_0$ expansion vectors (A.52) equal to zero and carrying out the same procedure, but now premultiplying with $\tilde{\mathbf{U}}_1$ rather than $\tilde{\mathbf{U}}_0$, we obtain the following differential equation for the transverse field variables:

$$\frac{d}{dZ}\begin{bmatrix} \hat{\mathbf{e}}_{x1}(Z) \\ \hat{\mathbf{h}}_{y1}(Z) \end{bmatrix} + j\begin{bmatrix} \mathbf{0} & \mathbf{1} - j\dfrac{k}{Z_0\sigma}(\tilde{\mathbf{U}}_1\mathbf{B}\mathbf{U}_1)^2 \\ -j\dfrac{Z_0}{k}\sigma & \mathbf{0} \end{bmatrix}\begin{bmatrix} \hat{\mathbf{e}}_{x1}(Z) \\ \hat{\mathbf{h}}_{y1}(Z) \end{bmatrix} = \mathbf{0} \qquad \text{(A.54)}$$

The longitudinal electric field $\hat{\mathbf{e}}_{z1}(Z)$ is directly related to $\hat{\mathbf{h}}_{y1}(Z)$ through the equation

$$\hat{\mathbf{e}}_{z1}(Z) = -j\frac{k}{Z_0\sigma}\tilde{\mathbf{U}}_1\mathbf{B}\mathbf{U}_1\,\hat{\mathbf{h}}_{y1}(Z) \qquad \text{(A.55)}$$

In these equations, relating to the set $\mathbf{U}_1$, we encounter the minor obstacle that the conductivity σ appearing as a parameter in the system matrix is very large, being infinite for a lossless metallic slit plate. But this particular feature does not present a major problem. The differential system is solved through an appropriate transformation scheme involving the normalized complex skin depth $\Delta = k\delta$, which is specified by the usual formula

$$\Delta = \sqrt{-j\frac{k}{Z_0\sigma}} \tag{A.56}$$

Furthermore, define new variables indicated by primes:

$$Z' = \frac{1}{\Delta}Z \qquad \hat{\mathbf{e}}'_{x1}(Z) = \frac{1}{\Delta}\hat{\mathbf{e}}_{x1}(Z) \qquad \hat{\mathbf{e}}'_{z1}(Z) = \frac{1}{\Delta}\hat{\mathbf{e}}_{z1}(Z) \tag{A.57}$$

Hereby, (A.54) and (A.55) are transformed to the simpler set

$$\frac{d}{dZ'}\begin{bmatrix} \hat{\mathbf{e}}'_{x1}(Z) \\ \hat{\mathbf{h}}_{y1}(Z) \end{bmatrix} + j\begin{bmatrix} 0 & 1 \\ -1 & 0 \end{bmatrix}\begin{bmatrix} \hat{\mathbf{e}}'_{x1}(Z) \\ \hat{\mathbf{h}}_{y1}(Z) \end{bmatrix} = \mathbf{0}+O(\Delta^2) \tag{A.58}$$

$$\hat{\mathbf{e}}'_{z1}(Z) = \mathbf{0}+O(\Delta) \tag{A.59}$$

where $O(\Delta)$ and $O(\Delta^2)$ are small terms of first and second order in the skin depth Δ. Correct to these orders the new system matrix in the transformed differential equation does not contain the conductivity σ. Therefore, it is readily solved for the transformed variables. The solutions are seen to be exponentially decaying wave solutions which are not shown explicitly here. Subsequent transformation back to the original variables yields the following solutions of (A.54):

$$\begin{aligned} \hat{\mathbf{h}}_{y1}(Z) &= e^{-(Z-Z_{21})/\Delta}\,\mathbf{a}_{11} - e^{(Z-Z_{22})/\Delta}\,\mathbf{a}_{12} \\ \hat{\mathbf{e}}_{x1}(Z) &= j\Delta\left[e^{-(Z-Z_{21})/\Delta}\,\mathbf{a}_{11} + e^{(Z-Z_{22})/\Delta}\,\mathbf{a}_{12}\right] \\ \hat{\mathbf{e}}_{z1}(Z) &= \mathbf{0}+O(\Delta^2) \end{aligned} \tag{A.60}$$

The coordinates Z_{21} and Z_{22} refer to the upper and lower slit plate surfaces, respectively. As expected, the two solutions $\mathbf{a}_{11}$ and $\mathbf{a}_{12}$ represent *surface* distributions, one at the upper surface, and one at the lower surface, with penetration depths equal to the skin depth Δ. The longitudinal electric field vector $\hat{\mathbf{e}}_{z1}(Z)$ is of order Δ^2 and therefore entirely negligible, whereas the X-component $\hat{\mathbf{e}}_{x1}(Z)$ is of order Δ and therefore also very small, being zero for a lossless slit plate. The associated current density does not vanish correspondingly, because $\hat{\imath}_{x1}(Z) = \sigma\hat{\mathbf{e}}_{x1}(Z)$ so that $\hat{\imath}_{x1}(Z)$ is proportional to $1/\Delta$, and therefore approaching infinity for the lossless plate. Of more interest than the current density is the X-directed *total* surface current $\mathbf{I}_x$ obtained by integrating the current density over the full surface penetration depth. At the upper and lower surfaces, respectively:

$$\mathbf{I}_{x1} = j\frac{\sigma}{k}\Delta \int_{Z=Z_{21}}^{\infty} e^{-(Z-Z_{21})/\Delta} dZ\ \mathbf{a}_{11} = \mathbf{a}_{11} = \frac{1}{Z_0}\hat{\mathbf{h}}_{y1}(Z_{21}) = \hat{\mathbf{H}}_y(Z_{21})$$

$$\mathbf{I}_{x2} = j\frac{\sigma}{k}\Delta \int_{-\infty}^{Z_{22}} e^{(Z-Z_{22})/\Delta} dZ\ \mathbf{a}_{12} = -\frac{1}{Z_0}\mathbf{a}_{12} = -\frac{1}{Z_0}\hat{\mathbf{h}}_{y1}(Z_{22}) = -\hat{\mathbf{H}}_y(Z_{22}) \tag{A.61}$$

The relations just confirm the well known fact that the surface currents $\mathbf{I}_x$ are equal to the corresponding tangential magnetic fields $\hat{\mathbf{H}}_y$ at the two surfaces. The exercise gone through in the development leading to these equations may appear as merely an adaptation of skin depth analysis to the slit plate configuration. But the real benefit is the linking of surface type field distributions to the $\mathbf{U}_1$ class solutions.

The surface type solutions belonging to the $\mathbf{U}_1$ class presented in (A.60) and (A.61) are contingent on the slit plate thickness $T = Z_{22} - Z_{21}$ being much larger than the skin depth $|\Delta|$; otherwise the two surface distributions would overlap, with obvious modifications needed in the stated formulae. But this is an utterly unrealistic possibility because the skin depth at optical frequencies in any metallic conductor is extremely small, so small in fact that a mechanically stable slit plate would have to be orders of magnitude thicker. On the other hand, for the $\mathbf{U}_0$ class fields in the slits the plate was assumed to be thin. In this case the statement means thin compared to the free space wavelength which is orders of magnitude larger than the skin depth. From inspection of (A.53) it is clear, however, that the assumption of a thin plate is merely a mathematical convenience allowing us to specify continuity of $\mathbf{U}_0$ class slit fields across the plate thickness t. If desirable the equation is readily solvable for any plate thickness.

A.1.8 Slit plate boundary conditions

The preceding analysis makes it quite clear that the slit plate boundary conditions are entirely different for fields of the $\mathbf{U}_0$ and $\mathbf{U}_1$ classes. The overall boundary conditions are summarized in Fig. A.3. The actual *inhomogeneous* slit plate with its open slits shown in (a) is represented by *homogeneous* virtual plates (b) and (c) having no slits at all. The fields of the $\mathbf{U}_0$ class shown in (b) are, within the assumption discussed above, constant across the plate thickness. For the $\mathbf{U}_1$ class fields depicted in (c) the situation is entirely different. Because all fields vanish identically *inside* the plate, the $\mathbf{U}_1$ class fields in free space above the plate is effectively insulated from the $\mathbf{U}_1$ class fields in free space below the plate. The coupling of $\mathbf{U}_1$ fields between the free space regions is completely missing.

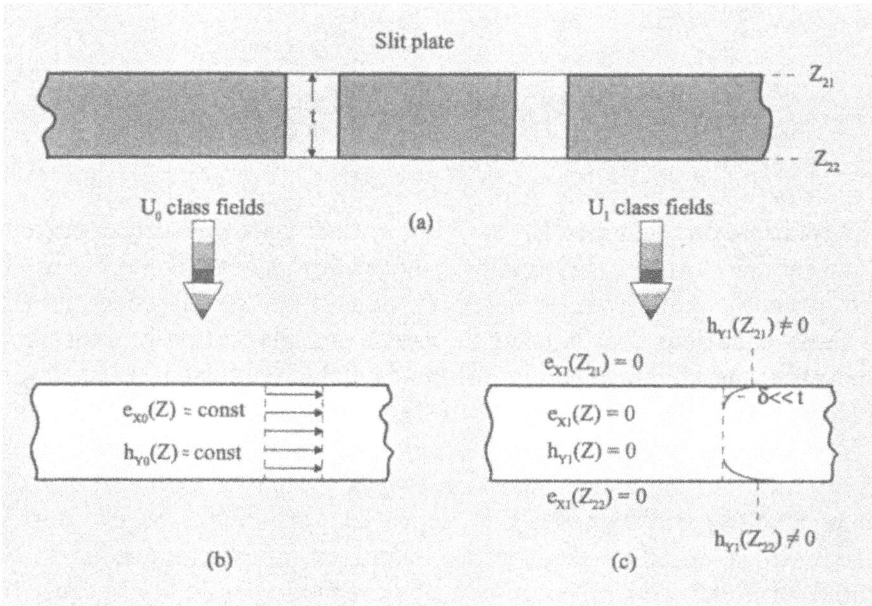

Figure A.3. Illustration of boundary conditions in the transformed virtual slit plate representations. (a) Original slit plate cross section with slits. (b) Equivalent homogeneous slit plate medium for $\mathbf{U}_0$ class electromagnetic fields. The fields are *continuous* across the plate. (c) Equivalent homogeneous slit plate medium for $\mathbf{U}_1$ class electromagnetic fields. The electric field components vanish everywhere. The magnetic field is non-zero at both surfaces Z_{21} and Z_{22}, decaying rapidly to zero inside the plate, with skin depth δ negligibly small compared to plate thickness t.

The fields in the free space regions 1 and region 3 are determined from additional boundary conditions having to do with transmitted and reflected wave components, to be specified in the following. The analysis is aimed at determining the *scattering matrix* of the slit plate. The full free space field solutions at the upper and lower surfaces of the slit plate are readily obtained from the general free space solutions (A.21). The plate thickness is negligible so that we can put $Z_{21} = Z_{22} = Z_2$.

Upper surface:

$$\begin{aligned} \mathbf{h}_y^1(Z_2) &= \eta^{-1/2}\,(\mathbf{a}_f^1 - \mathbf{a}_b^1) \\ \mathbf{e}_x^1(Z_2) &= \eta^{1/2}\,(\mathbf{a}_f^1 + \mathbf{a}_b^1) \end{aligned} \tag{A.62}$$

Lower surface:

$$\mathbf{h}_y^3(Z_2) = \eta^{-1/2}\,(\mathbf{a}_f^3 - \mathbf{a}_b^3)$$
$$\mathbf{e}_x^3(Z_2) = \eta^{1/2}\,(\mathbf{a}_f^3 + \mathbf{a}_b^3) \tag{A.63}$$

The $\mathbf{U}_0$ and $\mathbf{U}_1$ class field components are obtained from these general solutions by pre-multiplication with $\tilde{\mathbf{U}}_0$ and $\tilde{\mathbf{U}}_1$, respectively. The boundary conditions for the $\mathbf{U}_0$ and $\mathbf{U}_1$ class fields established above result in the following set of equations:

$$\begin{aligned}
\mathbf{e}_{x0} = \text{cont:} &\qquad \tilde{\mathbf{U}}_0\,\eta^{1/2}\,(\mathbf{a}_f^1 + \mathbf{a}_b^1 - \mathbf{a}_f^3 - \mathbf{a}_b^3) = 0\\
\mathbf{h}_{y0} = \text{cont:} &\qquad \tilde{\mathbf{U}}_0\,\eta^{-1/2}\,(\mathbf{a}_f^1 - \mathbf{a}_b^1 - \mathbf{a}_f^3 + \mathbf{a}_b^3) = 0\\
\mathbf{e}_{x1}^1 = 0: &\qquad \tilde{\mathbf{U}}_1\,\eta^{1/2}\,(\mathbf{a}_f^1 + \mathbf{a}_b^1) = 0\\
\mathbf{e}_{x1}^3 = 0: &\qquad \tilde{\mathbf{U}}_1\,\eta^{1/2}\,(\mathbf{a}_f^3 + \mathbf{a}_b^3) = 0
\end{aligned} \tag{A.64}$$

A.2 SLIT PLATE SCATTERING MATRIX

The set of equations (A.64) can be expressed in a form directly linking the relevant scatter variables

$$\mathbf{M}_2 \begin{bmatrix} \mathbf{a}_b^1 \\ \mathbf{a}_f^3 \end{bmatrix} = \mathbf{M}_1 \begin{bmatrix} \mathbf{a}_f^1 \\ \mathbf{a}_b^3 \end{bmatrix} \tag{A.65}$$

where the square matrices $\mathbf{M}_1$ and $\mathbf{M}_2$ are given by

$$\mathbf{M}_1 = \begin{bmatrix} \tilde{\mathbf{U}}_1\,\eta^{1/2} & 0 \\ \tilde{\mathbf{U}}_0\,\eta^{1/2} & -\tilde{\mathbf{U}}_0\,\eta^{1/2} \\ 0 & \tilde{\mathbf{U}}_1\,\eta^{1/2} \\ \tilde{\mathbf{U}}_0\,\eta^{-1/2} & \tilde{\mathbf{U}}_0\,\eta^{-1/2} \end{bmatrix} \tag{A.66}$$

$$\mathbf{M}_2 = \begin{bmatrix} \tilde{\mathbf{U}}_1\,\eta^{1/2} & 0 \\ \tilde{\mathbf{U}}_0\,\eta^{1/2} & -\tilde{\mathbf{U}}_0\,\eta^{1/2} \\ 0 & \tilde{\mathbf{U}}_1\,\eta^{1/2} \\ -\tilde{\mathbf{U}}_0\,\eta^{-1/2} & -\tilde{\mathbf{U}}_0\,\eta^{-1/2} \end{bmatrix} \tag{A.67}$$

It turns out that the matrix $\mathbf{M}_2$ is nonsingular so that (A.65) is transformed by inversion to

$$\begin{bmatrix} \mathbf{a}_b^1 \\ \mathbf{a}_f^3 \end{bmatrix} = \mathbf{M}_2^{-1}\mathbf{M}_1 \begin{bmatrix} \mathbf{a}_f^1 \\ \mathbf{a}_b^3 \end{bmatrix} \tag{A.68}$$

which is seen to comply with the standard scatter formulation $\mathbf{b}=\mathbf{S}\,\mathbf{a}$, with $\mathbf{a}_b^1$ and $\mathbf{a}_f^3$ representing the 'outgoing' components $\mathbf{b}$, and $\mathbf{a}_f^1$ and $\mathbf{a}_b^3$ the 'incoming' components $\mathbf{a}$. Accordingly, the scattering matrix $\mathbf{S}^p$ of the slit plate is given by the matrix product

$$\mathbf{S}^p = \begin{bmatrix} \mathbf{S}_{11}^p & \mathbf{S}_{12}^p \\ \mathbf{S}_{21}^p & \mathbf{S}_{22}^p \end{bmatrix} = \mathbf{M}_2^{-1}\mathbf{M}_1 \tag{A.69}$$

A.2.1 Symmetry properties of the slit plate

Due to geometrical symmetry properties of the slit plate itself the scattering matrix $\mathbf{S}^p$ satisfies a number of basic relations that are significant in the comprehensive treatment of photon behavior in the main text of Chapter 7. Lateral symmetry in the X-coordinate implies that $\mathbf{S}$ is symmetric.

$$\tilde{\mathbf{S}}^p = \mathbf{S}^p \tag{A.70}$$

The *unitary* condition, satisfied by all lossless scattering matrices, reduces to

$$\tilde{\mathbf{S}}^p\mathbf{S}^p = (\mathbf{S}^p)^2 = \mathbf{1} \tag{A.71}$$

The relation shows that

$$[(\mathbf{S}^p)^2]^2 = (\mathbf{S}^p)^2 = \mathbf{1} \tag{A.72}$$

A matrix whose square is equal to the matrix itself is *idempotent.* Such matrices are characterized by their eigenvalues being either unity or zero. Because $(\mathbf{S}^p)^2$ is seen to be idempotent the eigenvalues of $\mathbf{S}^p$ itself must belong to the set zero, plus one, and minus one. In the application of the scattering formulation in describing the double slit experiment in Chapter 7 the properties of the *submatrices* of $\mathbf{S}^p$ are more important. Symmetry with respect to Z implies

$$\mathbf{S}_{11}^p = \mathbf{S}_{22}^p \qquad \mathbf{S}_{12}^p = \mathbf{S}_{21}^p \tag{A.73}$$

which means that $\mathbf{S}^p$ is specified by only two independent submatrices $\mathbf{S}_{11}^p$ and $\mathbf{S}_{12}^p$. Because $\mathbf{S}^p$ is symmetric, so are the submatrices

$$\tilde{\mathbf{S}}_{11}^p = \mathbf{S}_{11}^p \qquad \tilde{\mathbf{S}}_{12}^p = \mathbf{S}_{12}^p \tag{A.74}$$

The two submatrices are related through the unitary condition $\tilde{\mathbf{S}}^p\mathbf{S}^p = \mathbf{1}$ which implies

$$(\mathbf{S}_{11}^p)^2 + (\mathbf{S}_{12}^p)^2 = \mathbf{1} \qquad \mathbf{S}_{11}^p\mathbf{S}_{12}^p + \mathbf{S}_{12}^p\mathbf{S}_{11}^p = \mathbf{0} \tag{A.75}$$

From the theory of linear algebra it can be shown that $\mathbf{S}^p_{11}$ and $\mathbf{S}^p_{12}$ are singular matrices, that the rank of $\mathbf{S}^p_{11}$ is equal to the nullity of $\mathbf{S}^p_{12}$ and vice versa, with the rank of $\mathbf{S}^p_{11}$ corresponding to the rank of the slit plate convolution matrix $\mathbf{\Phi}$. The nonzero eigenvalues of both $\mathbf{S}^p_{11}$ and $\mathbf{S}^p_{12}$ are equal to minus one. Both have a full set of orthogonal eigenvectors serving as appropriate expansion vectors in the analysis of photon behavior in the double slit experiment in Chapter 7.

A.2.2 Scattering matrix for the overall configuration

The scattering matrix $\mathbf{S}^p$ of (A.69) applies to the slit plate alone, relating forward and backward space harmonics referred to the plate position Z_2. The scattering matrix $\mathbf{S}$ of the full configuration referred to the input plane $Z = Z_1$ and the screen position $Z = Z_3$ is simply related to $\mathbf{S}^p$ by the transformation

$$\mathbf{S} = \begin{bmatrix} \mathbf{T}(Z_1-Z_2)^{-1} & \mathbf{0} \\ \mathbf{0} & \mathbf{T}(Z_3-Z_2) \end{bmatrix} \mathbf{S}^p \begin{bmatrix} \mathbf{T}(Z_1-Z_2)^{-1} & \mathbf{0} \\ \mathbf{0} & \mathbf{T}(Z_3-Z_2) \end{bmatrix} \tag{A.76}$$

where

$$\mathbf{T}(Z_1 - Z_2) = e^{-j\eta\,(Z_1-Z_2)} \qquad \mathbf{T}(Z_3 - Z_2) = e^{-j\eta\,(Z_3-Z_2)} \tag{A.77}$$

represents the complex *transmission matrices* from planes Z_1 and Z_3 to the slit plate at Z_2, η being the diagonal impedance matrix of the spatial harmonics. The scattering matrix $\mathbf{S}$ can be expressed by its submatrices in the form

$$\mathbf{S} = \begin{bmatrix} \mathbf{S}_{11} & \mathbf{S}_{12} \\ \mathbf{S}_{21} & \mathbf{S}_{22} \end{bmatrix} \tag{A.78}$$

where the complex submatrices are specified by the real slit plate scattering submatrices $\mathbf{S}^p_{11}$ and $\mathbf{S}^p_{12}$:

$$\begin{aligned} \mathbf{S}_{11} &= \mathbf{T}(Z_1 - Z_2)^{-1}\, \mathbf{S}^p_{11}\, \mathbf{T}(Z_1 - Z_2)^{-1} \\ \mathbf{S}_{12} &= \mathbf{T}(Z_1 - Z_2)^{-1}\, \mathbf{S}^p_{12}\, \mathbf{T}(Z_3 - Z_2) \\ \mathbf{S}_{21} &= \mathbf{T}(Z_3 - Z_2)\, \mathbf{S}^p_{12}\, \mathbf{T}(Z_1 - Z_2)^{-1} \\ \mathbf{S}_{22} &= \mathbf{T}(Z_3 - Z_2)\, \mathbf{S}^p_{11}\, \mathbf{T}(Z_3 - Z_2) \end{aligned} \tag{A.79}$$

The relations describe in a convenient form the complex scattering elements of the overall scattering matrix $\mathbf{S}$ by the *real* elements of the slit plate matrix $\mathbf{S}^p$ satisfying the simple relations (A.70) - (A.75). The overall scattering relation, linking the spatial harmonic vectors at the virtual input plane Z_1 and the screen plane Z_3, is then specified by

$$\begin{bmatrix} \mathbf{a}_b^1(Z_1) \\ \mathbf{a}_f^3(Z_3) \end{bmatrix} = \mathbf{S} \begin{bmatrix} \mathbf{a}_f^1(Z_1) \\ \mathbf{a}_b^3(Z_3) \end{bmatrix} \tag{A.80}$$

which serves as the basis for the analysis of macroscopic as well as single photon behavior in the double slit experiment in Chapter 7.

Chapter 8

THE DOUBLE SLIT EXPERIMENT FOR ELECTRONS

She went on, "Would you tell me please
which way I ought to go from here?"
"That depends a good deal on where
you want to get to," said the Cat.
"I don't much care where—" said Alice.
"Then it doesn't matter which way you go," said the Cat.
"—as long as I get somewhere," Alice added as an explanation.
"Oh, you are sure to do that," said the Cat,
"if you walk long enough."

Lewis Carroll
Alice in Wonderland

8.1 INTRODUCTION—THE PARADOXES

Together with the double slit experiment for photons the corresponding experiment for electrons represents one of the most celebrated and famous experiments in physics. Its fame is due more to the conceptual problems it creates rather than its contribution to understanding quantum physics. The observation of wavelike diffraction of electrons and their definite particlelike behavior in other circumstances has remained a conceptual mystery in quantum physics until this day. The basic experimental setup shown in Fig. 8.1 is simple enough. A stream of focused electrons from a hot cathode impinges on a plate with two narrow slits separated a small distance apart. The electrons transmitted through the slits are observed to form a typical diffraction pattern on the screen behind the slit plate. If the intensity of the electron source is reduced to the point when only one electron at a time is reaching the screen, it produces a pointlike spot located somewhere on the screen, not necessarily just below the slits. This behavior is certainly in accord with classical concepts of the electron as a particle, with the electron passing through *one* of the two slits. In the process it is deflected by some angle and finally hitting the screen at some localized point. The deflection has no classical explanation. But there are worse things to come. As more and more electrons are arriving at the screen the overall average macroscopic intensity builds up to a typical diffraction pattern as indicated in the figure. The similarity of the pattern with the familiar diffraction of an optical beam impinging on a double slit plate is immediately apparent.

In the absence of an immediate classical explanation quantum physics took refuge in the concept of duality. According to this principle the electron can be both a particle and a wave, depending on circumstances. When the single electron approaches the slit plate it mysteriously transforms itself to a wave encompassing both slits. The two wave components originating from the slits interfere behind the slit plate to produce the diffraction pattern, in accordance with regular wave principles. On hitting the screen the electron again transforms itself from a wave to a particle. The only virtue of such an explanation is that it explains the diffraction pattern. But it creates a number of questions regularly referred to as puzzles or paradoxes. One of the most famous is the 'which way' paradox. Classically, the electron must pass through *one* of the slits. But the diffraction pattern can not be explained unless the electron passes through *both* slits. If one tries to somehow measure which slit the electron passes through, the diffraction pattern disappears. It appears that nature refuses any attempt to gain information on its secrets, responding with severe penance in the form of experiment failure.

Lots of ink has been spilled in futile attempts to resolve the paradox. Richard Feynman, in his 'Lectures on Physics' [17] discusses at length the various aspects of the problem, and concludes that the phenomenon is impossible to explain in any classical way. David Mermin's hilarious article in Physics Today [34] is an amusing exposition of the quandary suffered by the poor electron in selecting 'which way' through the two slits. The article reflects in a humorous way the frustration apparently felt by the author himself and by others over the inability of quantum mechanics to provide a definite answer to this very simple question. His concluding remark on the 'which way' problem: the author knows which it is, but he's not telling, could well serve as a symptomatic description of the situation as projected in standard quantum interpretations, namely:

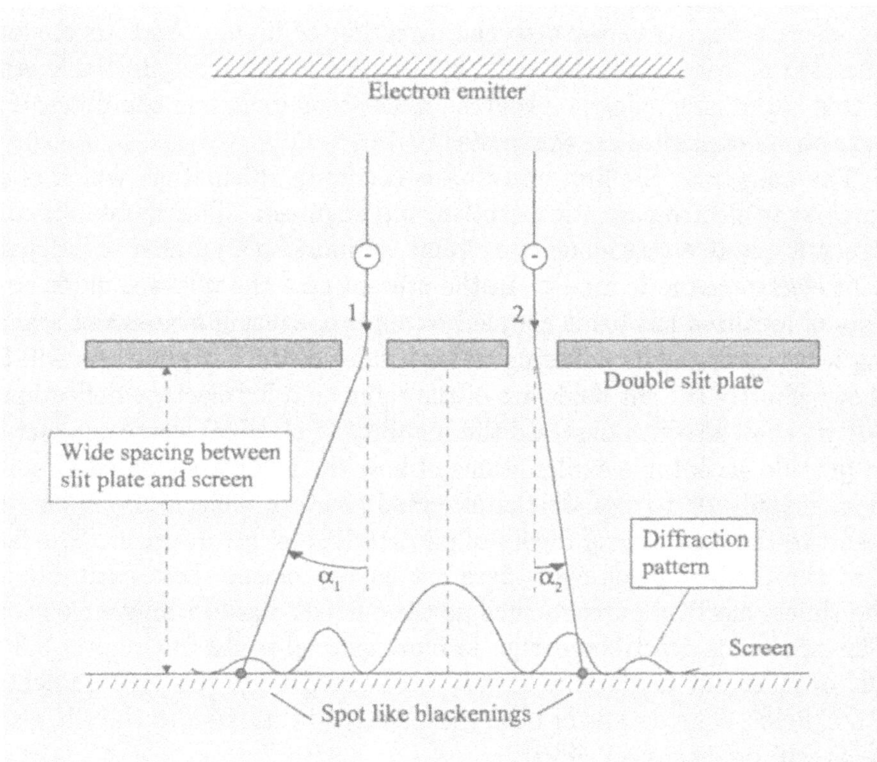

Figure 8.1. **Schematics of the double slit experimental set-up for electrons, showing electron emitter, slit plate, and screen. Two electrons labeled 1 and 2 are deflected by some angles α_1 and α_2 in the respective slits before hitting the screen at localized points, causing spot like blackening. The diffraction pattern represents the observed build up density of spots on the screen.**

'Nature knows but refuses to tell if we ask', that is, if we attempt to measure 'which way'.

In attempts to dig deeper into the two-slit paradoxes, scanning textbooks and literature, I am struck by the omission of discussions on how the electron is deflected in the slits. Quantum mechanics looks upon the deflection as a diffraction phenomenon associated with the electron as a wave. Retaining at least a cursory acceptance of the electron as a classical particle, it must be ascertained that it can be deflected only by forces, and the only forces acting on electrons are electromagnetic forces. In passing through the slits the electron is certainly not passing through 'nothing'—it is passing through 'nothing' surrounded by 'something', namely the solid slit plate. Ruling out any mysterious unknown force, the only possible deflection force on relatively low velocity electrons is an electric field transverse to the direction of flight. And an obvious question is, can such fields exist in the double slit configuration? And if this is the case, can the observed deflections including the diffraction pattern be explained on this basis?

The answer to the first question is certainly affirmative, which is no surprise to electromagnetic scientists and engineers. The double slit configuration is an electromagnetic circuit supporting a number of independent electromagnetic modes. In the present case the relevant mode consists of localized slit fields coupled to nonpropagating evanescent standing wave components adhering to both sides of the slit plate. As will be shown shortly the slit fields are of the right kind for electron deflections, but we shall also demonstrate the inability of classical electromagnetics to provide acceptable explanations of how the fields are excited in sufficient magnitude to explain the observed electron deflections, unless we resort to the neoclassical theory of photon doublet involvement. The fact that the doublet fields carry zero net energy makes them excitable by the tiniest electron current, such as the current due to a single electron. This process is described earlier in more general terms in Chapter 3.3.3. Or, doublet slit fields may simply be excited from the environmental PDE fields. The details of both mechanisms in relation to the slit plate fields will be discussed shortly.

The quantum description can be criticized on several grounds, not only for the nonphysical duality explanation but for the complete disregard of interaction between the electron system and the slit plate system, in particular with regard to momentum balance in the overall system. We shall find that proper consideration of lateral momentum transfer to the electrons leads to a theoretical diffraction pattern exactly as observed experimentally, for instance in [27]. In this reference the observed patterns were explained theoretically on a formal quantum mechanical basis as interfering electron waves originating at the two slits. It will be demonstrated that the diffraction pattern is equally well explained by the neoclassical theory, without incurring the concept of electron waves. Briefly, the main steps in the theoretical derivations presented in the following sections of this chapter are as follows.

The first step establishes the appropriate field distribution of the double slit configuration, expressed as a superposition of normal mode components. In the terminology of circuit theory the double slit configuration is more than the naive concept of just a plate with two slits. It is an electromagnetic circuit whose characteristic fields are reasonably well understood. The circuit field can be described in several equivalent ways. For the purpose of describing momentum exchange the preferred formulation is expressed in terms of localized slit fields matched to laterally traveling evanescent slow waves adhering to the upper and lower sides of the slit plate.

The next step describes the discrete momentum exchange between the slit fields and electrons, and the administration of this process by the *PDE* environmental fluctuations, a central theme in the neoclassical theory. It will be shown that this fundamental principle determines the routing of electrons to local spots on the screen and the associated build-up of a macroscopic diffraction pattern, in exact agreement with available experimental observations, in accordance with the sketch of Fig. 8.1.

The last step describes the local interaction taking place between the electron and the fields in the open slits, a process in which the electron acquires exact amounts of discrete lateral momentum to maintain momentum balance in the overall coupled system consisting of electrons and circuit fields.

8.2 ELECTROMAGNETIC FIELDS OF THE DOUBLE SLIT PLATE

The slit plate represents an electric circuit basically supporting a number of modes described by classical electromagnetic theory. As shown in earlier chapters the same modes apply to the doublet fields. For the purpose of exposing the fundamental processes giving rise to electron diffraction it seems a good point to avoid unessential generalities tending to obscure the basic processes. Adhering to this line of thought as a guidance, but still maintaining consistency, the slit plate fields will be defined as follows. With reference to Fig. 8.2, showing cross-sectional slit plate segments, the slit dimensions are specified by width L, plate thickness Δz, and length b (not shown in the figure). Within the rectangular slits themselves the fields are taken to be the fundamental Transverse Electric TE_{10} mode of rectangular waveguides. The electric field of the presumably undercut TE_{10} mode shown in Fig. 8.2a is given by

$$E_x(y,z) = E_x \sin(\pi \frac{y}{b}) \frac{\mathrm{Cosh}(\gamma_0 z)}{\mathrm{Cosh}(\gamma_0 \Delta z/2)} \tag{8.1}$$

which holds for $-\Delta z/2 < z < \Delta z/2$, and where

$$\gamma_0 = \left((\frac{\pi}{b})^2 - k^2 \right)^{1/2} \qquad k = \omega/c \tag{8.2}$$

The undercut condition requires γ_0 to be real, which is satisfied provided

$$k < k_{\max} = \frac{\pi}{b} \tag{8.3}$$

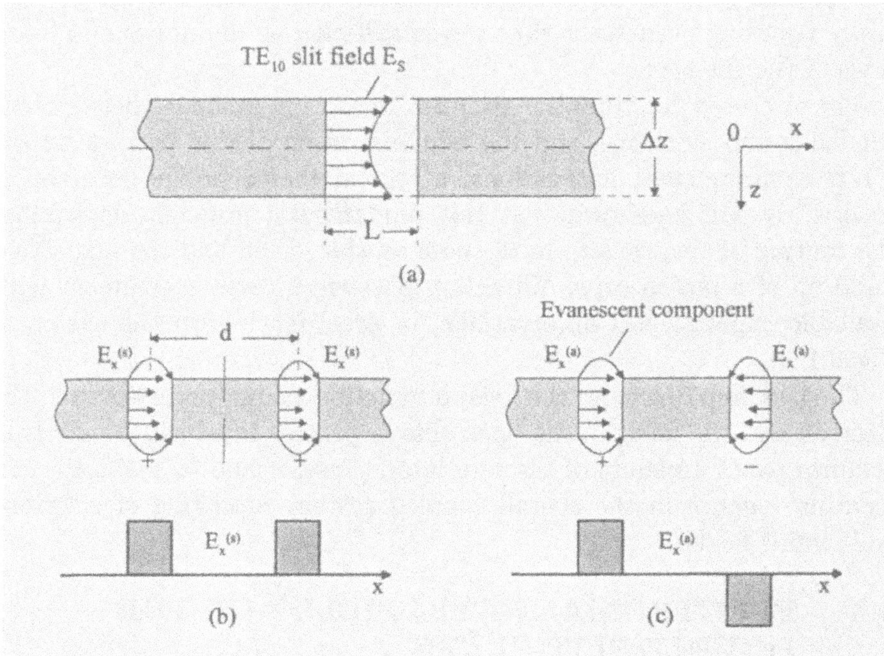

Figure 8.2. **Basic fields in the double slit configuration. (a) Fundamental undercut TE_{10} slit field. (b) Overall fields in the symmetric mode. (c) Overall fields in the antisymmetric mode.**

This condition is equivalent to the free space wavelength λ_0 being larger than twice the slit length b.

$$\lambda_0 > \lambda_{0,\min} = 2b \tag{8.4}$$

The magnetic field components of the TE_{10} mode are not needed. The TE_{10} waveguide mode satisfies all boundary conditions on the slit faces, thus eliminating the need for higher order TE_{mn} modes. For reasons soon to be explained the symmetric as well as the antisymmetric slit mode shown in Fig. 8.2 are involved in the deflection of electrons in the slits.

The stray fields above and below the slit plate surfaces shown in Figs. 8.2b and 8.2c are not part of the basic slit fields which are strictly the TE_{10} fields in Fig. 8.2a, but belong to the auxiliary evanescent fields in the free space immediately above and below the plate. The free space fields are evanescent Transverse Magnetic type (TM) solutions matching the TE_{10} fields at the two surfaces $z = \pm\Delta z/2$. Above the slit plate, where $z \leq -\Delta z/2$, the x-component of the electric field can be expressed by the Fourier integral

$$e_x(x,z) = \int_\beta e_x(\beta)\, e^{-j\beta x}\, e^{\gamma(z+\Delta z/2)}\, d\beta \tag{8.5}$$

where

$$\gamma = \sqrt{\beta^2 - k^2} \qquad \beta > k \tag{8.6}$$

Variations of $e_x(x,z)$ with regards to the coordinate y are disregarded as unimportant for understanding the essentials of the field configuration. Although the longitudinal electric field component e_z, normal to the slit plate, and the transverse magnetic component h_y are both nonzero, they are not needed in the matching process and therefore not specified. Below the slit plate, equation (8.5) is still applicable, with just the last factor replaced by $\exp[-\gamma(z-\Delta z/2)]$.

The symmetric and antisymmetric field components in Figs. 8.2b and 8.2c satisfy

$$\begin{aligned} e_x^{(s)}(x+d,z) &= e_x^{(s)}(x,z) \\ e_x^{(a)}(x+d,z) &= -e_x^{(a)}(x,z) \end{aligned} \tag{8.7}$$

These requirements put certain restrictions on the propagation factor β in the integral (8.5). In the symmetric mode, β must satisfy the condition:

$$e^{-j\beta d} = 1 \tag{8.8}$$

which requires β to be discrete.

$$\beta = \beta_n = 2\pi n \frac{1}{d} \qquad n = 0, \pm 1, \pm 2.... \tag{8.9}$$

Accordingly, in this symmetric case (8.5) reduces to the sum

$$e_x^{(s)}(x,z) = \sum_{n=-\infty}^{n=\infty} e_{xn}^{(s)}\, e^{-j\beta_n x}\, e^{\gamma_n(z+\Delta z/2)} \tag{8.10}$$

The expansion coefficients $e_{xn}^{(s)}$ are obtained through regular inversion, matching the evanescent sets with the TE_{10} slit fields at $z = -\Delta z/2$. This procedure results in:

$$e_{xn}^{(s)} = \frac{E_x^{(s)} L}{d} \frac{\sin(\pi n L/d)}{\pi n L/d} \tag{8.11}$$

In the antisymmetric mode the propagation factor β is restricted by the relation

$$e^{-j\beta d} = -1 \tag{8.12}$$

which again requires discrete β according to

$$\beta_m = \pi m \frac{1}{d} \qquad m = \pm 1, \pm 3, \pm 5..... \tag{8.13}$$

Accordingly, the antisymmetric fields are expressed by the sum

$$\begin{aligned} e_x^{(a)}(x,z) &= \sum_{m=odd} e_{xm}^{(a)}\, e^{-j\beta_m x}\, e^{\gamma_m (z+\Delta z/2)} \\ e_{xm}^{(a)} &= \frac{E_x^{(a)} L}{d} \frac{\sin(\pi m L/2d)}{\pi m L/2d} \end{aligned} \tag{8.14}$$

In the antisymmetric mode the even components, in particular the $m=0$ component, do not exist. It was stated that the field components represent evanescent waves, decaying exponentially away from the slit plate. This is certainly true provided

$$\beta_{n,m}^2 > k^2 \qquad \text{for all } m, n \tag{8.15}$$

Forgetting for a moment the symmetric $n=0$ component, this requirement is satisfied for relatively low frequencies, more specifically if the free space wavelength λ_0 satisfies

$$\lambda_0 > d \tag{8.16}$$

The slit spacing d is expected to be a few microns wide and there is ample reason to expect whatever frequency is occurring to be fairly low so that λ_0 is very much larger than d, say of the order of centimeters. Under these conditions it is legitimate to put $\gamma_n = \beta_n$ in the field equations (8.5)-(8.14). We shall return to the question of frequencies shortly.

Equations (8.11) and (8.14) describe the localized symmetric and antisymmetric slit fields as superpositions of slow wave components propagating laterally in both directions with propagation factors β_n and β_m. The components are evanescent in the sense that they decay exponentially away from the slit plate as $\exp[\gamma(z+\Delta z/2)]$. As already noted we must take exception for the $n=0$ component in the symmetric mode, because this mode is not evanescent in the $z-$direction but propagating, with $\beta_0 = 0$ and propagation factor γ_0 given by

$$\gamma_0 = \sqrt{\beta_0^2 - k^2} = \pm jk \tag{8.17}$$

These are plane uniform Transverse Electric Magnetic (TEM) components propagating with the velocity of light in the forward or backward z-directions. They provide a link between the confined evanescent mode of the slit plate and the environmental PDE fields in the sense that TEM environmental radiation can interact with and possibly excite the symmetric circuit mode $e_x^{(s)}$. In addition to this interaction the electrons themselves might be a source of excitation of the slit fields, a possibility considered next.

8.2.1 Can single electrons excite the slit fields?

The slit fields are presumably of the voltage doublet type defined in (3.44) so that the positive and negative frequency components are equal.

$$\mathbf{V}(\omega) = \mathbf{V}(-\omega) \tag{8.18}$$

In Chapter 3 we discussed excitation of doublet resonances and determined, as shown in (3.68), that they are indeed expected to be set up by the tiniest amount of driving current $\mathbf{I}_g(\omega,-\omega)$. Let us demonstrate that an electron traversing the slit represents a driving current that might excite the doublet slit fields. It suffices to consider the positive frequency part which is specified by the general formula (3.14)

$$\mathbf{I}_g(\omega) = \int\limits_V \vec{i}_g(\vec{r},\omega) \cdot \vec{\epsilon}(\vec{r},\omega)\, dV \tag{8.19}$$

Here, $\vec{i}_g(\vec{r},\omega)$ is the electron current density and $\vec{\epsilon}(\vec{r},\omega)$ is the *normalized* electric field of the circuit. We shall associate $\vec{i}_g(\vec{r},\omega)$ with the tiny current density from a single electron traversing one of the slits. The task at hand is to show that the integral (8.19) is indeed nonzero. With reference to Fig. 8.5, disregarding the small lateral deflection, the current density due to a single electron of charge $-q$ traversing the slit at the lateral position x_0, y_0 is specified by the delta type expression

$$\vec{i}_g(x,y,z,t) = -q\, v_0\, \delta(x-x_0)\, \delta(y-y_0)\, \delta[z - v_0(t-t_0)\,]\vec{a}_z \tag{8.20}$$

where the arrival time of the electron at the slit center $z=0$ is equal to t_0. The current is expressed as a function of time rather than frequency. Noting that the integrations over the lateral coordinates x and y are trivial, the driving current (8.19) is specified by

$$I_g(x_0,t) = -\,qv_0 \int\limits_{z=-\infty}^{\infty} \delta[z - v_0(t-t_0)]\, \epsilon_z[x_0, z - v_0(t-t_0)]\, dz \tag{8.21}$$

which evaluates to

$$I_g(x_0, t) = -qv_0\,\epsilon_z[x_0, v_0(t - t_0)] \tag{8.22}$$

Thus, the induced current due to a single electron traversing the slit at x_0 is proportional to the normalized longitudinal electric field component ϵ_z shown in Fig. 8.5b, causing a time variation of the current through the functional relation $z = v_0(t - t_0)$. The corresponding *frequency* components are specified by the Fourier transform of $I_g(x_0, t)$ which evaluates to

$$I_g(x_0, \omega) = j\,qe^{-j\omega t_0}\frac{1}{2\pi}\int_{-\infty}^{\infty} \epsilon_z(x_0, z)\,\sin(\omega z/v_0)\,dz \tag{8.23}$$

The omission of the cosine term follows from the fact that $\epsilon_z(x_0, z)$ is an odd function of z. The factor $\exp(-j\omega t_0)$ in front of the integral specifies the phase of $I_g(x_0, \omega)$ as a function of the traversing time t_0 of the electron. It follows, from the general shape of the function $\epsilon_z(x_0, z)$ in Fig. 8.5b, that $I_g(x_0, \omega)$ vanishes for zero frequency and has a maximum around $\omega \approx 2v_0/\Delta z$. However, it will be demonstrated shortly that the actual frequency range is fairly unimportant because the process of electron deflections in the slit and the build up of a diffraction pattern is completely independent of the actual frequencies of the slit fields. The essential point demonstrated by the analysis is that a single traversing electron can indeed induce a nonzero current in the slit circuit. The fact that the current is exceedingly small is of minor concern because it can still excite finite amplitude doublet slit fields as demonstrated earlier by (3.68) in Chapter 3. The physical reason is of course the absence of time average stored energy in the doublet fields so that the amplitude grows without any power expenditure. We are therefore led to the conclusion that the doublet fields of the slit plate indeed can be set up by the traversing electrons themselves. This process comes as a supplement to possible direct excitation from the environmental *PDE* fluctuations. Thus, there are two possible mechanisms available for slit field excitation.

8.2.2 Discrete momentum levels of the slit plate

At this point the treatment departs radically from classical electromagnetic theory. According to the neoclassical theory of Chapters 2, 5, and 6 the doublets of the environmental *PDE* fields are in equilibrium with the slit plate fields. Upper and lower slit plate faces are expected to be equally engaged in this process. The slit plate is an electric circuit on equal footing with any other electric circuit covered by the general theory of doublet excitation. Thus, further elaboration is not needed, except perhaps to justify the inclusion of the antisymmetric field configurations. These appear to be necessary when the circuit is interacting with an auxiliary perturbing system such as an electron passing through one of the slits upsetting the otherwise symmetric field pattern. The overall lateral momentum balance calls for independent and discrete levels of either slit plate field over a wide range from essentially zero to some maximum value. Apparently, this wide range can be achieved by appropriate superpositions of the symmetric and the antisymmetric modes. Graphically, the overall component picture is illustrated in the *momentum diagrams* in Fig. 8.3. The momentum levels do not refer to separate independent modes, but rather the momentum components of the laterally traveling evanescent slow wave components constituting the overall field patterns of the symmetric and the antisymmetric modes.

The laterally traveling evanescent wave components are photon doublets, but in the following we shall often refer to the components as photons. Each component n or m can be associated with a certain number of photons proportional to the amplitudes $e^{(s)}_{xn}$ and $e^{(a)}_{xm}$. A particular photon of the nth component carries the lateral electromagnetic momentum p_n specified by

$$p_n = \hbar\beta_n = \hbar 2\pi n \frac{1}{d} \tag{8.24}$$

where n can be positive as well as negative, corresponding to wave components traveling right or left. The neoclassical theory, describing the slit fields as a superposition of laterally traveling photons of discrete momenta, differs conceptually from the classical picture. In the classical theory, if we visualize the regular macroscopic slit fields set up by some unspecified external excitation, the set of space harmonic amplitudes $e^{(s)}_{xn}$ and $e^{(a)}_{xm}$ are fixed by the circuit configuration. In the neoclassical theory the macroscopic amplitudes are averages of a large number of laterally traveling photons having discrete momenta specified by (8.24). The total photon number in each category must be an integer. In order to satisfy (8.11) or (8.14) on a time average basis there must be photons engaged in transfers between levels, giving rise to spontaneous fluctuations in the slit fields. Following the theoretical concepts developed earlier in the book, the momentum fluctuations are expected to be governed by the principle of zero time average overall momentum—an adjunct of the principle of zero time average energy launched earlier.

Let us visualize the transfer process by assuming, for a moment, that the number of photons N_n in level n is a little high to satisfy the macroscopic amplitude $e^{(s)}_{xn}$ which represents the true time average level. This

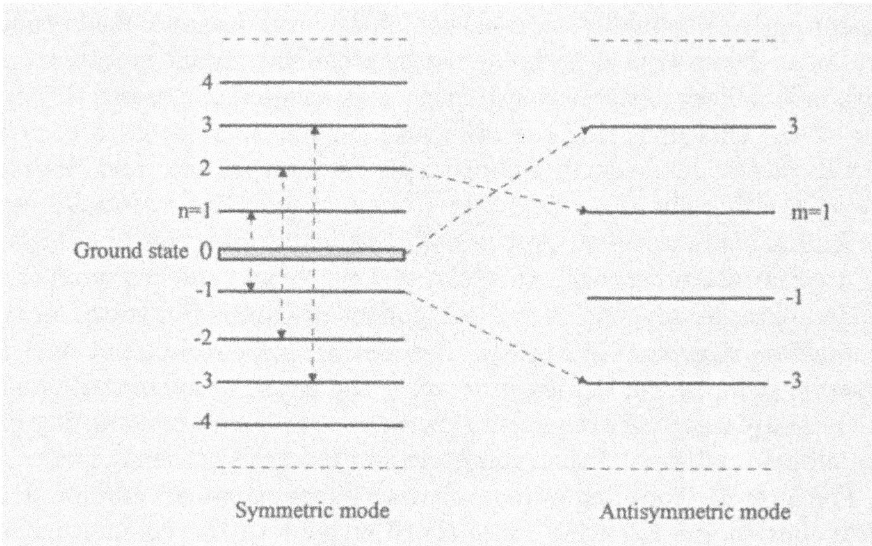

Figure 8.3. **Schematic momentum diagram showing the discrete lateral momentum levels $p_n = \beta_n \hbar$. The ground state $n=0$ has zero lateral momentum, but can interact with the environmental *PDE* fields.**

situation can not be endured by the system for a longer period of time. The photon is solving the problem by temporarily dropping down to level n–1, in the process giving up a certain amount of lateral momentum $\Delta p = p_n - p_{n-1} = \hbar/d$ as indicated in Fig. 8.4. This momentum must be absorbed by some extraneous system which might be either one of the following: It could be the mechanical system of the slit plate set into mechanical motion. Or, the momentum could be transferred to the $n=0$ component and 'radiated' into free space through the $z-$directed momentum $\hbar k$ carried by this component. This process would require a multistep procedure from level n to level zero. Or finally, the momentum could be transferred to an electron during its flight though the slits. It is this last process that appears as the most likely one, not necessarily excluding the others.

All the laterally traveling photons at the various momentum levels can take part in momentum exchange by jumping up and down the momentum ladder. The orchestration of this process is governed by a general principle of zero time average of momentum fluctuations, which is an adjunct to the principle of zero average energy fluctuations. Both are consequences of retaining zero time averages, one with respect to momentum, the other with respect to energy. The transition between two arbitrary levels $n1$ and $n2$ is accompanied by a lateral momentum transfer of

$$\Delta p_{n1,n2} = \hbar \frac{n2 - n1}{d} = \hbar \frac{\Delta n}{d} \tag{8.25}$$

The smallest momentum transfer occurs for transitions between adjacent levels, with $\Delta n = 1$ and $\Delta p_{n,n-1} = \hbar/d$. Thus, the permitted transfers of momentum are restricted to the following discrete quanta:

$$\Delta p = \pm\frac{\hbar}{d},\ \pm 2\frac{\hbar}{d},\ \pm 3\frac{\hbar}{d}, \ldots \tag{8.26}$$

Note that these levels are completely independent of the frequency ω which does not enter. Thus, a wide range of frequency components could be involved in the process (see discussion in connection with (8.23)). The probability of momentum transfer between any two levels $n1$ and $n2$ is expected to be proportional to the product of the photon numbers N_{n1} and N_{n2}, which is proportional to the product of the corresponding component amplitudes $e_{xn}^{(s)}$ or $e_{xm}^{(a)}$.

The interpretation emerging from this discussion can be summed up as follows. The slit plate circuit supports doublet resonances in equilibrium with the environmental *PDE* fields. A particular doublet resonance at some frequency ω gives rise to slit plate fields of a standing wave type, consisting of evanescent slow standing wave components. The components are interpreted as laterally traveling photon doublets with momentum corresponding to the lateral propagation factor $\beta_n = 2\pi n/d$ multiplied by $\hbar$. This depiction leads to a system interpretation in terms of a set of equally spaced photon momentum levels. The similarity with energy levels in atoms is immediately apparent. In this sense the slit plate circuit is a kind of 'macroatom', being in time average equilibrium with the environmental *PDE* fields, and interacting with other perturbing entities such as electrons passing the slits. The momentum transfer during interaction is governed strictly by the momentum levels of the system, expressed by (8.26).

Interaction between the traversing electron and discrete momentum states of the slit plate system is not a new concept. Alfred Landé in his book [29] is using similar arguments as those put forward above, but he is not pointing to specific physical mechanisms as responsible for the momentum transfer other than the mechanical slit plate itself. In the present bitemporal description the transfer involves the *PDE* environmental fields as a necessary ingredient in the process.

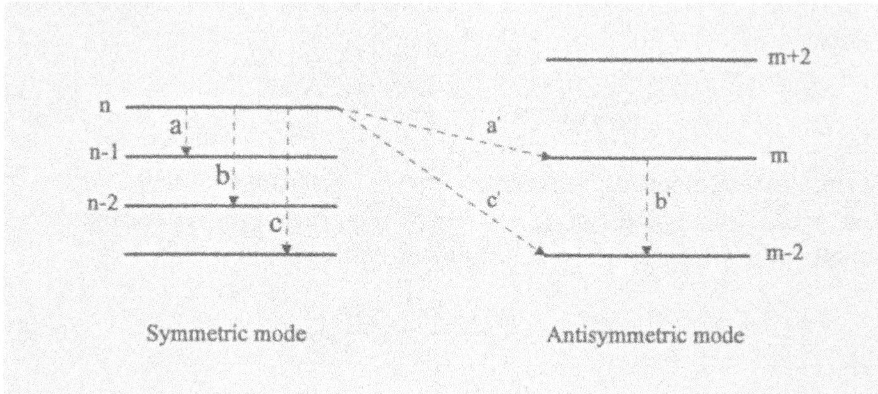

Figure 8.4. **Lateral momentum transitions. (a) Single-level transition: $\Delta p = \hbar/d$. (b) Two-level transition: $\Delta p = 2\hbar/d$. (c) Three-level transition: $\Delta p = 3\hbar/d$.**

8.2.3 Deflection of electrons traversing the slits

The restriction of momentum transitions to discrete levels implies that the electron, in traversing the slits, can receive only discrete lateral momentum changes Δp_e, specified by (8.25).

$$\Delta p_e = m v_{x,n} = \hbar \frac{\Delta n}{d} \qquad \Delta n = \pm 1, \pm, 2, \pm 3, ... \tag{8.27}$$

where m is the electron mass and $v_{x,n}$ the imparted transverse velocity. The associated lateral angular deflection $\alpha_n \cong \tan(\alpha_n)$ is then

$$\alpha_{\Delta n} = \frac{\hbar}{m v_0} \frac{\Delta n}{d} \qquad \Delta n = \pm 1, \pm, 2, \pm 3, ... \tag{8.28}$$

How can it be explained that the imparted deflection is limited to these specific values and does not form a continuum? Let us consider slit 1 shown in Fig. 8.5 and apply a regular classical approach to determine the deflection. The electric slit field is specified by the general expression

$$\vec{E}(x, z, t) = \vec{E}(x, z) e^{j\omega t} \tag{8.29}$$

An electron having z-velocity v_0, entering the slit at position x_0 relative to the center line, receives a lateral velocity impulse during the passage obtained from solution of the differential equation specifying momentum balance.

$$\frac{d v_x(x)}{dt} = -\frac{q}{m} E_x(x, z) e^{j\omega t} \tag{8.30}$$

For small deflections it is legitimate to put $x = x_0 =$ constant, and

$$z = v_0(t - t_0) \tag{8.31}$$

The reference time t_0 is the time of passage at $z = 0$, which is half way through the slit. With these specifications (8.30) reduces to

$$\frac{d}{dz}\alpha(x_0) = -\frac{1}{2V_0} E_x(x_0, z)\, e^{j\omega z/v_0}\, e^{j\omega t_0} \tag{8.32}$$

where V_0 is the accelerating potential and $\alpha(x_0)$ is the angular deflection. Direct integration results in the formula

$$\alpha(x_0) = \int_{slit} E_x(x_0, z) e^{j\omega z/v_0}\, dz \tag{8.33}$$

It suffices to approximate the integral by assuming constant electric field over the plate thickness Δz, in which case

$$\alpha(x_0) = -\frac{1}{2V_0}(E_x^{(s)} + E_x^{(a)})\, e^{j\omega t_0} \frac{\sin(\omega\Delta z/2)}{\omega\Delta z/2} \tag{8.34}$$

where the electric field is expressed as a sum of the symmetric and antisymmetric components. The factor $\sin(\omega\Delta z/2)/(\omega\Delta z/2)$ is a slit coupling coefficient, essentially the Fourier transform of the slit field, accounting for some reduction of deflection due to transit time phase delay. But the deciding factor in determining the deflection is the slit field $(E_x^{(s)} + E_x^{(a)})$ itself which must have the right magnitude to impart one of the permitted discrete angles $\alpha_{\Delta n}$ specified in (8.28). Is this stipulation possible? In light of the multiple momentum levels the answer is affirmative. In fact, this follows directly from the linking of the level intensities $e_{xn}^{(s)}$ and $e_{xm}^{(a)}$ with the slit field through (8.11) and (8.14), and their relations to the level occupancies N_n and N_m. The consequence of these interrelations is a quantization of the slit field into discrete levels, where a particular level corresponds to some combination of momentum level occupancies of the symmetric and antisymmetric modes. In light of the immensity of variables involved the particulars of such combinations are difficult to explain in detail, but under all circumstances momentum balance is guaranteed. And that is all what is needed. Accordingly, the imparted discrete velocity $v_{x,\Delta n}$ is specified by the momentum transfer $\Delta p_{\Delta n}$, in accordance with (8.27).

$$v_{x,\Delta n} = \frac{1}{m}\Delta p_{\Delta n} = \frac{\hbar}{md}\Delta n \qquad \Delta n = \pm 1, \pm 2, \pm 3, \ldots. \tag{8.35}$$

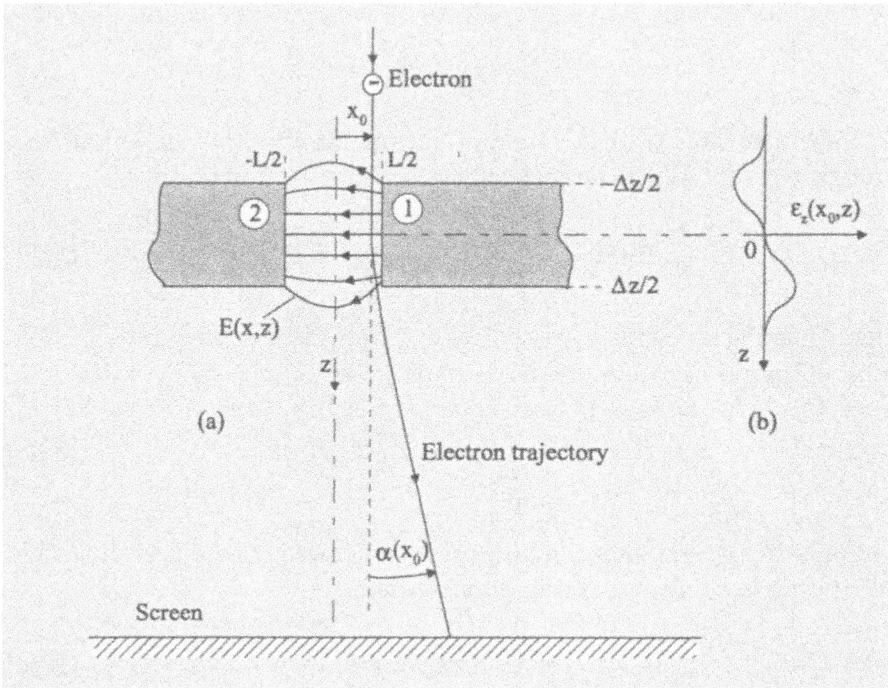

Figure 8.5. **Sketch showing one slit of thickness Δz and width L. (a) Projected electron deflection due to a slit field $E_x(x,z)$. The angular deflection is grossly exagerated. (b) Sketch showing the normalized ϵ_z-field which, through (8.23), specifies the induced current in the slit gap by a traversing electron at the position $x = x_0$.**

The corresponding discrete deflection angle is given by

$$\alpha_{\Delta n} \cong \tan(\alpha_{\Delta n}) = \frac{v_{x,\Delta n}}{v_0} = \frac{\hbar}{mv_0}\frac{1}{d}\Delta n = \alpha_1 \Delta n \qquad (8.36)$$

where

$$\alpha_1 = \frac{\hbar}{mv_0}\frac{1}{d} \qquad (8.37)$$

represents the smallest deflection angle, corresponding to $\Delta n = 1$.

8.2.4 Angular spread in the slits

At this stage we are almost at a point to discuss the overall diffraction pattern formed by electrons coming from both slits. But an important effect is still missing in the overall picture, namely *symmetric* angular spread in the slits. And why is the angular spread important? According to (8.34) the angular deflection $\alpha(x_0)$ is independent of the entrance coordinate x_0, at least in the approximation used to obtain this expression. This fact implies that a parallel beam of z-directed electrons filling the slit over the full width would continue as a parallel beam after deflection, with exactly the same width. This is an unlikely physical situation, in particular with regard to the expected diffraction pattern on a screen far away from the slit plate. In this so-called Frauenhofer approximation —which is the relevant one in discussing diffraction—the electron irradiation would be confined to tiny separated regions equal to the slit width L. This distribution does not satisfy the Frauenhofer criterion where the overall *shape* of the diffraction pattern is independent of distance. These deliberations call for an effect causing the beam to expand proportional with distance, rather than maintaining the parallel shape.

It follows then that we must look for an additional mechanism causing a symmetric spread $\Delta\alpha$ proportional to x_0, i.e., the distance from the slit center. And this mechanism must be entirely independent of the discrete angular deflections α_n specified in (8.34). The nonuniform field distribution in the slits, a small effect that was disregarded in arriving at (8.34), can be ruled out as a cause because the symmetric spread must also apply to nondeflected electrons corresponding to $\Delta n = 0$, in which case the electric deflection field vanishes. It appears that the self field of the electrons in the narrow slit location might be the missing mechanism causing angular spread, a process that indeed seems to satisfy the deflection criterion. With reference to Fig. 8.6 an electron entering the slit off center is attracted to the closest wall with a small but nevertheless finite self generated force causing a small deflection $\Delta\alpha(x_0)$, where x_0 is the offset. It will be shown that, to first order, $\Delta\alpha(x_0)$ is proportional to x_0, thereby explaining the angular spreading effect we were looking for. The force is evaluated on the basis of the equivalent image charges q_1 and q_2 induced on opposing slit walls. These are determined from the total normal flux at the two surfaces constituting the slit walls. The normal fluxes are obtained from a simple reciprocity procedure outlined in the following. Let the electron *self field* in the slit be $\vec{E}(\vec{r})$ and the corresponding potential $V(\vec{r})$. These are related by

$$\vec{E}(\vec{r}) = -\nabla V(\vec{r}) \tag{8.38}$$

where the boundary conditions are

$$V(\vec{r}) = 0 \qquad \text{on both walls} \tag{8.39}$$

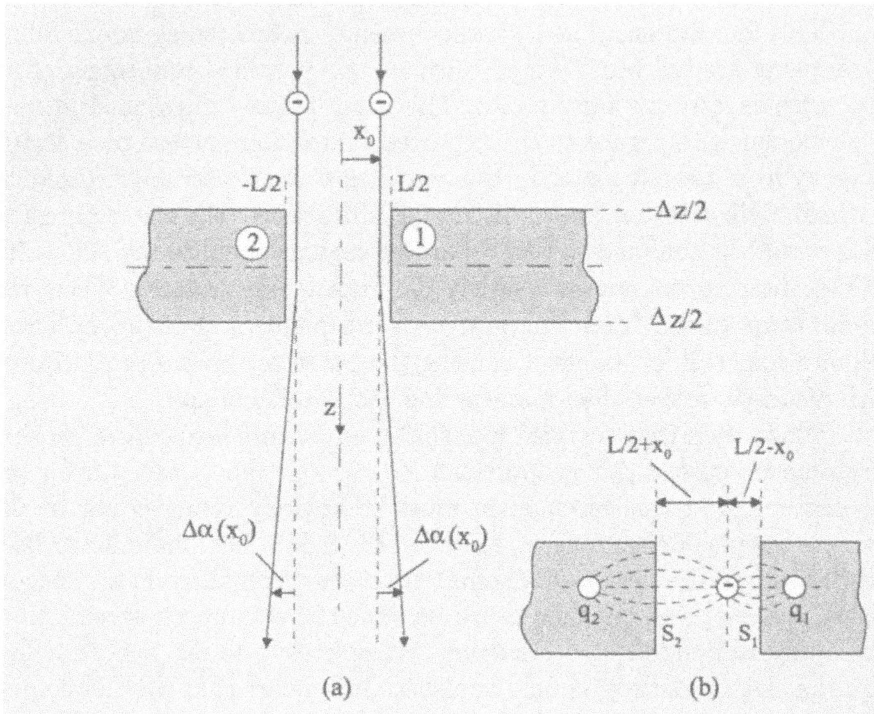

Figure 8.6. **Figure illustrating the self generated angular spread. (a) Basic configuration showing the symmetric nature of the angular spread. The spread is greatly exagerated. (b) Image charges q_1 and q_2 induced on the walls by a traversing electron.**

Let us define a purely mathematical 'test function' $\Phi(\vec{r})$ and its gradient $\vec{e}(\vec{r})$ by

$$\vec{e}(\vec{r}) = -\nabla\Phi(\vec{r}) \qquad \nabla\cdot\vec{e}(\vec{r}) = 0$$
$$\Phi(S_1) = 1 \qquad \Phi(S_2) = 0 \tag{8.40}$$

where S_1 and S_2 are the two faces of the slit. Let us form the expression

$$\nabla\cdot[\Phi\,\vec{E} - V\vec{e}] = \Phi\nabla\cdot\vec{E} - \vec{e}\cdot\vec{E} - V\,\nabla\cdot\vec{e} + \vec{E}\cdot\vec{e} = \Phi\,\nabla\cdot\vec{E} \tag{8.41}$$

Integration over the volume of the slit and use of Gauss' theorem results in

$$\int\limits_{S_1}\vec{E}\cdot\vec{n}\,dS = \int\limits_{V}\Phi\,\nabla\cdot\vec{E}\,dV = \frac{x_0'}{L}\int\limits_{V}\nabla\cdot\vec{E}\,dV = \frac{x_0'}{\varepsilon_0\,L}q \tag{8.42}$$

where $x_0'=L/2-x_0$. Because the image charge q_1 is given by

$$q_1 = -\varepsilon_0 \int\limits_{S_1} \vec{E}\cdot\vec{n}\,dS \tag{8.43}$$

the image charges q_1 and q_2 on the faces S_1 and S_2 due to the electron charge q are specified by

$$q_1 = -\frac{x_0'}{L}q \qquad\qquad q_2 = -q - q_1 \tag{8.44}$$

Disregarding edge effects due to the finite plate thickness Δz the net x-directed force F_x on the electron leaving the slit at x_0 is readily evaluated from the image charges.

$$F_x = \frac{1}{\varepsilon_0\pi}q^2\frac{1}{L^2}\frac{x_0}{L}\frac{1+(\frac{x_0}{L/2})^2}{\left(1-(\frac{x_0}{L/2})^2\right)^2} \tag{8.45}$$

which to first order is proportional to x_0/L. For electrons entering the slit close to the wall, where $x_0 \cong \pm L/2$, the last factor in (8.45) is quite large, so that such electrons are expected to be intercepted. However, detailed trajectory calculations including possible interception are beyond the scope of the discussion, and we shall be satisfied with the first order approximation retaining only the first order term. Assuming constant force during passage, the corresponding angular spread $\Delta\alpha(x_0)$ is readily evaluated and presented in the following convenient form:

$$\Delta\alpha(x_0) \cong \frac{q}{\pi\varepsilon_0}\frac{x_0\Delta z}{V_0L^3} = 240q^2\left(\frac{c}{v_0}\right)\left(\frac{\Delta z}{L}\right)\left(\frac{1}{mv_0L}\right)\left(\frac{x_0}{L}\right) \tag{8.46}$$

where c is the velocity of light.

Let us compare the angular spread $\Delta\alpha(x_0)$ with the *lowest* angular deflection α_1 specified by(8.37), by forming the ratio:

$$\frac{\Delta\alpha(x_0)}{\alpha_1} = 0.0092\left(\frac{\Delta z}{L}\right)\left(\frac{x_0}{L}\right)\left(\frac{d}{L}\right)\left(\frac{c}{v_0}\right) \tag{8.47}$$

Both equations (8.46) and (8.47) specify qualitative criteria that have to be satisfied in order to obtain well defined diffraction patterns. With a given slit plate geometry the parameters Δz, L, and d are fixed. The left part of (8.46) tells us that the angular spread is specified by the product of the voltage V_0 and slit width L to third power. The smaller the slit width the larger voltage must be used in order to retain a reasonably small angular spread. The ratio $\Delta\alpha(x_0)/\alpha_1$ in (8.47) is a more significant parameter which is seen to be specified exclusively by the ratio of electron velocity v_0 and the velocity of light, i.e., by the accelerating voltage. More specifically, if the ratio $\Delta\alpha(x_0)/\alpha_1$ approaches one half, two adjacent beamlets corresponding to n and $n+1$ begin overlapping, causing the pattern to blur out. This will happen if the accelerating voltage is too low. Let us see what voltage is needed to avoid this blurring effect. Solution of (8.47) with respect to v_0 and use of the formula $V_0 = mv_0^2/2$ results in the formula

$$V_0 = 5.4\left(\frac{\alpha_1}{\Delta\alpha(L/2)}\right)^2\left(\frac{2\Delta z}{L}\right)^2\left(\frac{d}{L}\right)^2 \tag{8.48}$$

Let us consider a typical set of slit plate parameters as follows:

$$2\Delta z/L = 1 \qquad d/L = 5 \tag{8.49}$$

and assume $\Delta\alpha(L/2)/\alpha_1$=0.1. In this case (8.48) specifies $V_0 = 13500$ volt. The calculation indicates a definite need for relatively high *DC* accelerating voltages as a precaution to avoid blurred out diffraction patterns, a condition apparently in agreement with experimental observations [27].

How can we explain that the symmetric electron deflection $\Delta\alpha(x_0)$, due to its self field, is a *continuous* phenomenon, particularly seen in light of the *discrete* deflections α_n? The discrete electron deflections are caused by interaction with the laterally traveling photon doublets of discrete momenta via the slit field. In contrast, the continuous deflection is due to the electron's *self field* which is independent of the slit field and therefore not discrete. A microwave tube scientist familiar with electron beam dynamics would conventionally refer to this field as the *space charge* field, to be distinguished from the *circuit field* of the slit configuration. With the emphasis on momentum balance in the system it is still fair to ask about momentum balance in the continuous deflection. What physical mechanism is taking up the imparted lateral momentum gained by the electron in the continuous deflection process? The external evanescent discrete field components were already ruled out as a possible explanation. The most likely candidate is a simple *mechanical* momentum transfer caused by the net difference in force on the two walls of the slit, essentially the force expressed by (8.45). The imparted lateral velocity of the slit plate would be entirely negligible due to the enormous ratio of its mass to the mass of the electron, but will still serve to satisfy momentum balance.

At this point, with the conditions for discrete and continuous electron deflections established, let us describe the expected diffraction patterns from a purely ballistic analysis of electron trajectories. We shall demonstrate that superpositions of discrete electron arrival states gives rise to an overall macroscopic diffraction pattern matching experimentally observed patterns.

8.3 DIFFRACTION PATTERN WITH NUMERICAL SIMULATIONS

If the deflections of electrons originating from the two slits are specified by indexes Δn and Δm, a particular cloud of Δn type electrons from slit 1 gives rise to a rectangular irradiation of width

$$\Delta w = L + 2z\Delta\alpha(L/2) \tag{8.50}$$

which is independent of the discrete deflection Δn. The center of the rectangular irradiation from slit 1 is located at

$$x_n(z) = -\frac{d}{2} + \alpha_1 z\,\Delta n \qquad \Delta n = 0, \pm1, \pm2, ... \tag{8.51}$$

Similarly, the Δm type electrons from slit 2 forms a rectangular irradiation strip of the same width as (8.50) but with center at

$$x_m(z) = \frac{d}{2} + \alpha_1 z\,\Delta m \qquad \Delta m = 0, \pm1, \pm2, ... \tag{8.52}$$

The beamlets from the to sets overlap to form the overall diffraction pattern on the screen. The pattern can be understood from the overlapping condition of beamlets from slit 1 and slit 2, putting

$$x_n(z) = x_m(z) \tag{8.53}$$

which solved for $z = z_{nm}$ results in the formula

$$z_{nm} = \frac{d}{\alpha_1|\Delta n - \Delta m|} \qquad n, m = \pm1, \pm2, \pm3, ... \tag{8.54}$$

The first three positions are

$$z_1 = \frac{d}{\alpha_1} \qquad z_2 = \frac{d}{2\alpha_1} \qquad z_3 = \frac{d}{3\alpha_1} \tag{8.55}$$

Figure 8.7 shows a typical computer generated diffraction pattern for screen locations fairly close to the slit plate. Each area illuminated by electrons is shown with constant blackening, disregarding any electron variation across. The first three positions specified in (8.55) are indicated in the figure. At each of these positions the electron irradiation pattern has the typical shape of a diffraction pattern, with areas of illumination separated by empty regions. These patterns correspond to Frenél diffraction in regular optics. For positions closer to the slit plate the pattern varies rapidly with z and tends to blur out.

8.3.1 Frauenhofer diffraction pattern

The special condition equivalent to the *Frauenhofer* diffraction in optics is not specified by the deflection angle α_1 through (8.54) but rather by the *angular spread* $\Delta\alpha(L/2)$. Electrons originating from both slits, deflected by the same angle $\alpha_{\Delta n}$, are merged together due to the angular spread $\Delta\alpha(L/2)$. This happens beyond a certain minimum distance $z_{\min}$ from the slit plate, where the diffraction pattern has the same relative shape provided $z >> z_{\min}$. The criterion specifying $z_{\min}$ is obtained from the two conditions

$$\Delta n = \Delta m \qquad x_n(z) + \frac{\Delta w(z)}{2} = x_m(z) - \frac{\Delta w(z)}{2} \tag{8.56}$$

where $\Delta w(z) = 2\Delta\,\alpha(L/2)\,z$ is the width of each beamlet. Under these circumstances the two beamlets of the same deflection $\Delta n = \Delta m$ just merges together and continues as a single beamlet. The equations define the Frauenhofer limit $z_{\min}$ by the relation

$$\Delta w(z) = 2\Delta\,\alpha(L/2)\,z_{\min} = x_m(z) - x_n(z) = d \tag{8.57}$$

which has the solution

$$z_{\min} = \frac{d}{2\Delta\,\alpha(L/2)} \tag{8.58}$$

The importance of the Frauenhofer limit $z_{\min}$ is readily recognized from the stationary state of the diffraction pattern beyond the Frauenhofer limit. The stationary ray width w and the center of ray location x_n are specified by the following equations:

$$\begin{aligned} &\lim\left(\frac{w(z)}{z}\right) = \frac{\Delta w(z)}{z} = 2\Delta\,\alpha(L/2) && z >> z_{\min} \\ &\lim\left(\frac{x_n}{z}\right) = \frac{x_n + x_n}{2z} = \alpha_1 \Delta n && \Delta n = 0, \pm 1, \pm 2, \ldots \end{aligned} \tag{8.59}$$

The separation $\Delta x(z)$ between maxima, i.e., the periodicity, is then given by the formula

$$\lim\left(\frac{\Delta x(z)}{z}\right) = \alpha_1 = \frac{\hbar}{m v_0}\frac{1}{d} \tag{8.60}$$

The ratio between ray width $w(z)$ and periodicity $\Delta x(z)$ is given by

$$\lim\left(\frac{w(z)}{\Delta x(z)}\right) = 2\frac{\Delta\alpha(L/2)}{\alpha_1} = \text{constant} \tag{8.61}$$

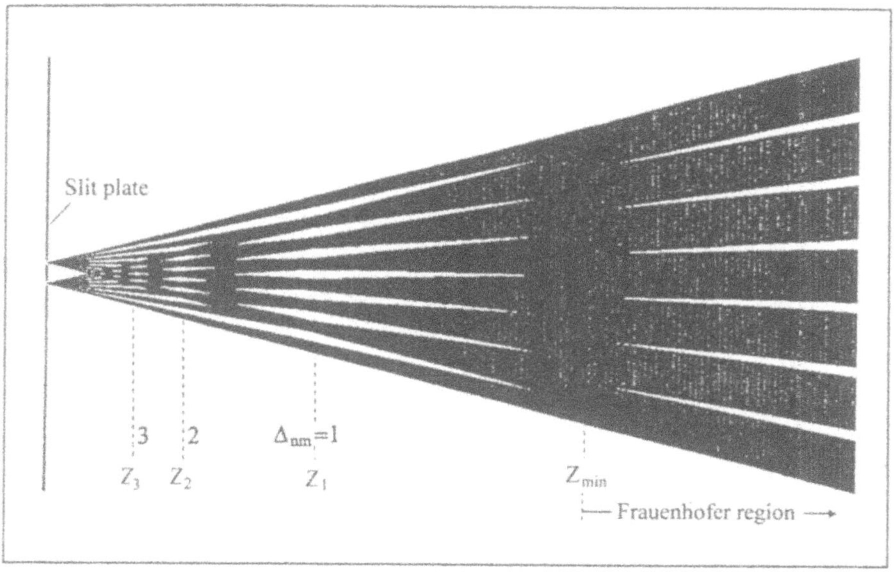

Figure 8.7. Computer generated electron diffraction pattern from a double slit plate configuration with selected parameters: $\alpha = 0.08$, $\Delta\alpha = 0.02$, $L/d = 0.3$. The transverse scale is greatly exagerated. The lower Frauenhofer limit is indicated by $z_{\min}$. Beyond this limit the stationary Frauenhofer diffraction prevails. The higher order diffraction patterns closer to the slit plate, corresponding to $\Delta_{nm} = 1, 2, 3 ...$, exist only over a limited range of z.

It is recalled, from (8.48), that the ratio $\Delta\alpha(L/2)/\alpha_1$ is specified by the *DC* accelerating voltage V_0. In the special case that the voltage is adjusted so that

$$\Delta\alpha(L/2)/\alpha_1 = 0.25, \tag{8.62}$$

the ray *widths* are equal to the separation between adjacent irradiated areas. This condition applies in the case illustrated in Fig. 8.7 where we must imagine the pattern continued to the right so that z becomes much larger than $z_{\min}$. Figure 8.8 shows the diffraction pattern at some distance $z >> z_{\min}$ where the strength $I(n)$ of irradiation is adjusted according to the reduced probability of higher order momentum transitions. This reduction is taken to be proportional to the square of the population density of the laterally traveling photon field components specified in (8.11).

$$I(n) = \frac{\sin^2(\pi n L/d)}{(\pi n L/d)^2} \tag{8.63}$$

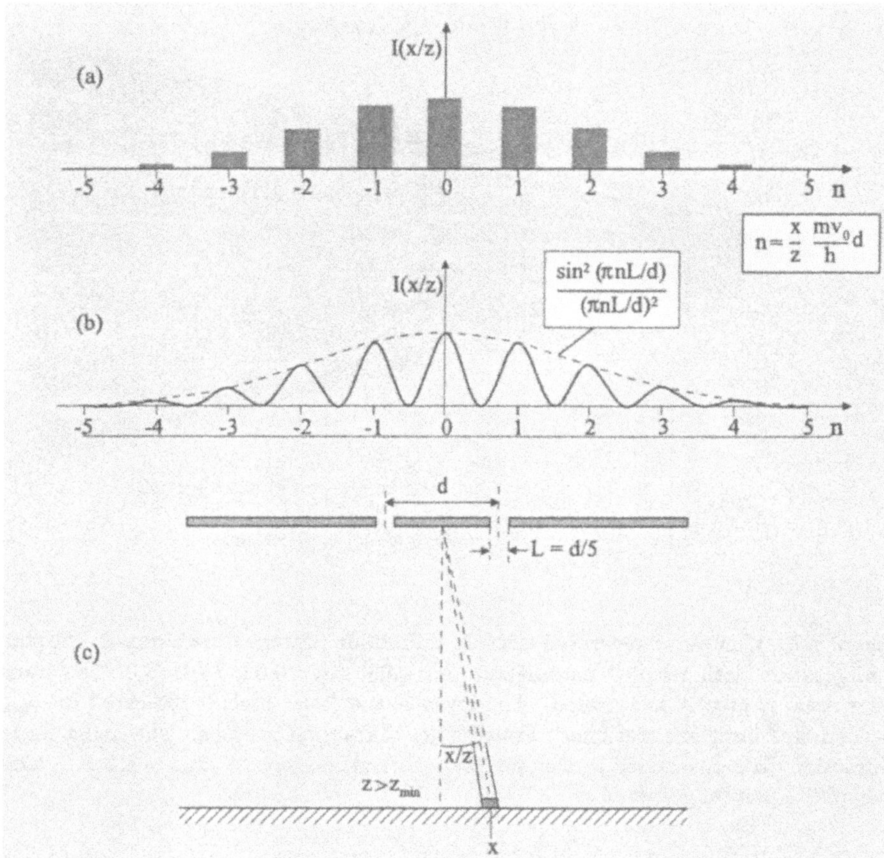

Figure 8.8. **Electron diffraction from a double slit plate. (a) Frauenhofer diffraction predicted by the neoclassical theory. (b) Diffraction pattern predicted by the quantum mechanical "electron wave" assumption based on wave particle duality. (c) Schematic diagram showing the double slit configuration and the Frauenhofer angle x/z.**

The relation can not be justified in a stringent manner because the detailed transitions taking place between the various symmetric and antisymmetric momentum levels are not easily described. The expression should be considered a likely but not necessarily rigorous prediction of reduced probability of higher order transitions.

8.3.2 Quantum mechanical electron waves

Quantum mechanics ascribe a wavelength λ_e to the electron, specified by

$$\lambda_e = \frac{\hbar}{mv_0} \tag{8.64}$$

which follows from equating the classical electron particle momentum mv_0 with the momentum $\hbar\beta = \hbar/\lambda_e$ of a hypothetical electron wave propagating in the direction of electron velocity v_0 according to $\exp(-j\beta z)$. What is gained by this construct is a description of the phenomenon of electron diffraction pattern as 'electron waves' originating from both slits and interfering to produce the diffraction pattern. The mathematics of the calculations is identical to familiar optical diffraction theory and will not be shown here. With the assumption of uniform slit 'electron wave' amplitudes, the diffraction pattern is specified by the formula:

$$I(\frac{x}{z}) = \frac{\sin^2\left(\pi\frac{x}{z}\frac{mv_0}{\hbar}L\right)}{\left(\pi\frac{x}{z}\frac{mv_0}{\hbar}L\right)^2}\cos^2\left(\pi\frac{x}{z}\frac{mv_0}{\hbar}d\right) \tag{8.65}$$

which is plotted in Fig. 8.8b. There is a remarkable similarity between the diffraction patterns described by the two theories. Not only is the predicted periodicity exactly the same, but the intensity amplitudes are specified by the same expression. The shape of both patterns agree well with experimental results of diffraction patterns shown in [27]. The agreement of the amplitude factors by the $[\sin(x)/x]^2$ type function is simply due to the assumption in both theories of uniform slit field distribution, but with entirely different physical substance—in one case uniform electromagnetic field—in the other uniform electron wave amplitude.

The fact that the periodicity of the patterns agree is basically a consequence of the acceptance of quantized momenta, but there is a profound conceptual difference. In the neoclassical theory the diffraction pattern comes about from discrete electron trajectories due to lateral momentum balance of an overall coupled system of photon doublets and free electrons. In this theory the wavelike diffraction pattern is not a consequence of any wave properties of the electron, but rather of the electromagnetic wave properties of the interacting electromagnetic system. The behavior is guided by the principle of minimization of the time average overall momentum, an adjunct of the corresponding energy principle. The particle aspect of the electron is maintained throughout, and full physical objectivity prevails.

In the electron wave theory of quantum physics the electron is considered a wave having the basic wavelength λ_e determined from its classical translational momentum mv_0. The electron wave is visualized as illuminating both slits, with the interference of electron wave radiation from the slits forming the diffraction pattern. In this dualistic wave-particle picture objectivity is lost. A number of problems familiar to the physics community are created in the process, such as: When is the electron a particle and when is it a wave? As a particle the electron must travel through one of the slits, as a wave it must travel through both. Moreover, experimental observations show that the diffraction pattern is made up of tiny localized spots, verifying the electron particle nature on arrival at the screen. The electrons are 'radiated' from both slits as waves, but arrive at the screen in a transformed costume as particles.

Besides these fundamental dilemmas with the electron wave picture itself there is the additional problem of wave coherence. In the corresponding double slit experiment for photons described in Chapter 7 the input stream of photons had to be coherent in order to form a distinct diffraction pattern on the screen (see paragraph following (7.74)). Should not the same criterium be expected to apply to the electrons? But the electron stream is certainly not coherent in any sense. The incident electrons are spread across the electron stream in a random lateral pattern with random arrival times and positions across the open slits. Thus, the individual electron waves radiating from the slits, with one wave component associated with each electron, can hardly be expected to form a coherent electron wave pattern at the screen, but rather a blurred out uniform illumination. The expected absence of coherence is in disagreement with the empirical observations and certainly detracts from the credibility of the electron wave proposition.

8.4 NEOCLASSICAL THEORY OF ELECTRON DIFFRACTION FROM A SINGLE SLIT

If the plate has just one single slit, the evanescent electromagnetic fields above and below can still be expressed by the Fourier transform (8.5), but we can not impose any periodicity as was done in (8.7) for the double slit plate. This difference implies that discrete momentum levels are nonexistent. Instead, the field is a continuum of evanescent lateral slow waves for which the amplitude factor $e_x(\beta)$ is readily obtained from inversion of (8.5).

$$e_x(\beta) = \frac{1}{2\pi} E_x L \frac{\sin(\beta L/2)}{\beta L/2} \tag{8.66}$$

A laterally traveling photon of propagation factor β posses the momentum

$$p(\beta) = \hbar\beta \tag{8.67}$$

In the interaction process between the evanescent modes and the electron, lateral momentum is expected to be conserved in the same manner as described for the double slit case. The electron deflection angle is therefore given by

$$\alpha(\beta) = \frac{1}{mv_0}\hbar\beta \tag{8.68}$$

which is the momentum change in the transition from level β to level $\beta=0$. The number of transitions is expected to be proportional to the square of the population density obtained from (8.66). If $x = z\,\alpha(\beta)$ is the point of arrival at the screen, then the corresponding β is given by

$$\beta = (\frac{x}{z})(\frac{mv_0}{\hbar}) \tag{8.69}$$

which inserted into (8.66) specifies the following screen irradiation density:

$$I(\frac{x}{z}) = \frac{\sin^2\left(\pi\frac{x}{z}\frac{mv_o}{\hbar}L\right)^2}{\left(\pi\frac{x}{z}\frac{mv_o}{\hbar}L\right)^2} \tag{8.70}$$

The resulting Frauenhofer pattern is shown in Fig. 8.9. The single slit has no additional diffraction zones closer to the slit plate. The Frauenhofer zone is the only one, extending all the way from the slit plate to infinity. According to [27] the expected *electron wave* diffraction pattern of quantum theory is specified by an expression identical to (8.70). Thus, the two theories give identical answers also for the single slit case, but for entirely different physical reasons. In the neoclassical theory the continuous momentum levels of the laterally traveling photon doublets, i.e., the evanescent surface modes matching the slit field, are transferred to continuous electron deflections. The number of doublets contributing to a particular deflection angle is proportional to the square of the population level, i.e., to the expression in (8.70). As a typical example, the absence of irradiation at point $\beta(x/z)=1.0$ in Fig. 8.9 is due to the absence of laterally traveling photons with this particular value of β. The corresponding momentum is simply not available for transfer to the electron. In the electron wave theory the irradiation pattern is a result of interference of wave components originating from different points in the single slit, which results in an overall pattern identical to (8.70). In particular, the zero at $\beta(x/z)=1.0$ in Fig. 8.9 is a result of exact phase cancellations from rays originating at $x=-L/2$ and $x=L/2$, i.e., from both ends of the slit.

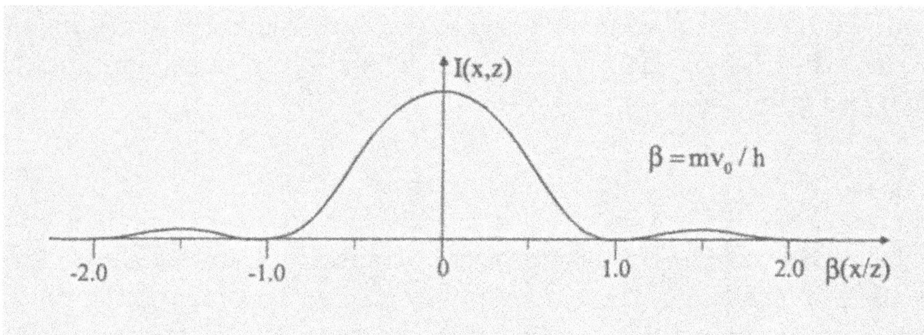

Figure 8.9. **Neoclassical electron diffraction pattern from a single slit plate. The pattern is identical to the prediction from the electron wave theory of quantum mechanics.**

The demonstrations in this chapter of electron diffraction due to interaction with periodic doublet type electromagnetic fields is in sharp contrast with the dualistic electron wave picture. The two theories predict essentially the same diffraction patterns, although for entirely different physical reasons. The neoclassical theory has the advantage of retaining the particle nature of the electron, thereby maintaining full objectivity of the diffraction phenomena taking place in the double slit experiment, avoiding the infamous problems and paradoxes associated with the wave particle duality proposition.

Chapter 9

THE ENIGMATIC 1/F NOISE

'Real knowledge is to know the extent of ones ignorance.'

Confucius

9.1 WHAT IS 1/F NOISE?

In this chapter the general formulations of the neoclassical circuit theory will be adapted to the well known phenomenon of $1/f$ noise in active electronic devices and material samples under nonequilibrium conditions. It will be demonstrated that the theory predicts low frequency noise excitations with power spectral density inversely proportional to the frequency. The $1/f$ noise comes in addition to the ever present thermal noise which is well understood and has no relation to $1/f$ noise. But first a brief description of $1/f$ noise which is observed experimentally in essentially all electronic materials and devices, such as simple resistors, semiconductors, metallic thin films, diodes, transistors, electrolytic solutions, Josephson junctions and many more. The universal presence of $1/f$ noise in such a diverse group of materials and devices points to the existence of some deeper principle or law of nature that applies to all nonequilibrium electronic systems. Lots of ink has flown in efforts to pin down and understand its origin, with numerous specific models of the noise mechanism proposed. But none of these have satisfied the requirement of being universally applicable to a large number of diverse cases and systems.

The $1/f$ noise can be observed experimentally under quite simple conditions where a *DC* (Direct Current) voltage V_0 is applied across a resistor or whatever electronic component is involved, drawing current through the sample. As indicated in Fig. 9.1 the observed noise in the form of voltage or current fluctuations in the sample shows a typical $1/f$ noise power spectrum $\overline{S(\omega)}$, giving rise to the name $1/f$ noise. In a more general environment, with a nonzero input frequency ω_0, the corresponding noise spectrum is of the form $1/|\omega_0 - \omega|$. In this case the generalized $1/f$ noise appears in two symmetrically located bands centered at $\omega = \omega_0$.

In spite of the ubiquitous nature of the $1/f$ noise and the enormous amount of experimental data that has accumulated since this noise phenomenon was first recognized and described around 1925 [26], its physical origin has stubbornly refused to be revealed. And this in spite of a colossal amount of theoretical and experimental work on the subject from a considerable number of researchers in the field. Essentially all the search for the source of the $1/f$ fluctuations has been directed towards the charge carriers in the electronic components, looking for a possible $1/f$ noise generating mechanism. Many such mechanisms have been proposed but it is a fair assessment that none has been sufficiently broad in scope to gain general acceptance.

9.2 HOOGE'S EMPIRICAL 1/F HYPOTHESIS

Earlier it was believed that $1/f$ noise showed up in semiconductor materials but not in homogeneous metals. But then Hooge and Hoppenbrowers [24] reported $1/f$ noise in thin gold films and Hooge[25] put forward the hypothesis that $1/f$ fluctuations in homogeneous materials can be represented by the *empirical* formula

$$\frac{\overline{S(\omega)}}{P_0} = \frac{\alpha_H}{N_0 f} \tag{9.1}$$

where $\overline{S(\omega)}$ is the noise power spectral density, P_0 is the *DC* power dissipated in the sample, N_0 is the *total* number of carriers in the sample, and $\alpha_H \approx 2 \times 10^{-3}$ is a universal constant, the *Hooge constant.*

Hooge's empirical law had considerable success in predicting $1/f$ noise behavior in various conducting homogeneous materials, although later research did not corroborate the 'universality' of Hooge's constant α_H. In fact, α_H was found to vary over a large range, being considerably smaller for short samples, and larger for materials having low mobility such as metallic whiskers [13],[14]. The reason for quoting these comments is that the theory of $1/f$ noise based on the neoclassical theory, as presented in the following sections, ends up with a *theoretical* derivation

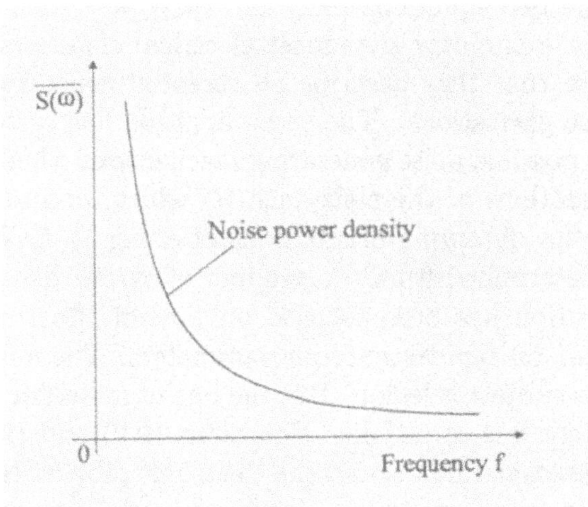

Figure 9.1. Typical $1/f$ noise spectrum characterized by noise power density $\overline{S(\omega)}$ inversely proportional to frequency.

of Hooge's law, in which α_H contains geometrical and physical parameters in qualitative agreement with the quoted observations.

9.3 PHOTON EXCITATION OF 1/F NOISE

The analysis of noise mechanisms to be presented shifts the attention from the charge carriers to electromagnetic excitation caused by single photon absorption and reflection aided by environmental photon doublets. The noise theory is based on the theory of photons and doublets in general networks presented earlier in Chapter 6. The sample under study is a simple cylindrical resistor carrying DC current I_0 with applied DC voltage V_0 across the resistor terminals. The sample resistance R_0 is related to V_0 and I_0 through the familiar $Ohm's$ law:

$$R_0 = \frac{V_0}{I_0} \tag{9.2}$$

The dissipated DC power in the resistor is given by

$$P_0 = V_0 I_0 \tag{9.3}$$

If the DC power supply has negligible inner resistance, the voltage V_0 is constant, in which case the $1/f$ noise is observed as a fluctuating component of current superimposed on the DC current I_0. Similarly, if the DC current I_0 is kept constant from a power supply of very high inner resistance, the $1/f$ noise is observed as fluctuations in the voltage across the terminals. The two equations (9.2) and (9.3) play such a fundamental role in electronic circuitry that most electrical engineers and scientists tend to believe that they describe all essential aspects of the resistor, including noise excitations. The same attitude has certainly pervaded the search for possible noise generating mechanisms which has been centered on fluctuations of the resistance R_0 which, according to (9.2), is a prerequisite for observing fluctuations of either V_0 or I_0. Because the resistance is determined from the product of carrier density and carrier mobility, attention has been focused on possible fluctuations in these parameters, i.e., on typical material parameters. The number of papers written on this subject is legion. But the line of reasoning described here is not absolutely free of pitfalls. Equations (9.2) and (9.3) do not tell everything there is to know about the electronic properties of the resistor because they are based on an oversimplified model disregarding the electromagnetics of the problem, in particular the role of discrete photons. The power dissipated in the sample is not *basically* specified by (9.3) but rather by the electromagnetic influx of energy through the resistor surface, described by Poynting's theorem. For the *DC* case Poynting's theorem has the form

$$\nabla \cdot (\vec{E} \times \vec{H}) + \vec{E} \cdot \vec{i} = 0 \tag{9.4}$$

Integration over the resistor volume V, followed by transformation of the volume integral to a surface integral using Gauss' theorem, converts the equation to

$$\begin{aligned} &\int_S \vec{n} \cdot (\vec{E} \times \vec{H}) dS = -\int_V \vec{E} \cdot \vec{i} dV \\ \text{or} \quad &P_{em} = \int_S E_z H_\varphi(r_1) dS = E_z L \, 2\pi r_1 H_\varphi(r_1) = V_0 I_0 \end{aligned} \tag{9.5}$$

In these equations r_1 is the resistor radius, and the electric field E_z is assumed constant over the resistor cross section. The surface integral is the electromagnetic *DC* power P_{em} expressed as the integrated Poynting flux into the resistor.

$$P_{em} = V_0 I_0 = P_0 \tag{9.6}$$

With these equations as a base let us consider possible origins of the noise. In the usual search for $1/f$ noise mechanisms—the materials approach—one takes fluctuations in the resistance R_0, i.e., in V_0 or I_0 to be the basic mechanism, assuming that the electromagnetic power P_{em} and associated field variables E_z and $H_\varphi(r_1)$ *adjust* themselves accordingly, to satisfy (9.6). With its emphasis on photon involvement, the neoclassical theory takes the point of view that the basic noise is caused by fluctuations in the electromagnetic fields due to the discrete nature of photons. The fluctuations occur in the electromagnetic flux P_{em} with $P_0 = I_0 V_0$ *adjusting* itself to satisfy (9.6). If this tentative supposition is correct, any search for a general $1/f$ noise mechanism in the materials themselves is doomed to failure. The lack of much success along this line is certainly an indication that this might be the case.

In the neoclassical description of the noise generating mechanism the power influx P_{em} is due to photons of 'zero frequency' impinging radially on the resistor surface, with associated noise excitations caused by the unavoidable photon doublet admixture aiding the process. Therefore, in describing the resistor and its environment it is essential to establish an adequate model reflecting the emphasis on the electromagnetic aspects, in particular the photon involvement. As already noted, the familiar but oversimplified concept of a resistor component fed by the voltage V_0 and current I_0 does not satisfy this criterion. The local electromagnetic fields in the immediate neighborhood of the resistor play a role in the process and must somehow be included. The model must also express the *DC* state in a more general form than expressed in classical electromagnetics, so that it can be applied to *DC* photons as well. The concept of the *DC* photon state is discussed in the next section.

9.3.1 The DC photon state–definition and interpretation

It is informative first to consider the *DC* state in classical electromagnetics. In the spherical coordinate system of Fig. 9.2 the fields inside and outside the resistor satisfy

$$\nabla \times \vec{e}(r,\theta,\varphi) = 0 \qquad \text{and} \qquad \nabla \times \vec{H}(r,\theta,\varphi) = \vec{\imath}(r,\theta,\varphi) \tag{9.7}$$

Because the current density $\vec{\imath}(r,\theta,\varphi)$ is zero in free space external to the resistor, the electric and magnetic fields outside the resistor are basically uncoupled. Therefore, assuming rotational symmetry, with no fields depending on θ, the magnetic field $H_\theta(r,\varphi)$ *outside* the resistor satisfies

$$\frac{\partial[rH_\theta(r,\varphi)]}{\partial r} = 0 \qquad \text{and} \qquad \frac{\partial[H_\theta(r,\varphi)\sin(\varphi)]}{\partial \varphi} = 0 \tag{9.8}$$

The overall outside solution, through matching with the inside field at the resistor surface, is then given by

$$2\pi r \sin(\varphi) H_\theta(r,\varphi) = I_0 \tag{9.9}$$

where I_0 is the DC current in the resistor. Furthermore, from the equation $\nabla \times \vec{e}(r,\varphi) = 0$:

$$\frac{\partial\left[re_\varphi(r,\varphi)\right]}{\partial r} - \frac{\partial\left[e_r(r,\varphi)\right]}{\partial\varphi} = 0 \tag{9.10}$$

Integration with respect to the spherical angle φ from δ to $\pi-\delta$, keeping r constant, converts the equation to a form that is readily recognized. With reference to the insert in Fig. 9.2 the angle δ is specified by

$$\sin(\delta) = \frac{r_0}{r} \tag{9.11}$$

where r_0 is the radius of the feeder wire. If the wire is thin, the last term in the integration of (9.10) can be disregarded, because

$$\left|\frac{e_r(r,\delta)}{e_\varphi(r,\delta)}\right| = \left|\frac{e_r(r,\pi-\delta)}{e_\varphi(r,\pi-\delta)}\right| = \tan(\delta) \cong 0 \tag{9.12}$$

Thus, integration of (9.10) results in

$$\frac{\partial}{\partial r}\int_{\delta}^{\pi-\delta} re_\varphi(r,\varphi)\,d\varphi = 0 \qquad \text{or} \qquad \int_{\delta}^{\pi-\delta} re_\varphi(r,\varphi)\,d\varphi = -V_0 \tag{9.13}$$

In this equation the line integral of $e_\varphi(r,\varphi)$ is the DC resistor voltage V_0 which does not depend on the selected radius r. The two components I_0 and V_0 are independent solutions of Maxwell's equations. According to (9.2) they are coupled through the boundary condition imposed by the resistor.

From the platform of classical electromagnetism the fields (9.9) and (9.13) represent static solutions of Maxwell's equations. But conceptually it is difficult to reconcile the static picture with the continuous feeding of DC power into the resistor, as required by (9.5). The radially directed power flux is expected to contain electromagnetic entities propagating with the velocity of light. And the photons constituting the DC fields are supposedly going to be identified with these entities. Thus, we are looking for field solutions that overcome the apparent contradiction of simultaneously describing static fields and propagating fields. Do such solutions exist in classical electromagnetics? As demonstrated in the following the answer is affirmative. But in order to arrive at these solutions we must start from a base of *time-dependent* field equations.

Referring again to the spherical model shown in Fig. 9.2, and assuming the same field symmetries as in the DC case already discussed, the time-dependent Maxwell's equations outside the resistor are specified by the

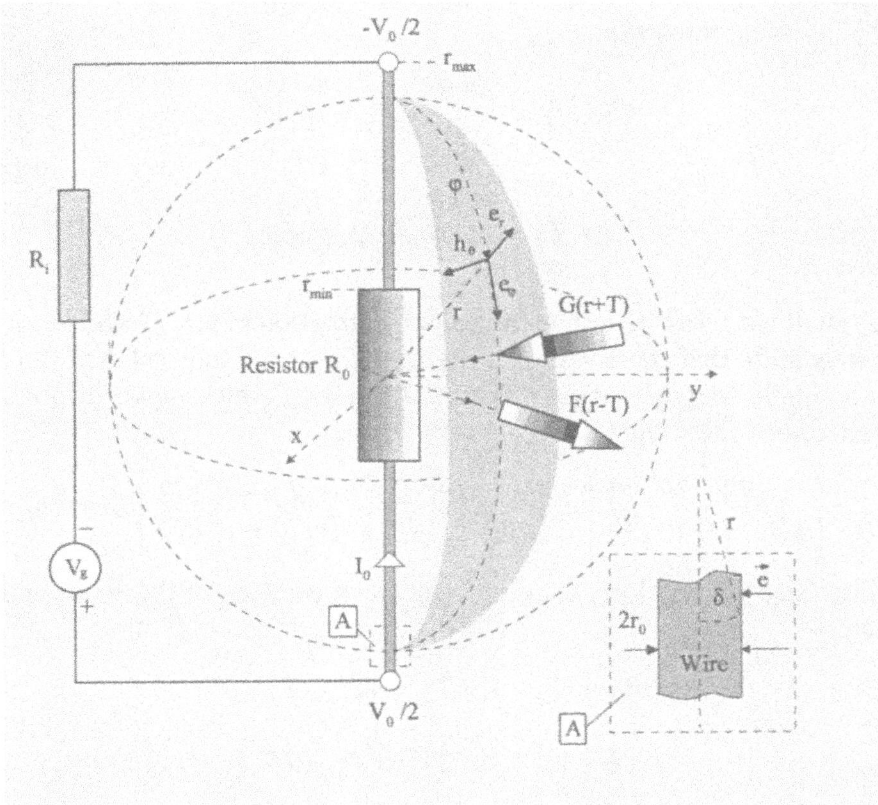

Figure 9.2. **Sperical model of the resistor and its environment. The approriate field components are e_r, e_φ, and h_θ. The two DC pulses $F(r-T)$ and $G(r+T)$ indicated by heavy arrows are the transmission line representation of the staticDC fields.**

two partial differential equations:

$$\frac{\partial\left[re_\varphi(r,\varphi,T)\right]}{\partial T}+\frac{\partial\left[rh_\theta(r,\varphi,T)\right]}{\partial r}=0$$

$$\frac{\partial\left[rh_\theta(r,\varphi,T)\right]}{\partial T}+\frac{\partial\left[re_\varphi(r,\varphi,T)\right]}{\partial r}-\frac{\partial\left[e_r(r,\varphi,T)\right]}{\partial\varphi}=0 \tag{9.14}$$

As in earlier chapters the normalized variables appearing in these equations are defined by $h_\theta = Z_0 H_\theta$ and $T = ct$, where $Z_0 = \sqrt{\mu_0/\varepsilon_0}$ is the impedance and $c = 1/\sqrt{\mu_0\varepsilon_0}$ the velocity of light, both referred to free space. Let us keep the radius r constant and again integrate both equations with respect to the cone angle φ from δ to $\pi-\delta$. The integral of $\partial e_r(r,\varphi,T)/\partial\varphi$ is vanishingly small for the same reason as explained in connection with (9.10). Let us define the variables $v(r,T)$ and $i(r,T)$ by the following integrals:

$$v(r,T) = -\int_{\delta}^{\pi-\delta} re_{\varphi}(r,\varphi,T)\,d\varphi$$
$$i(r,T) = \int_{\delta}^{\pi-\delta} rh_{\theta}(r,\varphi,T)\,d\varphi \tag{9.15}$$

We shall not dwell with the physical interpretation of $v(r,T)$ and $i(r,T)$, merely state that these variables are appropriate for conversion of (9.14) to a simple coupled partial differential equation which, upon diagonalization, has the form:

$$\frac{\partial}{\partial T}\begin{bmatrix} v(r,T)+i(r,T) \\ v(r,T)-i(r,T) \end{bmatrix} + \begin{bmatrix} -1 & 0 \\ 0 & 1 \end{bmatrix} \frac{\partial}{\partial r}\begin{bmatrix} v(r,T)+i(r,T) \\ v(r,T)-i(r,T) \end{bmatrix} = \mathbf{0} \tag{9.16}$$

In its most general form the time-dependent solution of the equation is simply

$$v(r,T) - i(r,T) = F(r-T)$$
$$v(r,T) + i(r,T) = G(r+T) \tag{9.17}$$

where $F(r\text{–}T)$ and $G(r\text{+}T)$ are pulses of *arbitrary shape* propagating with the velocity of light in opposite radial directions with no change of pulse shapes. Because (9.17) is the general solution of (9.16), it must contain *all* possible solutions, and therefore the time-independent *DC* state as a special stationary case. This particular condition is seen to be satisfied when the pulses $F(r\text{–}T)$ and $G(r\text{+}T)$ are both of constant shape. In this special case the solutions are certainly stationary, because neither $v_0 = v(r,T)$ nor $i_0 = i(r,T)$ depend on r and T.

$$v_0 - i_0 = F(r-T) = \text{constant}$$
$$v_0 + i_0 = G(r+T) = \text{constant} \tag{9.18}$$

Solved with respect to v_0 and i_0:

$$v_0 = \tfrac{1}{2}[G(r+T) + F(r-T)]$$
$$i_0 = \tfrac{1}{2}[G(r+T) - F(r-T)] \tag{9.19}$$

The pulses $F(r-T)$ and $G(r+T)$, both of constant shape, represent *DC* pulses propagating with the velocity of light, thus combining static fields with propagating fields.

How do the time-independent variables v_0 and i_0 relate to the *DC* voltage V_0 and current I_0? From the definitions in (9.15) it follows that

$$V_0 = v_0$$

$$I_0 = i_0 \frac{2\pi/Z_0}{\int_\delta^{\pi-\delta} d\varphi/\sin(\varphi)} = i_0 \frac{\pi/Z_0}{|\log(\delta)|} \tag{9.20}$$

The variables v_0 and i_0 of (9.19) are coupled through Ohm's law $V_0 = R_0 I_0$ which, together with (9.20), result in the equivalent relation:

$$\frac{v_0}{i_0} = z = \frac{\pi R_0/Z_0}{|\log(\delta)|} \tag{9.21}$$

The normalized load impedance z, appropriate to the variables v_0 and i_0, is of course proportional to the sample resistance R_0 but depends in addition on the feeder wire thickness r_0 through the slowly varying logaritmic term $|\log(\delta)| \cong |\log(r_0/r)|$. Equation (9.21) together with (9.19) specify the ratio ρ of the outward and inward *DC* pulses $F(r-T)$ and $G(r+T)$:

$$\rho = \frac{F(r-T)}{G(r+T)} = \frac{z-1}{z+1} \tag{9.22}$$

Readers familiar with transmission line theory will recognize the equation as defining the *reflection coefficient* ρ of a transmission line loaded with the normalized impedance z. The fact that the outward traveling pulse $F(r-T)$ is less than the inward traveling pulse $G(r+T)$ reflects the fact that net power is dissipated in the resistor.

The field solutions of the spherical model presented here are valid for radii between $r_{\min}$ and $r_{\max}$, where $r_{\min}$ and $r_{\max}$ are defined in Fig. 9.2. For $r > r_{\max}$ the fields are decaying quite rapidly. For $r < r_{\min}$ the spherical model suffers the inconvenience of not possessing the right symmetry for field matching at the surface of a presumably cylindrical resistor. For this reason the actual noise analysis to be carried out a little further on, in 9.3.3, is done on a cylindrical model matching more closely the circular shape of the resistor. The spherical model has the virtue of projecting a mental image of a field distribution that closely matches the actual physical fields, except close to the resistor. Moreover, it serves to introduce the somewhat unfamiliar concept of static fields expressed by *DC* pulses propagating with the velocity of light. The formulation allows us to interpret the static fields in transmission line terminology as propagating *DC* pulses, subject to transmission and reflection according to transmission line theory. As demonstrated in the subsequent treatment this particular formulation can first be appreciated fully when applied to single photons.

The inward and outward *DC* pulses $G(r+T)$ and $F(r-T)$ are expected to contain a very large number of individual *DC* photons, each satisfying the same basic propagation characteristics as the macroscopic pulses. Still the photons behave differently due to their energy integrity. The absorption and reflection of photons at the resistor surface is a discontinuous phenomenon in stark contrast to the continuous macroscopic process described in the preceding analysis. An inward propagating photon in the form of a pulse $G(r+T)$ can not have *part* of its energy absorbed and part of it reflected, but is subject to the option of *either* dissipating all its energy in the resistor *or* reflecting all energy back into the pulse $F(r-T)$. In order to gain some insight into this discontinuous process we shall leave the spherical model, which has served to provide the conceptual background, and apply a somewhat simpler but in all major respects equivalent model. The discussion serves as an intermediate step before the final noise analysis of section 9.3.3.

9.3.2 Behavior of single DC photon pulses

This simpler model shown in Fig. 9.3 refers to a finite section of a two-wire *TEM* transmission line charged to the *DC* voltage V_0. A switch permits the line to be either open or discharged through a resistor R_0 matched to the characteristic impedance of the line. The propagation of pulses along the transmission line is described by the familiar *transmission line equation* described in any textbook on the subject.

$$\frac{\partial}{\partial T}\begin{bmatrix} V(Z,T)+I(Z,T) \\ V(Z,T)-I(Z,T) \end{bmatrix}+\begin{bmatrix} -1 & 0 \\ 0 & 1 \end{bmatrix}\frac{\partial}{\partial Z}\begin{bmatrix} V(Z,T)+I(Z,T) \\ V(Z,T)-I(Z,T) \end{bmatrix}=0 \tag{9.23}$$

where $T=ct$ is normalized time. The variables $V(Z,T)$ and $I(Z,T)$ are the line voltage and line current, respectively, with line current normalized to the characteristic transmission line impedance Z_0. The fact that the differential equation is identical to the corresponding spherical equation (9.16) merely confirms the common features of the two models and the legitimacy of the simple model. Therefore, the general solution is the same as (9.17).

$$V(Z,T) = G(Z+T) + F(Z-T)$$
$$I(Z,T) = G(Z+T) - F(Z-T) \tag{9.24}$$

In these expressions $F(Z–T)$ and $G(Z+T)$ are pulses of *arbitrary shape* propagating with the velocity of light in opposite directions along the line with no changes of pulse shape. The special case of *DC* requires these to be stationary, a condition implying that

$$F(Z-T) = G(Z+T) = \text{constant} \tag{9.25}$$

yielding the following two equations:

$$V(Z,T) = 2F(Z-T) = V_0 = \text{ constant}$$
$$I(Z,T) = 0 \tag{9.26}$$

The relations specify a pure static *DC* condition expressed as a superposition of constant amplitude pulses propagating with the velocity of light along the line.

The two traveling pulses (9.25) are shown in the two lower sketches of Fig. 9.3a, illustrating how this alternate formulation of *DC* state is expressed as a standing wave of constant amplitude pulses $F(Z–T)$ and $G(Z+T)$ traveling in opposite directions along the line, and being subject to total reflections at both ends. If no energy is extracted, this description appears to be entirely equivalent to the simple static condition (9.7), but conceptually it is quite different. The difference becomes crucial in the discharge process illustrated in Fig. 9.3b showing how the standing wave pulses react upon closing the switch at time $T=0$. As shown in the upper sketch the forward traveling pulse $F(Z–T)$, with wave impedance Z_0, suddenly sees a load offering a perfect match $R_0 = Z_0$ for discharging its energy. Instead of being reflected it propagates into the resistor, thereby failing to excite the backward pulse $G(Z+T)$ at $Z=L$. In turn, the curtailed backward pulse, after a time interval L/c corresponding to a full transit along the line, fails to excite the forward pulse $F(Z–T)$ at $Z=0$, which subsequently is fully discharged after the lapse of another time interval L/c. The resulting voltage over the resistor is a perfect rectangular pulse of constant amplitude $V = F(Z–T) = V_0/2$ and duration $2L/c$, i.e., twice the transit time along the line. The configuration discussed here is by no means a theoretical construct or curiosity; the finite length transmission line has found numerous applications in

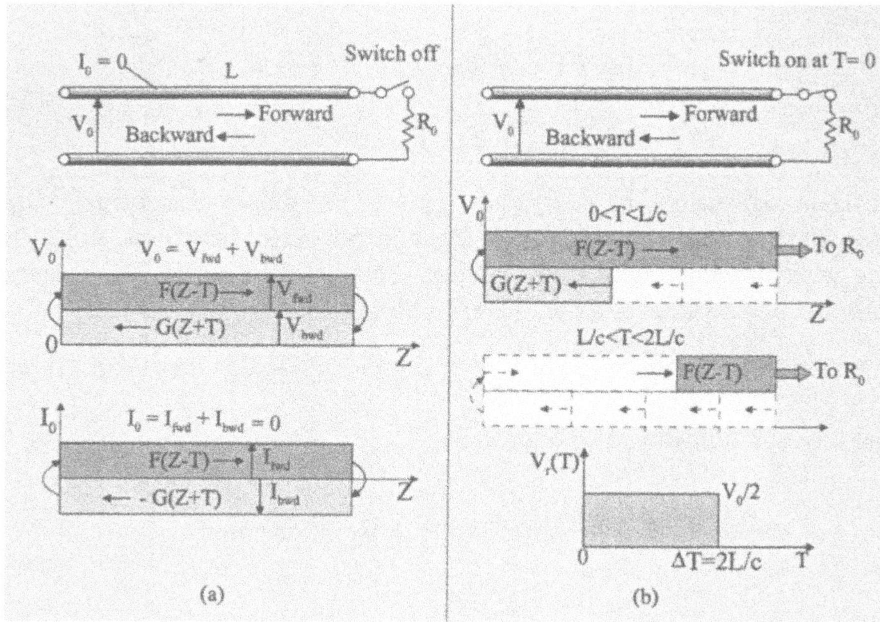

Figure 9.3. **Illustration of the discharge of a transmission line section with** ***DC*** **voltage V_0. (a) The** ***DC*** **state of the line is fully described by two oppositely traveling constant pulses F(Z–T) and G(Z+T) of same amplitude, the pulses being reflected into each other at the line ends. (b) With the switch closing at time T=0 the line discharges according to the two intermediate sketches. The lower sketch shows the rectangular pulse voltage $V_r(T)$ across the resistor. The amplitude is equal to** ***one half*** **the line voltage V_0, and the pulse duration is** ***twice*** **the transit time L/c along the transmission line.**

experimental work in plasma and solid state, serving as a simple one shot pulse generator, providing extremely short pulses of perfect rectangular shape and relatively high power. A typical device consists of a roll of flexible large size coaxial line several meters long which is charged and discharged through mercury switches.

The discharge process shown in Fig. 9.3 has been described from classical electromagnetics alone, but it applies equally well to the discharge of single photons in the resistor system, with the important difference that the switching is administered and executed by the environmental photon doublets rather than by the mechanical switch in the figure. What has been learned from the particular example discussed here are twofold. First, the zero frequency photons can be interpreted as a superposition of two oppositely traveling constant amplitude pulses, forming a standing wave pattern. Second, the discharge of the photon into the resistor is a pulse of *finite* duration. In the macroscopic situation described in Fig. 9.3 the switch was operated manually. In the actual resistor system the switching is done through the intermediary of photon doublets. The similarity of the two processes is immediately apparent. In both cases the photons are either transferred to the resistor or reflected back. In the model of Fig. 9.3 the pulse duration Δt is specified by the line length L. In the actual physical configuration Δt is expected to bear a relation to the radial extension $r_{\max}$ of the static field around the resistor, which is proportional to the resistor length L. A simple approximate calculation estimates the static fields beyond $r = 2L$ to be negligible. Therefore, in the actual physical configuration let us put

$$r_{\max} \simeq 2L \tag{9.27}$$

so that the pulse duration can be approximated by

$$\Delta t \simeq 4L/c \tag{9.28}$$

Based on this comparison the expected situation is further illustrated in Fig. 9.4 showing schematic diagrams of the switching operation affected by photon doublets. In the upper sketch the doublets have adjusted the photon impedance to match the resistor impedance; the gate is open, causing photon excitation of the resistor. In the lower sketch the doublet admixture puts the photon impedance at zero or infinite; the gate is closed, causing complete photon reflection. The sequence of pulses, each of duration Δt, shown to the right in Fig. 9.4a, is merely meant to indicate the general type of photon input to the resistor. The fact that each *DC* pulse is of finite duration implies that it can not be described as a pure *DC* condition at zero frequency, but rather as a spectrum $a(\omega)_{phot}$ of nonzero frequency components:

$$a(\omega)_{phot} = \frac{1}{2\pi} \int_{-\Delta t/2}^{\Delta t/2} a_{phot}\, e^{-j\omega t}\, d\omega t = \frac{1}{2\pi} a_{phot} \Delta t \frac{\sin(\omega \Delta t/2)}{\omega \Delta t/2} \tag{9.29}$$

The frequency spectrum is essentially constant up to the frequency $1/(2\Delta t)$ which is very high. If the pulse is identified as the '*DC* photon' its energy is not zero but spread out over the large frequency spectrum of (9.29). Even if the energy of each '*DC* photon' is not zero, it may still be quite small and therefore expected to be compensated by a correspondingly large number of photons per second. In the analysis to be shown shortly we shall indeed find that there is a very large photon number N_Δ inside the time interval Δt. The frequency spectrum of this sequence of overlapping pulses is obtained as follows.

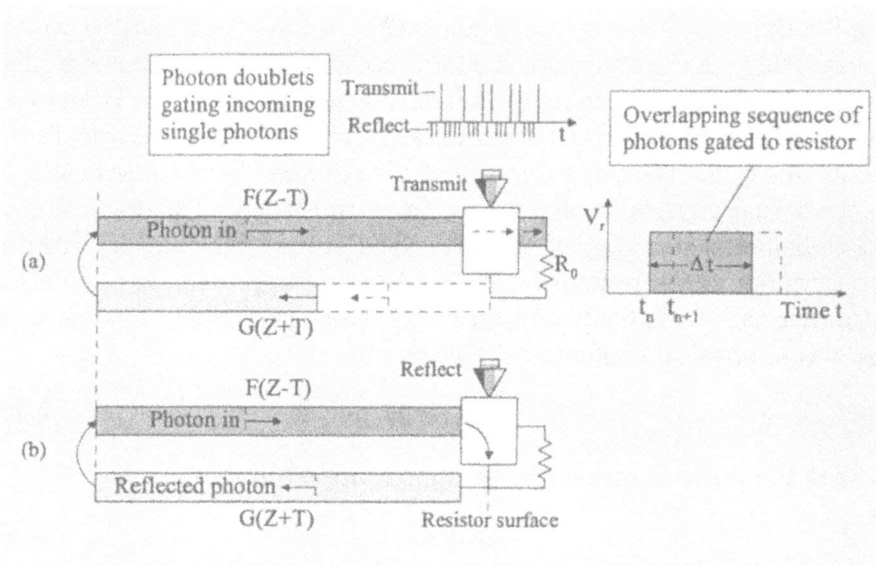

Figure 9.4. Schematic illustration of the switching operation affected by photon doublets. (**a**) *Transmit:* The photon wave impedance is matched to the resistor impedance through appropriate doublet supplement. The sequence to the right shows two pulses out of an infinite set of overlapping pulses. (**b**) *Reflect*: The doublet supplement renders the photon wave impedance either zero or infinite, causing complete reflection.

$$a(\omega) = \frac{1}{2\pi} a_{phot} \sum_{n=1}^{N_\Delta} \int_{t_n}^{t_n+\Delta t} e^{-j\omega t}\, d\omega t \tag{9.30}$$

where t_n is the random launch time of the nth pulse, Δt is the pulse duration, and a_{phot} is the identical height of each of the N_Δ pulses. The expression is readily converted to

$$a(\omega) = \frac{1}{2\pi} a_{phot} \Delta t \frac{\sin(\omega \Delta t/2)}{\omega \Delta t/2} e^{-j\omega \Delta t/2} \sum_{n=1}^{N_\Delta} e^{-j\omega t_n} \tag{9.31}$$

Assuming *random* distribution of launch times t_n over the interval Δt, the sum averages out to zero unless $\omega = 0$ in which case it is equal to N_Δ. Accordingly

$$a(\omega) = \begin{cases} 0 & \text{if } \omega > 0 \\ \frac{1}{2\pi} a_{phot} N_\Delta \, \Delta t & \text{if } \omega = 0 \end{cases} \tag{9.32}$$

The random spreading in time of the N_Δ superimposed pulses has effectively canceled the harmonic frequency components (9.29) appearing in the single pulse discharge. The only nonzero component $a(0)$ is simply the sum of the individual pulses, and as such it represents a true *DC* state in spite if the finite duration Δt of the superimposed pulses. It will be shown shortly that $a(0)$ forms the basis for the specification of *DC power* P_0 dissipated in the resistor.

From these deliberations it appears that the excitation of the resistor by the infinite sequence of photons randomly localized in time does not give rise to any frequency components of noise, just the *DC* component $a(0)$. So, where does the $1/f$ noise come from? The present theory holds that the noise comes from the switching process itself, with photon doublets organizing the transmission and reflection of the photons at the resistor surface.

9.3.3 The radial transmission line–appropriate model

For detailed analysis of the doublet noise the radial transmission line model shown in Fig. 9.5 is selected. In this model, which is an approximate representation of the resistor constellation, the electromagnetic fields are contained within the two virtual conducting planes at $Z = \pm L/2$. The fundamental Transverse Magnetic (TM) mode of the line is assumed to adequately represent the basic noise processes. Why is the radial line model preferable to the spherical model of Fig. 9.2 and the two-wire line of Fig. 9.3? We have seen that both these models describe the *DC* state in transmission line terminology as two oppositely propagating *DC* pulses of constant amplitude. But they suffer from certain inadequacies in describing the actual resistor system. The spherical model does not properly match the fields at the resistor surface, and the constant characteristic impedance of the two-wire model does not describe the actual impedance variation with frequency. The radial transmission line configuration represents a more adequate model in that it satisfies these basic requirements. In this model the electromagnetic fields near the resistor surface are modeled more accurately. The modeling of the far fields is of little consequence, allowing us to terminate the radial line at the outer radius r_2, beyond which the actual fields are negligible.

The fundamental mode of the radial transmission line equation, specified by Maxwell's equations in cylindrical coordinates, has the simple form

$$\frac{\partial}{\partial T}\begin{bmatrix} e_z \\ rh_\theta \end{bmatrix} - \begin{bmatrix} 0 & 1/r \\ r & 0 \end{bmatrix} \frac{\partial}{\partial r}\begin{bmatrix} e_z \\ rh_\theta \end{bmatrix} = \mathbf{0} \tag{9.33}$$

The form of the differential equation is similar to (9.16) and (9.23), but unlike these equations the 2×2 system matrix is not constant but depends on r. In contrast to these equations the exact solution is not readily obtained, but it is still expected to contain two oppositely traveling pulses, albeit of modified shapes as compared to the earlier simple solutions (9.17) and (9.24). However, the explicit form of the solution of (9.33) with respect to T and R in the form of detailed pulse shape is of minor interest, simply because the specification of noise in terms of power spectral density calls for the analysis of (9.33) to be done in the frequency domain, which is readily available. Thus, the radial line model appears to contain the essential features for describing and understanding the noise phenomena, at the same time being sufficiently simple to be mathematically tractable.

Following the same approach as in earlier chapters, the first step in the analysis is to establish the *macroscopic* electromagnetic properties of the configuration. These serve as the basis for the subsequent analysis of *single photon* behavior. According to general electromagnetic theory, applied to the assumed fundamental TM mode, all the field components outside the resistor are derived from the longitudinal electric field $e_z(r)$ satisfying the wave equation:

$$\nabla^2 e_z(r) + k^2 e_z(r) = 0 \tag{9.34}$$

where $k = \omega/c$. In the fundamental mode there are no variations with regard to θ and z. In cylindrical coordinates the wave equation has the explicit form

$$R\frac{d}{dR}(R\frac{de_z(R)}{dR}) + R^2 e_z(R) = 0 \tag{9.35}$$

where the normalized radius R is given by

$$R = kr \tag{9.36}$$

and therefore implicitly contains the frequency through $k = \omega/c$. The general solution of (9.35) is the sum of two independent Hankel function components.

$$e_z(R) = \sqrt{\frac{kZ_0}{2L}}\left[f_1 H_0^{(1)}(R) + f_2 H_0^{(2)}(R)\right] \tag{9.37}$$

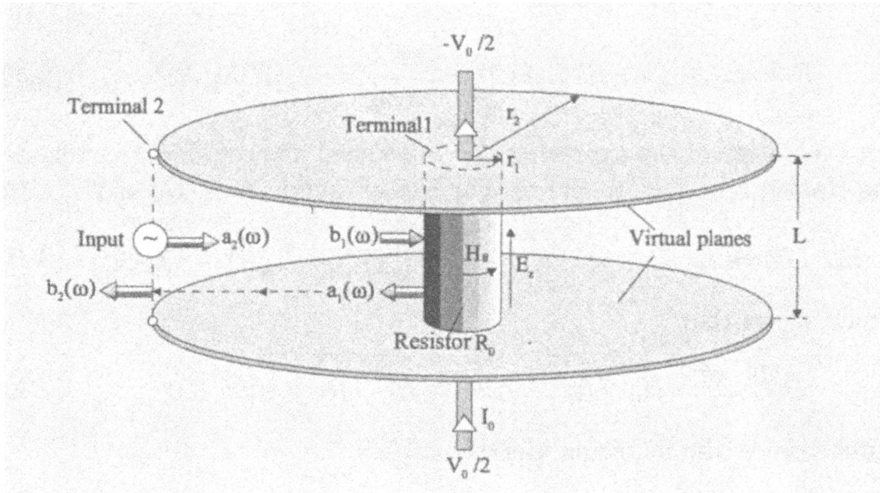

Figure 9.5. **Radial line model of the resistor configuration analyzed in the text. The cylindrical resistor is of radius r_1, length L, and resitance R_0.**

where $Z_0 = \sqrt{\mu_0/\varepsilon_0}$ is the impedance of free space, and $H_0^{(1)}(R)$ and $H_0^{(2)}(R)$ are zero order Hankel function of first and second kind. The rotational symmetry of the field model assures that the longitudinal electric field $e_z(R)$ and the angular magnetic field $h_\theta(R) = Z_0\, H_\theta(R)$ are the only nonzero field components. They are related through Maxwell's first curl equation

$$h_\theta(R) = -j\frac{dE_z(R)}{dR} \tag{9.38}$$

yielding

$$h_\theta(R) = j\sqrt{\frac{kZ_0}{2L}}\left[f_1 H_1^{(1)}(R) + f_2 H_1^{(2)}(R)\right] \tag{9.39}$$

The Hankel functions $H_1^{(1)}(R)$ and $H_1^{(2)}(R)$ appearing in the equation are of first order and first and second kind, respectively. In (9.37) and (9.39) the two components f_1 and f_2 represent the amplitudes of the radially propagating waves in the *inward* and *outward* directions. The factor $(kZ_0/2L)^{1/2}$ appearing in both equations serves to normalize the amplitudes f_1 and f_2 to satisfy appropriate scatter formulations. This is readily demonstrated from evaluation of the net inward power in the line, specified by the formula

$$P = \pi r L \,\mathrm{Re}\,\left[e_z(R)\, H_\theta(R)^*\right] = \frac{\pi R L}{k Z_0}\,\mathrm{Re}\left[e_z(R) h_\theta(R)^*\right] \qquad (9.40)$$

For evaluation of the expression for P we need a set of identities relating the Hankel functions to the regular Bessel functions $J_n(R)$ and $Y_n(R)$:

$$H_n^{(1)}(R) = J_n(R) + jY_n(R) \qquad H_n^{(2)}(R) = J_n(R) - jY_n(R) \qquad (9.41)$$

which shows that

$$H_n^{(2)}(R) = H_n^{(1)}(R)^* \qquad H_n^{(1)}(R) = H_n^{(2)}(R)^* \qquad (9.42)$$

Furthermore, the following identity applies:

$$J_1(R)Y_0(R) - J_0(R)Y_1(R) = \frac{2}{\pi R} \qquad (9.43)$$

Evaluation of (9.40) using these identities results in the simple formula

$$P = f_1 f_1^* - f_2 f_2^* \qquad (9.44)$$

which is seen to be compatible with the usual scatter formulation in which the product $f_1 f_1^*$ is interpreted as the *inward* power and $f_2 f_2^*$ as the *outward* power, with net power specified by the difference.

The normalized characteristic impedance $z(R)=e_z(R)/h_\theta(R)$ of the two wave components follows from (9.37) and (9.39).

$$z(R)_{inw} = \frac{H_0^{(1)}(R)}{jH_1^{(1)}(R)} \qquad z(R)_{outw} = z(R)_{inw}^* \qquad (9.45)$$

which show that the impedances in the general case are complex and functions of $R = kr$, i.e., of frequency. The frequency dependence of $z(R)_{inw}$ and the corresponding mismatch at the resistor is a basic clue to understanding the noise excitation problem. A photon is discharged into the resistor as a rectangular pulse, or reflected, according to the general scheme of Fig. 9.4. To accommodate the two alternatives open to the photon—full absorption or full reflection—the frequency dependent impedance must be adjusted over the full frequency range by appropriate photon doublet supplements, in strict adherence to the neoclassical theory in earlier chapters. It will be demonstrated shortly that the frequency dependent doublet supplement corresponds to an $1/f$ power spectral density variation which is expected to occur as noise across the resistor terminal.

From a macroscopic viewpoint the overall impedance $e_z(R)/h_\theta(R)$ of the combined waves must be matched to the normalized resistor surface impedance z_L specified by

$$z_L = R_0 \frac{2\pi r_1}{Z_0 L} \tag{9.46}$$

where R_0 is the sample resistance, r_1 the radius, and L the resistor length. The matching condition is expressed by the formula

$$z_L = \frac{e_z(R_1)}{h_\theta(R_1)} = \frac{f_1 H_0^{(1)}(R_1) + f_2 H_0^{(2)}(R_1)}{j\left[f_1 H_1^{(1)}(R_1) + f_2 H_1^{(2)}(R_1)\right]} \tag{9.47}$$

As usual, the *reflection coefficient* $\rho(R_1)$ at the resistor surface is defined by the ratio of the two amplitudes f_2 and f_1.

$$\rho(R_1) = \frac{f_2}{f_1} \tag{9.48}$$

which can be expressed in terms of z_L in the following form:

$$\rho(R_1) = \frac{j\, z_L\, H_1^{(1)}(R_1) - H_0^{(1)}(R_1)}{H_0^{(2)}(R_1) - j\, z_L\, H_1^{(2)}(R_1)} \tag{9.49}$$

The *transmission coefficient* $\tau(R_1)$ at $r = r_1$, describing the transmitted component, is expressed in terms of the reflection coefficient by

$$\tau(R_1) = [1 - \rho(R_1)\rho(R_1)^*]^{1/2} \tag{9.50}$$

Although the specified macroscopic equations (9.45)-(9.50) are valid at any frequency through the implicit relation $R_1 = \omega r_1/c$, the domination of low frequencies in the $1/f$ noise process allows first order expansions in R_1. In order to form these expansions we need the first order representations of the Bessel functions at $R_1 = \omega r_1/c$:

$$H_0^{(1)}(R_1) = 1 - j\frac{2}{\pi}\log(R_1)$$

$$H_0^{(2)}(R_1) = 1 + j\frac{2}{\pi}\log(R_1) \tag{9.51}$$

$$H_1^{(1)}(R_1) = -H_1^{(2)}(R_1) = -j\frac{2}{\pi}\frac{1}{R_1}$$

Insertion of the expansions into (9.45), (9.49), and (9.50) results in the following first order approximations of the scattering parameters.

The inward characteristic impedance:

$$z(R_1)_{inw} = R_1 \left[\frac{\pi}{2} - j\log(R_1)\right] \to jR_1 \log(R_1) \tag{9.52}$$

The reflection coefficient and transmission coefficient:

$$\rho(R_1) = 1 - \pi R_1/z_L \qquad\qquad \tau(R_1) = (2\pi R_1/z_L)^{1/2} \tag{9.53}$$

If the frequency becomes exactly zero, the limits are:

$$z(0)_{inw} = 0 \qquad \rho(0) = 1 \qquad \tau(0) = 0 \tag{9.54}$$

These first order expansions form the basis for the noise analysis of the following sections. The fact that the characteristic impedance $z(R)_{inw}$ approaches zero with frequency clearly reflects the increasing mismatch between the radial line and the resistor, a mismatch that also is reflected in the reflection and transmission coefficients $\rho(R_1)$ and $\tau(R_1)$. Thus, inward traveling low frequency components are experiencing increased difficulty in dumping their energies at the resistor. The functional dependence of the listed parameters with respect to frequency is the clue to understanding the $1/f$ noise problem, to be demonstrated shortly. The analysis relies somewhat on the general theory of Chapter 6, which the reader might consult for details.

9.4 MACROSCOPIC SCATTERING IN THE RADIAL RESISTOR SYSTEM

Following the general theory of Chapter 6 by first developing the macroscopic scatter relations, the response of the radial line configuration of Fig. 9.5 is specified by the scatter relation (6.18). As shown below the scattering matrix satisfies $\tilde{\mathbf{S}}^* = \mathbf{S}$ so that the equation takes the form

$$\mathbf{c}(\omega) = \boldsymbol{\tau}(\omega)[\mathbf{S} - \boldsymbol{\rho}(\omega)]^{-1}\mathbf{a}_g(\omega) \tag{9.55}$$

In accordance with the sketch in Fig. 9.5 the input $a_{g2}(\omega)$, representing a constant amplitude generator at frequency ω, is located at terminal 2, with no input at terminal 1. Therefore, the input column $\mathbf{a}_g(\omega)$ and the response column $\mathbf{c}(\omega)$ are specified by

$$\mathbf{a}_g(\omega) = \begin{bmatrix} 0 \\ a_{g2}(\omega) \end{bmatrix} \qquad\qquad \mathbf{c}(\omega) = \begin{bmatrix} c_1(\omega) \\ c_2(\omega) \end{bmatrix} \tag{9.56}$$

The validity of the scatter relation (9.55) rests on the condition (6.12) that $\mathbf{a}_g(\omega)$ is orthogonal to the reflected column $\mathbf{a}(\omega)$, which does not appear explicitly in (9.55). This requirement is satisfied by the model of Fig. 9.5 by the assumption that no part of the column $\mathbf{b}_2(\omega)$ at terminal 2 is reflected into $\mathbf{a}_2(\omega)$, a condition implying perfect match at this terminal, for instance through radiation into free space. The radial line parameters appearing in the scatter relation (9.55) are specified by:

$$\mathbf{S}=\begin{bmatrix} 0 & 1 \\ 1 & 0 \end{bmatrix} \quad \boldsymbol{\rho}(R_1)=\begin{bmatrix} \rho_1(R_1) & 0 \\ 0 & 0 \end{bmatrix} \quad \boldsymbol{\tau}_1(R_1)=\begin{bmatrix} \tau_1(R_1) & 0 \\ 0 & 1 \end{bmatrix} \tag{9.57}$$

where $\mathbf{S}$ is the *scattering matrix*, $\boldsymbol{\rho}(R_1)$ the *reflection matrix*, and $\boldsymbol{\tau}_1(R_1)$ the *transmission matrix*. With these relations the scatter relation (9.55) reduces to the simpler form

$$\mathbf{c}(\omega) = \begin{bmatrix} \tau_1(R_1) \\ \rho_1(R_1) \end{bmatrix} a_{g2}(\omega) \tag{9.58}$$

If we assume an input component $a_{g2}(\omega)$ at frequency ω, the product $c_1(\omega)^* c_1(\omega)$ is the macroscopic power absorbed in the resistor, and $c_2(\omega)^* c_2(\omega)$ is the macroscopic power reflected at terminal 2, i.e., back into whatever matching impedance is provided, be it free space or the signal source itself. It is readily verified that the macroscopic power balance condition is satisfied

$$|c_1(\omega)|^2+|c_2(\omega)|^2=\{|\tau_1(R_1)|^2+|\rho_1(R_1)|^2\}\,|a_{g2}(\omega)|^2=|a_{g2}(\omega)|^2 \tag{9.59}$$

Thus, the fraction of the incoming macroscopic power absorbed in the resistor is given by

$$\frac{P(\omega)_{resistor}}{P(\omega)_{input}} = \frac{|c_1(\omega)|^2}{|a_{g2}(\omega)|^2} = |\tau_1(R_1)|^2 = 2\pi k r_1/z_L \tag{9.60}$$

which is proportional to k, i.e., the frequency. The reflected part:

$$\frac{P(\omega)_{reflected}}{P(\omega)_{input}} = \frac{|c_2(\omega)|^2}{|a_{g2}(\omega)|^2} = |\rho_1(R_1)|^2 = 1 - \pi k r_1/z_L \tag{9.61}$$

If we visualize the frequency approaching zero, the equations clearly demonstrate the increasing difficulty of feeding power into the resistor, because most of it is reflected back. The physical explanation for the large reflection is the mismatch between the load impedance z_L provided by the resistor and the characteristic impedance $z(\omega)$ evaluated in (9.52).

$$z(\omega) = z(R_1)_{inw} \cong k r_1 \log(k r_1) \tag{9.62}$$

It is emphasized that (9.60) and (9.61) specify the subdivision of *macroscopic* power between terminals 1 and 2, i.e., between the power absorbed by the resistor and power reflected. With the assumption of incoherent photon flux the corresponding relative photon number is given by

$$\frac{N_{trans}}{N_{refl}} = \frac{|c_1(\omega)|^2}{|c_2(\omega)|^2} = \frac{|\tau_1(R_1)|^2}{|\rho_1(R_1)|^2} \cong 2\pi k r_1/z_L \tag{9.63}$$

In this macroscopic description the resistor is bombarded with a swarm of photons of which a very small number is absorbed compared to the number of photons reflected, the ratio approaching zero as the frequency goes to zero. It follows then that the true *DC* condition, i.e., zero frequency, represents a limiting case with the transmission line impedance being zero and therefore all the power expected to be reflected back. But this is a situation that can not be sustained. The *DC* power dissipated in the resistor somehow *has* to be fed into the sample in spite of this at first glance impossible situation. The resolution to the problem follows from the earlier discussion of the *DC* photon state in the dissipative resistor system. The *DC* photons are identified as the rectangular pulses of Fig. 9.3b, having a finite and very short duration Δt, and therefore a corresponding frequency spectrum. To resolve the problem we must depart from the macroscopic description and apply the neoclassical theory of Chapter 6 to photon routing and absorption.

9.5 SINGLE PHOTON ABSORPTION AND REFLECTION AT THE RESISTOR

At the resistor surface $r = r_1$ in the radial line the behavior of a photon is similar to the description in Figs. 9.3 and 9.4 in the sense that the photon pulses are discharged into the resistor or reflected as one shot processes. Whatever difference that might exist between these models and the present radial line model is due to the frequency dependence of the scattering parameters of the latter, requiring the analysis to be done in the frequency domain. The first step in this direction is to identify the input $a_{g2}(\omega)$ in the radial line of Fig. 9.5 with the frequency spectrum $a(\omega)$ of a single *DC* photon pulse a_{phot} of rectangular shape and duration Δt_0 as specified in Fig. 9.3. The spectrum was already evaluated in (9.29) as:

$$a_{g2}(\omega) = a(\omega)_{phot} = \frac{1}{2\pi} a_{phot} \Delta t \, \frac{\sin(\omega \Delta t/2)}{\omega \Delta t/2} \tag{9.64}$$

In determining the $1/f$ noise spectrum only the lowest frequencies need to be considered, in which case the factor $\sin(\omega\Delta t/2)/(\omega\Delta t/2)$ is essentially unity, so that

$$a_{g2}(\omega) = a_{g2} = \frac{1}{2\pi} a_{phot} \Delta t \tag{9.65}$$

The equation shows that the low frequency spectrum $a_{g2}(\omega)$ of the rectangular photon pulse a_{phot}, serving as input to the radial line, is constant. The pulse represents one photon; therefore its energy is specified by

$$w_{phot} = a_{phot}^2 \Delta t \tag{9.66}$$

Because $a_{g2}(\omega)^2 \sim w_{phot}\Delta t$ the dimension of $a_{g2}(\omega)^2$ is *energy spectral density*. The routing of the photon to either full absorption or full reflection at the resistor is accomplished by means of appropriate doublet supplements $\boldsymbol{\alpha}_g(\omega, -\omega)$ added to the regular input $\mathbf{a}_g(\omega)$, so that the macroscopic scatter relation (9.55) is modified to

$$\mathbf{c}(\omega) = \boldsymbol{\tau}(\omega)[\mathbf{S} - \boldsymbol{\rho}(\omega)]^{-1}[\mathbf{a}_g(\omega) + \boldsymbol{\alpha}_g(\omega, -\omega)] \tag{9.67}$$

where now $\mathbf{c}(\omega)$ presumably has the appropriate form reflecting the particulars of the routing condition, to be explained shortly. Through inversion of the equation the doublet admixture $\boldsymbol{\alpha}_g(\omega, -\omega)$ is specified by:

$$\boldsymbol{\alpha}_g(\omega, -\omega) = [\mathbf{S} - \boldsymbol{\rho}(\omega)]\,\boldsymbol{\tau}(\omega)^{-1}\mathbf{c}(\omega) - \mathbf{a}_g(\omega) \tag{9.68}$$

According to (9.65) the photon pulse input $\mathbf{a}_g(\omega)$ is independent of frequency so that

$$\mathbf{a}_g(\omega) = \begin{bmatrix} 0 \\ a_{g2} \end{bmatrix} \tag{9.69}$$

In view of the power balance condition (9.59) it follows that also the response column $\mathbf{c}(\omega)$ is frequency independent.

$$\mathbf{c}(\omega) = \begin{bmatrix} c_1 \\ c_2 \end{bmatrix} \tag{9.70}$$

Thus, in the present case (9.68) reduces to

$$\boldsymbol{\alpha}_g(\omega, -\omega) = \begin{bmatrix} -\dfrac{\rho_1(R_1)}{\tau_1(R_1)} & 1 \\ \dfrac{1}{\tau_1(R_1)} & 0 \end{bmatrix} \begin{bmatrix} c_1 \\ c_2 \end{bmatrix} - \begin{bmatrix} 0 \\ a_{g2} \end{bmatrix} \tag{9.71}$$

The appropriate power columns $\mathbf{c}_{trans}$ and $\mathbf{c}_{refl}$ for full transmission or full reflection are simply:

$$\mathbf{c}_{trans} = \begin{bmatrix} c_1 \\ 0 \end{bmatrix} = \begin{bmatrix} a_{g2} \\ 0 \end{bmatrix} \qquad \mathbf{c}_{refl} = \begin{bmatrix} 0 \\ c_2 \end{bmatrix} = \begin{bmatrix} 0 \\ a_{g2} \end{bmatrix} \tag{9.72}$$

Let us first consider the case of full transmission, with $\mathbf{c} = \mathbf{c}_{trans}$ inserted into (9.71). The corresponding doublet supplement is then:

$$\boldsymbol{\alpha}_g(\omega, -\omega)_{trans} = \begin{bmatrix} -\dfrac{\rho_1(R_1)}{\tau_1(R_1)} \\ \dfrac{1}{\tau_1(R_1)} - 1 \end{bmatrix} a_{g2} \tag{9.73}$$

As far as resistor noise is concerned the first element is the relevant component.

$$\alpha_{g1}(\omega, -\omega)_{trans} = -\frac{\rho_1(R_1)}{\tau_1(R_1)} a_{g2} \cong \frac{-1}{\tau_1(R_1)} a_{g2} = \frac{-a_{g2}}{(2\pi k r_1/z_L)^{1/2}} \tag{9.74}$$

The equation shows that the absorption of a photon in the resistor requires a doublet supplement inversely proportional to the square root of frequency. Note that $\boldsymbol{\alpha}_g(\omega, -\omega)$ was defined in (9.67) as pointing in the same direction as the regular input $\mathbf{a}_g(\omega)$, which is *into* the transmission line section. Therefore, the negative sign in (9.74) indicates propagation out of the section and *into* the sample, giving rise to a corresponding noise component in the resistor.

Let us proceed with the process of full photon *reflection* at the resistor surface by substituting the column $\mathbf{c} = \mathbf{c}_{refl}$ into (9.71). The corresponding doublet supplement is then

$$\boldsymbol{\alpha}_g(\omega, -\omega)_{refl} = \begin{bmatrix} 1 \\ -1 \end{bmatrix} a_{g2} \tag{9.75}$$

The signs of the elements in the column $\boldsymbol{\alpha}_g(\omega, -\omega)_{refl}$ are such that both specify propagation in the *outward* radial direction. Therefore, they are not expected to give rise to noise in the resistor. It is seen that the components do not depend on frequency and are very much smaller than the noise component (9.74) associated with absorption of a photon.

However, the large difference in their relative magnitudes is offset by the much larger number N_{refl} compared to N_{trans}. In fact, from (9.63) the following relation is obtained:

$$\frac{N_{trans}\, \alpha^2_{g1,trans}}{N_{refl}\, \alpha^2_{g2,refl}} = 1 \tag{9.76}$$

showing that the doublet power radiated into the resistor as noise equals the doublet power radiated into the environment, i.e., into free space. These relations are illustrated schematically in Fig. 9.6.

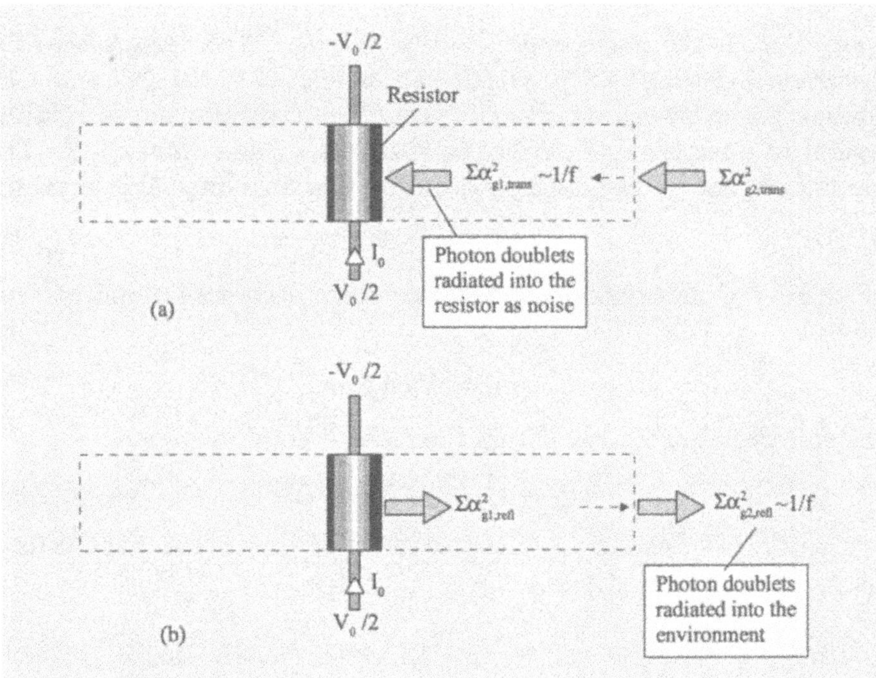

Figure 9.6. **Figures illustrating the flow of photon doublet power in the resistor system. (a) Arrows indicate direction and magnitude of doublet noise power *into* the resitor. (b) Outward radiation of doublet power into the environment. Both components are inversely proportional to frequency.**

9.5.1 Theoretical noise power spectrum

The component $\alpha_{g1}(\omega,-\omega)_{trans}$ specified in (9.74) is the doublet supplement associated with the routing of one *single* photon to the resistor. It follows, from the discussion in connection with (9.66), that $\alpha_{g1}(\omega,-\omega)^2_{trans}$ is equal to the noise energy spectral density (energy per unit frequency) per photon. Because the switching process administered by the environmental photon doublets is presumably random, there is no reason to expect coherence between the various noise constituents arising from absorption of different photon pulses. Thus, the *rms* noise *power* density generated across the resistor is obtained by summing over the number N of absorbed photons per second.

$$\overline{S(\omega)} = \sum_{n=1}^{N} \alpha_{g1}(\omega, -\omega)^2_{trans} = \frac{1}{|\tau_1(R_1)|^2} \sum_{n=1}^{N} |a_{g2n}|^2 = \frac{N\,|a_{g2n}|^2}{|\tau_1(R_1)|^2} \qquad (9.77)$$

where $\overline{S(\omega)}$ is the *noise power spectral density.* With reference to the discussion following (9.32) the sum can be related to the *DC* power P_0 through the following arguments. Each photon was discharged into the resistor as a rectangular pulse of amplitude a_{phot} and duration Δt. The number of photons N_Δ discharged during this time interval is equal to

$$N_\Delta = N\,\Delta t \qquad (9.78)$$

All these N_Δ photons add together so that the overall amplitude becomes

$$a_\Delta = N_\Delta\, a_{phot} \qquad (9.79)$$

The corresponding energy within the interval Δt:

$$w_\Delta = a_\Delta^2\,\Delta t = N_\Delta^2\, a_\Delta^2\, \Delta t \qquad (9.80)$$

Each separate interval Δt contributes the same amount, so that the total energy per second is given by

$$w_{total} = w_\Delta \frac{N}{N_\Delta} = \frac{w_\Delta}{\Delta t} = N_\Delta^2\, a_{photon}^2 = N\,N_\Delta\,\Delta t\, a_{photon}^2 \qquad (9.81)$$

According to (9.65)

$$a_{photon}^2 = 4\pi^2 |a_{g2}|^2 \frac{1}{\Delta t^2} \qquad (9.82)$$

so that (9.81) becomes

$$P_0 = w_{total} = 4\pi^2 N_\Delta N\,|a_{g2}|^2 \frac{1}{\Delta t} \qquad (9.83)$$

where the *DC* power P_0 dissipated in the resistor is identified with w_{total}, i.e., the dissipated energy per second. The equation provides the desired relation between $N\,|a_{g2}|^2$ and P_0:

$$N\,|a_{g2}|^2 = \frac{1}{4\pi^2}\frac{1}{N_\Delta} P_0 \Delta t \qquad (9.84)$$

This expression, together with the transmission coefficient $\tau_1(R_1)$ from (9.53), are substituted into (9.77) resulting in the following *relative* spectral density of noise power:

$$\frac{\overline{S(\omega)}}{P_0} = \frac{z_L}{4\pi^2 \omega r_1} \frac{c\Delta t}{N_\Delta} \qquad (9.85)$$

Insertion of normalized surface impedance z_L and pulse duration Δt from (9.46) and (9.28), respectively, results in the formula

$$\frac{\overline{S(\omega)}}{P_0} = \frac{1}{\omega}\frac{2R_0}{\pi Z_0}\frac{1}{N_\Delta} \tag{9.86}$$

where R_0 is the sample resistance and $Z_0 = \sqrt{\mu_0/\varepsilon_0}$ the impedance of free space, both measured in ohm.

9.5.2 A theoretical 'Hooge' formula

The formula (9.86) demonstrates that noise generated by the photon doublet admixture in the routing of *DC* photons is of the $1/f$ type. The sample resistance R_0 appearing in the equation can be expressed by the material parameters.

$$R_0 = \frac{1}{\mu_{00}\, q\, n_0}\frac{L}{\pi r_1^2} = \frac{L^2}{\mu_{00}\, q\, N_0} \tag{9.87}$$

where μ_{00} is the carrier *mobility*, q is the carrier *charge*, n_0 is the carrier *density*, and $N_0 = n_0 L \pi r_1^2$ is the *total* number of charge carriers in the sample. Hereby formula (9.86) is converted to

$$\frac{\overline{S(\omega)}}{P_0} = \left(\frac{L^2}{\pi^2 Z_0\, \mu_{00}\, q\, N_\Delta}\right)\frac{1}{N_0\, f} = \alpha_H \frac{1}{N_0\, f} \tag{9.88}$$

This is Hooge's empirical equation (9.1) discussed earlier, with his empirical 'constant' α_H specified by the now *theoretical* expression

$$\alpha_H = \left(\frac{1}{\pi^2 Z_0\, q\, N_\Delta}\right)\left(\frac{L^2}{\mu_{00}}\right) = \frac{1.68\times 10^{15}}{N_\Delta}\left(\frac{L^2}{\mu_{00}}\right) \tag{9.89}$$

It is quite remarkable that (9.88), obtained by a direct application of the neoclassical theory, is in *formal* agreement with the purely *empirical* Hooge's law (9.1). The main theoretical result emerging from the analysis is the prediction of the inverse frequency dependence of the noise spectral density $\overline{S(\omega)}$, thus confirming the general empirical observance of $1/f$ noise characteristics. In particular, the theory predicts no lower frequency limit for the validity of the formula, in agreement with an abundance of empirical data. Not only does the formula specify the $1/f$ noise behavior but also the inverse dependence on the total number of carriers N_0. Moreover, the *theoretical* ' Hooge's constant' α_H in (9.89) is dependent on the sample properties through the factor L^2/μ_{00}, a dependence that at least superficially appears to be in qualitative agreement with several experimentally observed and reported deviations from the originally claimed value of $\alpha_H = 0.002$. For instance van der Ziel [44], commenting on Hooge's constant, states that it is now generally conceded that α_H is not a universal constant but can be several orders of magnitude smaller than 0.002 for short devices. Dutta et al [14] reported on $1/f$ noise in copper whiskers and found values of α_H as much as 2×10^3 times larger than Hooge's α_H. Noting that the mobility is generally smaller in metallic conductors than in semiconductors and resistive samples, both these observations are in qualitative agreement with the theoretical predictions in (9.89) of α_H being proportional to the square of sample length L and inversely proportional to the mobility μ_{00}. Although these predictions appear to be in fair agreement with empirical results, no attempts of extensive comparison have been made. Therefore, not too much significance should be attached to the comments on detailed variation of α_H.

The only unknown parameter in the formula (9.89) is the number of photons N_Δ discharged into the resistor during the time interval Δt. I have no good theory for providing estimates of its numerical value. It could well depend on some of the resistor parameters. Empirically, N_Δ will have to be of the order of 10^{12} in order for α_H in (9.89) to match the original Hooge's value of $\alpha_H \cong 0.002$. In the absence of adequate theoretical foundation, a value of N_Δ equal to 10^{12} can be taken as an order of magnitude approximation. In assessing the present theoretical derivations it is felt that the numerical uncertainty is less important than the theoretical predictions of the functional dependence on the main physical parameters, in particular the $1/f$ frequency dependence.

9.5.3 A modified 'Hooge' formula

It is interesting to speculate that Hooge could have formulated his empirical law according to (9.86), thereby getting a corresponding $\hat{\alpha}_H$ that would really be a constant, at least estimated from the theoretical results of the present chapter, namely in the form:

$$\frac{\overline{S(\omega)}}{P_0} = \hat{\alpha}_H \frac{R_0}{f} \tag{9.90}$$

where the new $\hat{\alpha}_H$ would be given by

$$\hat{\alpha}_H = \frac{1}{\pi^2 Z_0 N_\Delta} \tag{9.91}$$

which is constant, assuming N_Δ is a constant. The *sample* variable in this alternative formulation is the resistance R_0 rather than the number of carriers N_0 in the original Hooge law.

The two alternative formulations clearly reflect the different viewpoints concerning noise origin. The original Hooge formula associates the $1/f$ noise with some kind of material phenomena having to do with the charge carriers. This has been the predominant thinking amongst researchers; therefore the emphasis on the *total carriers* N_0 in the sample. The present theory holds that the noise is caused by the routing process of photons organized by the environmental photon doublets. In this process the $1/f$ frequency dependence is contained in the *transmission coefficient* from free space to the resistor, which is merely functioning as an impedance load. Therefore, the dependence on resistance R_0.

9.5.4 The 1/f noise phenomenon—more than just noise

It was concluded, in Chapters 3 and 6, that the environmental doublet fields carried no energy and were not measurable in *passive circuits*, which is equivalent to saying that the $1/f$ noise is not observable under such circumstances, in agreement with experimental observations. Apparently, it takes dissipation of energy in the form of distinct *DC* or *RF* photon pulses to achieve coupling of the regular photon fields to the doublet fields, giving rise to the $1/f$ noise fluctuations. In a more general context discussed in Chapter 3.3.3, and also shown schematically in Fig. 9.6, the phenomenon contributes to the doublet spectral energy of free space. The excitation of $1/f$ noise in the dissipative resistor system is associated with a corresponding radiation into the environment of photon doublets with the characteristic $1/f$ equilibrium energy spectrum. Rather than being an isolated process in the dissipative system itself, the $1/f$ noise appears to be part of a universal physical phenomenon affecting the energy balance of the environmental sea of *PDE* fluctuations. It should be emphasized, however, that the $1/f$ noise theory presented in this chapter is not in a any way based on the prerequisite of $1/f$ frequency variation of the environmental *PDE* fluctuations described in Chapter 3. The fact that both phenomena appear to be somehow interconnected, giving the same frequency variation, tends to strengthen the credence of both theories.

From this viewpoint the radiated doublet noise is a mirror representation of the frequency variation in the device. This argument must apply to all kind of devices and gadgets dissipating electromagnetic power. Following this line of thought it is tempting to speculate whether the human brain belongs to the category of electric circuits exhibiting $1/f$ noise phenomena. We saw that the only requirement for $1/f$ noise was dissipation of electromagnetic energy, and the brain certainly satisfies this criterion. Conscious thought processes require electric energy supplied by the blood stream through electrophysiological effects. The associated electromagnetic energy is expected to be dissipated through exactly the same photon absorption processes as described in the previous modeling. The propagation of nerve impulses and their interactions in the neural network of the brain is of course a vast and extremely complex phenomenon which nonetheless might be described by the overall Fourier representation containing a wide band of frequencies f_0. According to the theoretical proposition of the present chapter, these extremely complex time varying processes must involve the absorption of photons accompanied by $1/|f_0 - f|$ noise as well as corresponding photon doublet noise radiation. These processes accompany the dissipation of electromagnetic energy in the brain and must exist regardless of whatever phenomena lie behind conscious thinking, which in this context is an irrelevant issue.

According to this somewhat vague description, the brain is not completely isolated from the environment, but in equilibrium with the environmental sea of photon doublets modulated with frequency components f_0 somehow mirroring whatever thought processes goes on in the brain. In accordance with the theory presented in Chapters 2 and 3 the environmental photon doublets (PDE) consist of entangled positive and negative time elements balancing each other and filling all space with *apparent* infinite velocity. We also saw that these fluctuations are observable only in dissipative circuits, of which the human brain qualifies as a special case. It can be speculated then, that the $1/f$ doublet noise absorbed from the environment by a particular brain might contain frequency components f_0 originating from another brain. If such effects are real, further speculation along the same lines might link the processes to phenomena such as extrasensory perception (ESP). In spite of this interesting rational perspective on the paranormal, I hasten to emphasize the extremely speculative nature of the subject. It certainly falls outside the more realistic material presented in the book, but might nonetheless deserve attention in different kind of media addressing such phenomena.

9.6 CONCLUDING REMARKS

Briefly, the theory of $1/f$ noise presented in this chapter can be summed up as follows. The analysis is based on the simple model of a circularly symmetric homogeneous resistor shown in Fig. 9.5, which is simple enough to be analytically tractable and sufficiently complex for demonstration of the proposed basic origin of $1/f$ noise. Resistive samples of more complex shape and composition, as well as active components such as transistors and electron beam devices, would certainly call for more complex models implying more advanced analytical methods of treatment with supplements of higher order components in addition to the lowest order component of rotational symmetry assumed in this text. Regardless of geometrical details of more complex configurations, the fundamental component is expected to dominate the noise properties.

As stated in the introduction, if the input source is *RF* at some frequency ω_0 rather than *DC*, the noise spectrum manifests itself as two symmetrically located noise bands of the form $1/|\omega_0-\omega|$ centered at ω_0. This process has all the flair of a parametric interaction between the two frequency components. It is unlikely that the additional parametric interaction should lead to interpretations of the basic noise origin which in any major way would be different from the one described in the present text. Further analysis of the more general cases with regard to geometrical configuration and *RF* input source was not attempted.

The theory proposes that the origin of the noise is basically photonic, caused by fluctuations in the environmental photon doublets in the process of administering the absorption and reflection of *DC* photons at the resistor surface. Any electronic device dissipating *DC* or *RF* power is subject to a corresponding electromagnetic influx in the form of a stream of photon pulses of finite duration, and therefore of a wide frequency spectrum. Their energy integrity requires individual photons to be either fully absorbed or fully reflected. These are requirements that can not be satisfied by the regular macroscopic matching conditions. The boundary conditions can only be satisfied through the intermediary of supplementary wave components in the form of photon doublets from the environmental pool of doublets (*PDE*). The low frequency components of the photon pulses experience an increasingly worse transmission mismatch at the resistor surface, culminating at zero frequency. Thus, the process of getting the photon pulse across the resistor surface requires a correspondingly large doublet supplement for providing a perfect match of resistor surface impedance and the impedance of the combined entity of the photon plus its doublet admixture. The difficult task faced by the photon in overcoming a progressively smaller transmission coefficient proportional to the square root of frequency calls for a correspondingly larger doublet supplement in order to accomplish perfect match and photon absorption. Thus, the required energy supplement is inversely proportional to frequency, giving rise to $1/f$ type noise in the resistor. In particular, the theory predicts no lower limit to the $1/f$ frequency behavior, in agreement with a vast pool of empirical observations. Noise effects similar to those in a simple resistor are expected in any electronic device in which power is dissipated.

Chapter 10

QUESTIONING STERN-GERLACH

'But whether this be true or no
the Devil any of you know.'

Samuel Butler

10.1 BASIS OF THE STERN-GERLACH EXPERIMENT

The Stern-Gerlach experiment [18] is hailed as a dramatic demonstration of the necessity of a radical departure from the concepts of classical mechanics. The basic postulates of quantum mechanics are often formulated in axiomatic manners with the example of the Stern-Gerlach experiment in the back of the reader's mind. In fact, several textbooks on quantum mechanics introduce the subject by devoting the very first chapter to the experiment [41], [43] thereby intending to attune the reader from the very beginning to the 'quantum mechanical way' of thinking.

The theoretical interpretations of the Stern-Gerlach experiment are not entirely without flaws. In fact, some of these appear to be nontrivial and might possibly cause some reassessment of the familiar conclusions based on the experiment. The analysis and discussions presented in this chapter, together with supporting computer simulations, are meant as a contribution to the interpretation of the experiment itself rather than a discussion of the background theory. This chapter distinguishes itself from all the other chapters of the book in that bitemporal effects are not at all a consideration; the issue at hand is a pure classical versus quantum interpretation. The main reason for inclusion of the Stern-Gerlach experiment into a book basically concerned with bitemporal phenomena is its association with fundamental dogma in the orthodox Copenhagen Interpretation such as nonobjective quantum entities and collapse of wave functions, concepts that are subject to scrutiny in the book. Contrary to current quantum theory we shall find that there is insufficient basis for the Stern-Gerlach claim of experimental verification of two-state phenomena.

The experiment, discussed in almost any text on modern physics, is conceptually quite simple. Before discussing the interpretation, let us give a brief description of the experiment as it might be presented in textbooks on quantum mechanics. This account is henceforth referred to as 'the standard *SG* description'.

10.1.1 Standard description of the SG experiment

The experiment measures the intrinsic magnetic moment of electrons by an indirect method involving silver atoms. Silver atoms are composed of 47 electrons and a nucleus. The total orbital and total spin angular momentum of 46 of the electrons add up to zero, leaving a surplus from the 47*th* electron. The magnetic moment of the silver atom is therefore due to the magnetic moment of one single electron alone. The basic experimental set-up is shown in Fig. 10.1. The silver atoms are heated in an oven provided with a small hole from which some atoms escape with presumably thermal velocity distribution. The beam of silver atoms is collimated by means of a slit and then subjected to an inhomogeneous magnetic field $\vec{B}$ produced by an asymmetric set of magnetic pole pieces as shown in Fig. 10.1b. The silver atoms are electrically neutral but posses a magnetic moment $\vec{\mu}$ due to the single 47*th* electron. The atom experiences a force from the interaction energy W between magnetic dipoles and the magnetic field, where W is given by

$$W = -\vec{\mu} \cdot \vec{B} \tag{10.1}$$

The force $\vec{F} = -\nabla W = \nabla(\vec{\mu} \cdot \vec{B})$ so that the force in the direction of strong magnetic field is given by

$$F_x = \frac{\partial}{\partial x}(\vec{\mu} \cdot \vec{B}) = \mu_x \frac{\partial B_x}{\partial x} \tag{10.2}$$

where components of $\vec{B}$ in directions other than the x-direction are ignored. Classically, the magnetic moment in the x-direction is given by

$$\mu_x = |\mu| \cos(\theta_\mu) \tag{10.3}$$

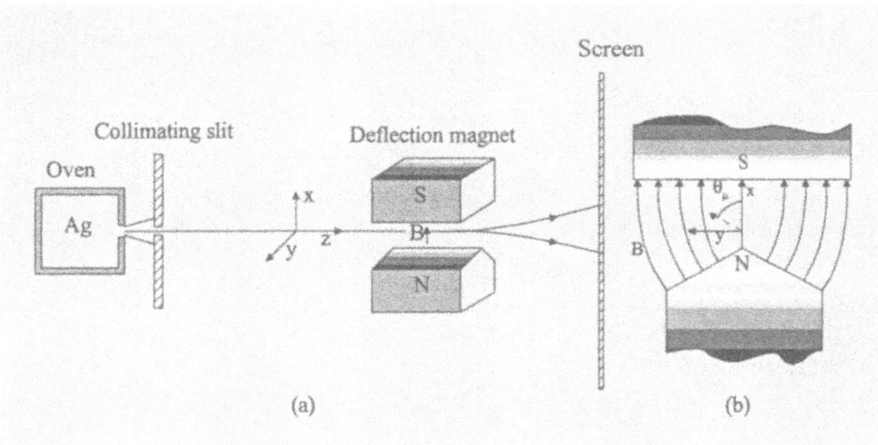

Figure 10.1. (**a**) Schematic diagram of the Stern-Gerlach experiment. Collimated silver atoms Ag escaping from the hot oven are subject to a nonuniform magnetic field deflecting the atoms up and down. (**b**) Cross sectional view of the magnet pole pieces.

where θ_μ is the angle that the magnetic moment makes with the x-axis. With the direction of field shown in Fig. 10.1b atoms with $\mu_x > 0$ experience a downward force, whereas atoms with $\mu_x < 0$ are subject to an upward force, both proportional to μ_x. Therefore, from a naive point of view the beam is expected to get split according to the presumably random values of μ_x. The set-up is expected to measure the x-component of $\vec{\mu}$, provided the atom behaves like a classical object. The atoms leaving the oven are certainly randomly oriented, and classically we would expect random orientation between $|\vec{\mu}|$ and $-|\vec{\mu}|$ and therefore a continuum of deflections. In the original Stern-Gerlach experiment the deflected silver atoms were registered through deposition on a glass plate. The expected deflection is indicated in Fig. 10.2a, whereas the observed deposition is shown in Fig. 10.2b. The actual observed cross section of the deflected beam, according to [18], is shown in Fig. 10.2c. The fussiness of the cross section is explained by the velocity spread and by the width of the incoming beam.

The experimental observations are interpreted as a quantum effect that has no classical explanation. Apparently, the *SG* apparatus splits the silver beam into two components, one with its spin, or magnetic

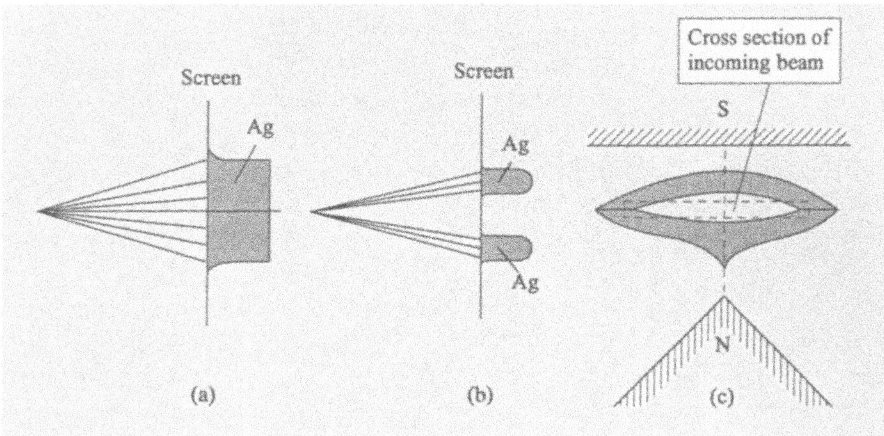

Figure 10.2. **(a) Continuous deflection expected from the Stern-Gerlach classical argument. (b) Deposition into two constituents as observed in the actual experiment. (c) Actual cross section of the deposited beam by Gerlach and Stern (see [18]).**

moment, pointing up, and one with spin pointing down.[1] The randomly oriented beam of silver atoms has somewhat mysteriously 'collapsed' to orientations *up* and *down* during the deflection process. This picture fits into the prevalent quantum philosophy of 'state function collapse' to the macroscopic level. It is argued that the possible quantum spin states are only two, somewhat arbitrarily referred to as 'up' and 'down' and that the actual polarizations measured are determined by the measuring apparatus.

This is essentially the standard description of the Stern-Gerlach experiment, with interpretation as a quantum mechanical effect and corresponding conclusions. The background quantum theory of magnetic moment is not a central issue addressed in the present text, which is concerned with a nontrivial inconsistency in the magnetic interaction argument in the standard *SG* experiment, to be described in the next section.

[1]The observed cross section in Fig. 10.2c is not shown in the original Stern-Gerlach publication [18], but on a postcard from Walther Gerlach to Niels Bohr dated February 8, 1922 reproduced in [43]. The communication refers to their experimental verification of directional quantization and congratulates Bohr on the confirmation of his theory.

10.2 THE FLAW IN THE STERN-GERLACH THEORY

A close scrutiny of the deflection argument in the *SG* experiment reveals the flaw. Whereas (10.2) correctly specifies the force in the vertical x-direction, it fails to account for force components in other directions. The sentence immediately following (10.2), stating that components of $\vec{B}$ in directions other than the x-direction are ignored, is certainly an acceptable approximation, but is entirely irrelevant. The deflection force is not specified by the magnetic field but by its gradient. And the gradients in directions other than the x-direction are not negligible. In fact, it is readily shown that the gradients of $\vec{B}$ in the y-direction and the x-direction are the same except for a sign difference. The proof is straight forward. In view of the solenoidal nature of the magnetic field distribution, satisfying $\nabla \cdot \vec{B} = 0$, and of the fact that the longitudinal component B_z is entirely negligible, the following equation holds everywhere:

$$\frac{\partial B_x}{\partial x} = -\frac{\partial B_y}{\partial y} \tag{10.4}$$

which proves the statement above, that the forces in the x- and y-directions are equal in magnitude. But the argument can be carried quite a bit further by proving that the magnitude of the *radial force* on *all* particles polarized in the x-y plane are the same, regardless of the *direction* of their magnetic moment vector $\vec{\mu}$. The proof goes as follows. Instead of the force being specified by (10.2) it is given by the more complete expression

$$\vec{F} = \nabla(\vec{\mu} \cdot \vec{B}) = \nabla(\mu_x B_x + \mu_y B_y) \tag{10.5}$$

from which

$$F(\theta_\mu)_x = \mu_x \frac{\partial B_x}{\partial x} + \mu_y \frac{\partial B_y}{\partial x} \tag{10.6}$$

$$F(\theta_\mu)_y = \mu_x \frac{\partial B_x}{\partial y} + \mu_y \frac{\partial B_y}{\partial y} \tag{10.7}$$

In addition to being solenoidal as in (10.4) the magnetic field is also irrotational, satisfying $\nabla \times \vec{B} = 0$, so that

$$\frac{\partial B_y}{\partial x} = \frac{\partial B_x}{\partial y} \tag{10.8}$$

At the symmetry axis $y = 0$ the y-component of $\vec{B}$ vanishes for all x. Therefore, if the beam enters at the symmetry axis, the following holds:

$$B_y(x, 0) \equiv 0 \qquad \text{for all } x \tag{10.9}$$

and therefore also

$$\frac{\partial B_y(x,0)}{\partial x} = \frac{\partial B_x(x,0)}{\partial y} \equiv 0 \tag{10.10}$$

Therefore, at the symmetry axis the general force components (10.6) and (10.7) reduce to

$$\begin{aligned} F(\theta_\mu)_x &= \mu_x \frac{\partial B_x(x,0)}{\partial x} = \mu \cos(\theta_\mu) \frac{\partial B_x(x,0)}{\partial x} \\ F(\theta_\mu)_y &= -\mu_y \frac{\partial B_x(x,0)}{\partial x} = -\mu \sin(\theta_\mu) \frac{\partial B_x(x,0)}{\partial x} \end{aligned} \tag{10.11}$$

Thus, the *magnitude* of the force on an atom with polarization θ_μ is given by

$$F(\theta_\mu) = \sqrt{[F(\theta_\mu)_x]^2 + [F(\theta_\mu)_y]^2} = \mu \frac{\partial B_x(x,0)}{\partial x} \tag{10.12}$$

The equation shows that the *magnitude* of the force is constant and independent of the polarization θ_μ of the silver atoms. The *direction* of the force, and therefore the deflection, is in the radial direction, not generally at the polarization angle θ_μ but rather at the angle θ specified by

$$\tan(\theta) = \frac{F_y(\theta_\mu)}{F_x(\theta_\mu)} = -\tan(\theta_\mu) \tag{10.13}$$

These simple analytical results follow directly from the inherent properties of the magnetic field expressed in (10.4) and (10.8). They show that there is no basis for the claim made in the standard description of the *SG* experiment that the atoms *classically* are expected to be spread over a continuous band along the vertical axis as depicted in Fig. 10.2a. Instead, the expected classical distribution of the deflected atoms is the cylindrical shell of Fig. 10.3, telling an entirely different story. With allowance made for the differences in the incident beam shapes, the observed lip shaped pattern in Fig. 10.2c bears a certain resemblance to the cylindrical deflection pattern of Fig. 10.3. At this point it can only be speculated that the difference between the two might be due to the fact that the *SG* beam was laterally quite wide, so that the circular pattern of Fig.10.3 is necessarily widened laterally, and moreover, the nonuniformity of the magnetic field might provide further distortion of the deposited pattern. These points will be elaborated further at the end of the chapter together with presentation of computer simulations of the deflection pattern based on the present revised analysis.

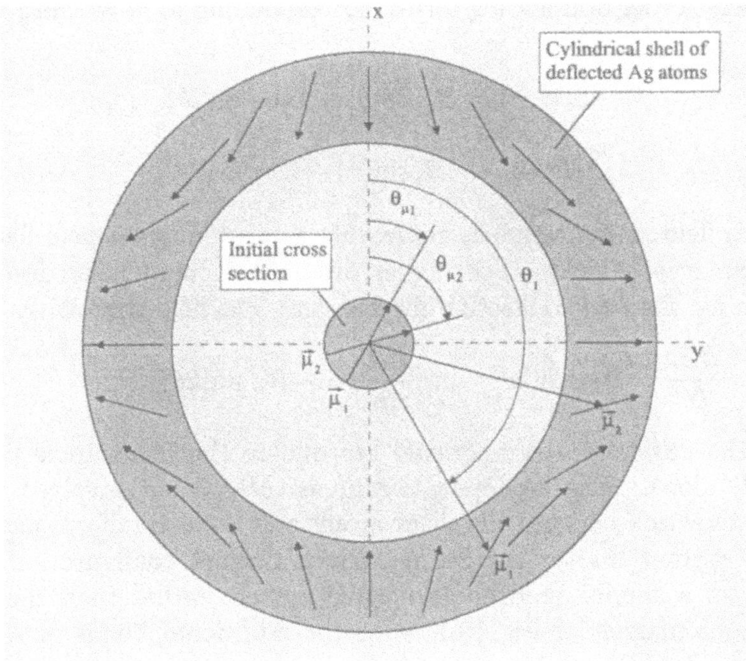

Figure 10.3. **Sketch showing the basic deflection pattern of a thin circular beam of randomly polarized silver atoms traversing a non-uniform magnetic field, assuming the atoms treated as *classical* entities. The deflected atoms are all located on a cylindrical shell rather than in the continuous band claimed by the standard SG theory illustrated in Fig. 10.2a. Arrows refer to the direction of magnetic polarization of the silver atoms.**

10.2.1 Multipole expansion of the magnetic field

Multipole expansion of the magnetic field components provides further insight into the basic deflection mechanism and its dependence on the magnetic field pattern. Referring to Fig. 10.3, with the origin located at the beam center and $\theta = 0$ along the vertical x-axis, the following multipole expansions expressed in polar coordinates r, θ apply:

$$B_r(r,\theta) = \sum_{m=1}^{\infty} B_m \, r^{m-1} \cos(m\theta)$$

$$B_\theta(r,\theta) = -\sum_{m=1}^{\infty} B_m \, r^{m-1} \sin(m\theta) \tag{10.14}$$

The lowest order dominating terms, corresponding to $m = 1$ and $m = 2$, are

$$\begin{aligned} B_r(r,\theta) &= B_1 \cos(\theta) + B_2 r \cos(2\theta) \\ B_\theta(r,\theta) &= -B_1 \sin(\theta) - B_2\, r \sin(2\theta) \end{aligned} \tag{10.15}$$

Of these field components B_1 represents a constant *monopole* field and B_2 a *quadrupole* field. The higher order field components are of no concern for the basic deflection mechanism. The field derivatives are

$$\frac{\partial B_r}{\partial r} = B_2 \cos(2\theta) \qquad \frac{\partial B_\theta}{\partial r} = -B_2 \sin(2\theta) \tag{10.16}$$

Thus, the basic force components are due to the quadrupole component B_2 alone. The monopole component B_1 is not involved in the basic interaction between the silver atoms and the nonuniform magnetic field. For that matter the Stern-Gerlach magnet configuration could have been a simple quadrupole magnet system rather than the inhomogeneous magnet of Fig. 10.1, with the additional bonus of a much 'cleaner' experiment.

The complete absence of any participation of the monopole component B_1 in the basic interaction comes as a bit of a surprise. But it is an unavoidable conclusion provided we stick to the central theme, that the force is specified by the classical expression $\vec{F} = \nabla(\vec{\mu} \cdot \vec{B})$ as is indeed the assumption in both the standard *SG* theory and in the present revised one. More serious is the second unavoidable conclusion. Since there is no preferred direction of deflection by the quadrupole field, it is quite unfounded to interpret the deflection in the Stern-Gerlach apparatus as a quantum mechanical effect signifying polarization in the 'up' or 'down' directions. Because all polarizations are deflected equally it is impossible to single out preferred directions. According to these deliberations the *SG* effect is pure classical, showing no special quantum effects. In particular, interpretations suggesting two-state quantum effects with experimental support of a nonobjective quantum world appear quite unfounded.

10.2.2 Is the silver atom model adequate?

This being said, we must bear in mind that these conclusions relate to the very simplest model of the silver atom, which may or may not be adequate. But it is a fact that this model is being used in the standard *SG* description as well as in the present revised theory. In fairness to both, comparison of theoretical predictions should be based on the same model. And this comparison definitely introduces considerable doubt about interpretations, not only of the basic *SG* experiment, but also of the more far reaching conclusions drawn from *sequential* Stern-Gerlach experiments (see for instance [41] or [43]). In these experiments the silver atoms are made to traverse two or more Stern-Gerlach magnets tilted with respect to each other. They appear to be exclusively 'Gedanken' experiments meant to provide additional insight into the peculiarities of two-state systems of quantum mechanics. The conclusions drawn from these are valid only if the basic *SG* experiment is valid. And the revision of the theory presented here does not support the standard *SG* interpretation. At least not in the context of the semiclassical model of the silver atom used in both the original *SG* experiment and in the present revised theory.

The consistency of the standard *SG* description is not questioned in the many textbook presentations as one of the cornerstones of quantum physics. It is somewhat surprising that the Stern-Gerlach claim has managed to survive the long period of time since the inception of the theory, without being exposed by quantum physicists or even by clever students at the elementary level. Why did it not happen? It can only be speculated that it might be due to a tendency in quantum physics to promote specific objectives or causes that happen to support the prevailing quantum wisdom at the time. The abundance of paradoxes, weird behavior, and unsettled interpretations that characterize quantum mechanics is certainly conducive to such attitudes. Perhaps one more added to the list did not make much difference in the overall picture. The Stern-Gerlach effect, with its alleged experimental observations of two-state quantum effects, predicting a nonobjective quantum world, fitted snugly into the trend and culture of the 'new physics' developing around the time of the experiment.

In spite of the obvious flaw in the interpretation of the *SG* experiment, it is legitimate to ask whether a more refined modeling of the silver particles and the overall interaction with the nonuniform magnetic field might possibly restore the interpretation as a two-state quantum effect. The electron *spin* associated with its magnetic polarization could possibly be involved in the interaction process. From a simple minded point of view the electron spin causes a precession of its axis about the magnetic field lines represented by the *monopole* component B_1. The precession is expected to effectively cancel the net lateral deflection caused by the lateral force component $F(\theta_\mu)_y$ in (10.11) whereas the vertical deflection would be proportional to $\cos(\theta_\mu)$. Therefore, this model predicts exactly the continuous deflection shown in Fig. 10.2a. But the *SG* experimental results contradicts this prediction, forcing us to reject the model of spin precession about the field line.

Another dubious point about the electron's magnetic moment is its relation to the silver atom itself. Is the magnetic moment of the *47th* electron associated with the electron alone or is it an attribute of the whole atom? Questions such as these would be of significance in a more refined analysis beyond the zero order *SG* deflection model, which is characterized by no rotation of the silver atom and its electron during the passage through the nonuniform magnetic field. The rotational force $\vec{\mu} \times \vec{B}$ in addition to the deflection force $\nabla(\vec{\mu} \cdot B)$ would seek to align the atom system with the magnetic field much in the same way as a magnetic needle is aligned towards the north pole. Efforts of improvements of the model in this direction is bound to encounter a number of apparently unsettled questions on the quantum mechanical modeling of the silver atom and its adaptation to the present modified theory of interaction. Problems of this nature go way beyond the primary goal of the theoretical material presented in this chapter, which is centered on demonstrating the basic flaw in the interpretation of the standard Stern-Gerlach experiment.

10.2.3 Numerical simulations of deposition pattern

In order to throw light on some of the questions posed above a set of numerical computer simulations were carried out. Still based on the original silver atomic model, the analysis went beyond the first-order approximations presented in the first part of the chapter. The computations, based on a nonlinear analysis, were set up to simulate deposition patterns of deflected silver atoms for a wide range of operational parameters. The inhomogeneous Stern-Gerlach magnetic field was represented by a computer generated true field analysis accounting properly for nonuniformities in the field. Figure 10.4 shows the results of a typical simulation referred to the exit plane of the magnet deflection system, depicting the deposition pattern of a total of 5000 randomly polarized atoms. The incident flat beam of silver atoms, shaded in the figure, was assumed to have constant density with no longitudinal or lateral velocity spread. A comparison with the experimental deposition in the original Stern-Gerlach paper [18], reproduced in Fig. 10.2c, reveals the same lip-shaped pattern. Some of the minor differences might be ascribed to velocity spread in the incident stream of silver atoms. To my best knowledge the data from the Stern-Gerlach experiment [18] are the only ones available at the present time, but its lack of details on magnet and stream parameters prevents comparison through computer simulation.

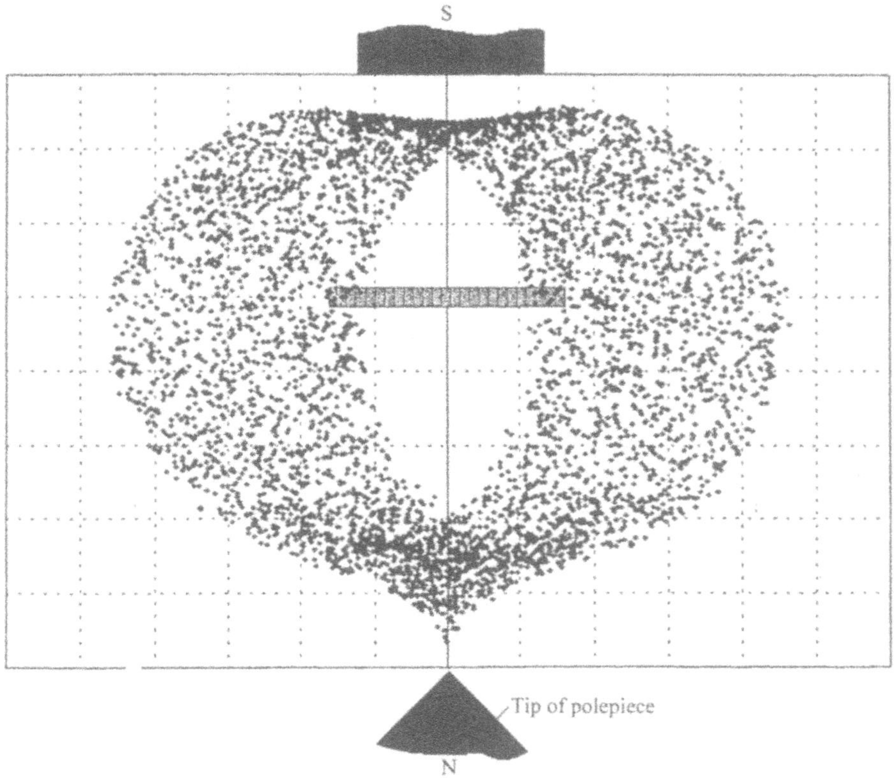

Figure 10.4. Computer generated deposition pattern of 5000 randomly polarized atoms in a simulated Stern-Gerlach experiment based on the present revised theory. The density of the incident Ag beam of rectangular cross section (shaded) is constant, with no longitudinal or lateral velocity spread. Only the tips of the magnetic pole pieces are shown in the figure.

Resolutions of the issues discussed in this chapter would require further theoretical modeling of the *Ag* atom with its electron spin, and above all, more experimental evidence on the deposition pattern under controlled conditions. Regardless of the outcome of such future investigations, it is clear that the standard background theory of the Stern-Gerlach experiment contains serious flaws—flaws that introduce serious doubts about the interpretation presented in quantum texts. Based on the present exposure, and in lack of further experimental evidence, the experiment must be interpreted as a pure classical phenomenon not supporting the *SG* claim of experimental verification of two-state quantum effects.

Chapter 11

PHOTON TUNNELING–SUPERLUMINAL VELOCITY?

'In nature's infinite book of Secrecy
a little I can read.'

The soothsayer in Shakespeare's
Anthony and Cleopatra

11.1 PHOTON TUNNELING IN A MICROWAVE CONFIGURATION

Recently, there has been a great deal of interest in tunneling times. In particular, there seems to be experimental evidence that the tunneling velocity of photons exceed the velocity of light [11],[42]. Attempts of explaining these observations have been based on quantum theory. In an effort to find out if bitemporal effects can possibly contribute to the understanding of these puzzling effects the present chapter analyzes photon tunneling from the viewpoint of the neoclassical theory. As photon tunneling is a universal effect not limited to optical photons, we shall consider a microwave version of tunneling in the waveguide configuration shown in Fig. 11.1. The transmission system consists of an undercut Transverse Electric TE_{10} waveguide section B of length L_0 and width a inserted into a regular propagating waveguide transmission path. For convenience the rectangular dimensions are the same throughout the entire configuration. By appropriate choice of frequency below the TE_{10} cut-off frequency ω_c, section B will be nonpropagating, with cut-off frequency given by

$$\omega_c = c\frac{\pi}{a} \tag{11.1}$$

The adjoining waveguide sections, the input section A and the output section C, contain lossless dielectric material with dielectric constant $\varepsilon > \varepsilon_0$. Their common cut-off frequency is then

$$\omega_{c,A} = \omega_{c,C} = c\frac{\pi}{a}\sqrt{\frac{\varepsilon_0}{\varepsilon}} \tag{11.2}$$

Provided the frequency satisfies the inequality

$$\frac{\pi}{a}\sqrt{\frac{\varepsilon_0}{\varepsilon}} < \frac{\omega}{c} < \frac{\pi}{a} \tag{11.3}$$

the configuration operates in the described manner, with sections A and C supporting unattenuated TE_{10} waves and section B attenuated TE_{10} waves. The real propagation factor β in the propagating sections A and C is given by

$$\beta = \sqrt{(\frac{\omega}{c})^2\frac{\varepsilon}{\varepsilon_0} - (\frac{\pi}{a})^2} \tag{11.4}$$

In the nonpropagating section B the propagation factor is imaginary:

$$\beta_0 = -j\gamma_0 = -j\sqrt{(\frac{\pi}{a})^2 - (\frac{\omega}{c})^2} \tag{11.5}$$

meaning that the attenuation factor $\gamma_0 = j\beta_0$ is real. Because the configuration is a special case of the general electromagnetic circuit configuration analyzed in Chapter 6, it must necessarily comply with the general theory of that chapter. The circuit properties were there expressed in terms of the macroscopic scattering matrix $\mathbf{S}$, which is derived in Appendix A of this chapter for the configuration of Fig. 11.1.

$$\mathbf{S} \cong \begin{bmatrix} \rho & (1-\rho^2)\,e^{-\gamma_0 L_0} \\ (1-\rho^2)\,e^{-\gamma_0 L_0} & \rho \end{bmatrix} \tag{11.6}$$

The inverse scattering matrix $\mathbf{S}^{-1}$ is evaluated to

$$\mathbf{S}^{-1} \cong \begin{bmatrix} 1/\rho & -e^{-\gamma_0 L_0}(1-\rho^2)/\rho \\ -e^{-\gamma_0 L_0}(1-\rho^2)/\rho & 1/\rho \end{bmatrix} \tag{11.7}$$

The equations hold for large attenuation $\gamma_0 L_0$, certainly an adequate approximation in discussing tunneling. The variable ρ is the regular

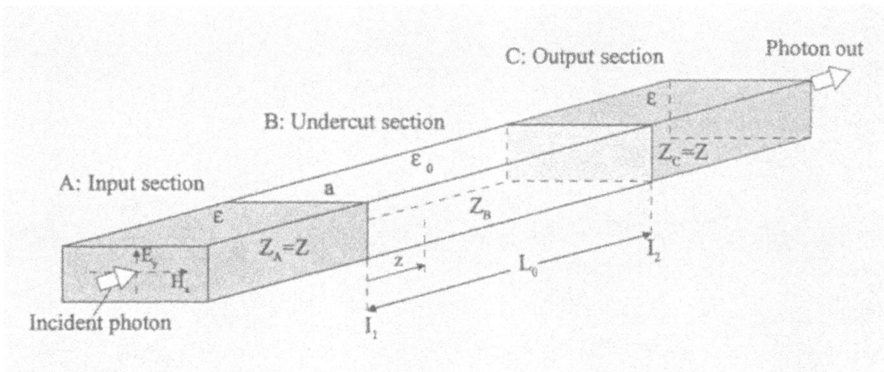

Figure 11.1. **Illustration of a microwave tunnelling system consisting of an undercut waveguide section B inserted into a regular transmission system $A - C$.**

reflection coefficient at the interface I_1 between the input section A and the undercut section B.

$$\rho = \frac{Z_B - Z_A}{Z_B + Z_A} \tag{11.8}$$

where Z_A and Z_B are the characteristic impedances of the sections. Because the impedance Z_B of the undercut section is pure imaginary, the reflection coefficient ρ satisfies the relation

$$\rho^* \rho = 1 \tag{11.9}$$

showing that ρ is of unit magnitude. The scattering relation referred to interfaces I_1 and I_2 is expressed in the standard form

$$\begin{bmatrix} b_1 \\ b_2 \end{bmatrix} = \mathbf{S} \begin{bmatrix} a_1 \\ a_2 \end{bmatrix} \tag{11.10}$$

where the $a's$ and $b's$ represent components traveling *in* and *out* of section B, respectively. Because there is no reason why the output section C should not be matched at its termination, the component a_2 can be taken to be zero so that (11.10) reduces to the two equations

$$\begin{aligned} b_1 &\cong \rho a_1 \\ b_2 &\cong (1 - \rho^2)\, e^{-\gamma_0 L_0}\, a_1 \end{aligned} \tag{11.11}$$

again valid for large attenuation $\gamma_0 L_0$. The analysis leading to these relations is strictly classical. Let us proceed by first discussing the macroscopic behavior and then continue with single photon characteristics based on the neoclassical theory.

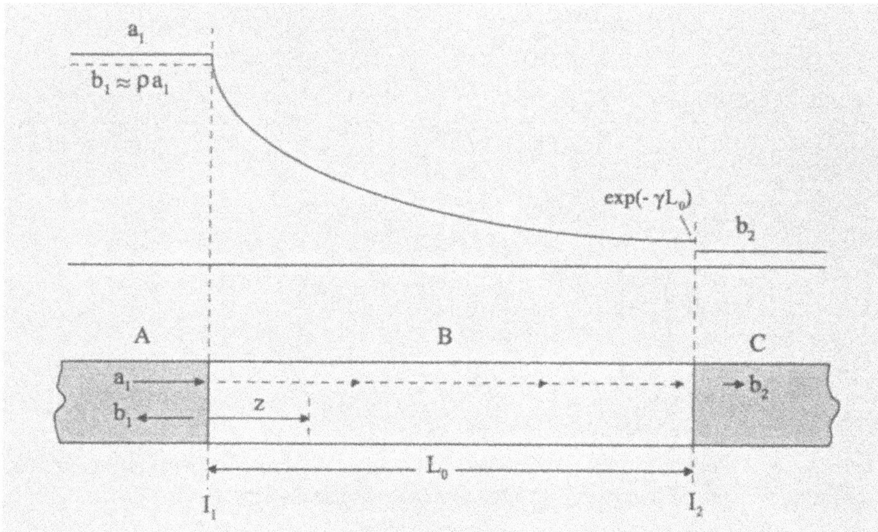

Figure 11.2. **Sketch showing 'macroscopic tunneling'. The undercut section B of length L_0 is excited with the incident component a_1. The reflected component $b_1 = \rho a_1$, and the transmitted component is b_2.**

11.1.1 Macroscopic 'tunneling'

The interpretation of (11.11) is straight forward. The reflected wave component $b_1 \approx \rho a_1$ is caused by the mismatch at terminal I_1 expressed by the reflection coefficient ρ. Because $|\rho| = 1$ it follows that $|b_1| \approx |a_1|$ so that practically all the power is reflected, provided the attenuation $\gamma_0 L_0$ is very large. The transmitted component b_2 is damped by the factor $\exp(-\gamma_0 L_0)$ and reduced further by the transmission loss in the two interfaces I_1 and I_2. The transmission coefficient for each interface is specified by

$$\tau = (1 - \rho^2)^{1/2} \tag{11.12}$$

so that the overall transmission coefficient for the two interfaces I_1 and I_2 is $1-\rho^2$, the first factor in the lower equation of (11.11). The macroscopic relations specified by (11.11) are illustrated in Fig.11.2. It should be noted that the difference between the components b_1 and a_1 at the input plane I_1 is exaggerated. The expected time average number ratio N_{trans}/N_{refl} of photons is extracted from the macroscopic relations (11.11).

$$\lim_{N\to\infty} \frac{N_{trans}}{N_{refl}} = \frac{|b_2|}{|b_1|} = \frac{|1 - \rho^2)|}{|\rho|} e^{-\gamma_0 L_0} \tag{11.13}$$

which is of order $\exp(-\gamma_0 L_0)$. This equation is the only information that can be extracted from the *macroscopic* relations (11.11) as far as single photons are concerned. The fact that a certain number N_{trans} out of a total of N photons are transmitted through the undercut region is a macroscopic electromagnetic 'tunneling' effect which follows directly from classical electromagnetism. No quantum effects are involved in this macroscopic description.

11.1.2 Neoclassical theory of single photon tunneling

Let us next discuss the behavior of single photons constituting the electromagnetic wave in the undercut section B. A single photon, out of a total of N photons per second, entering at terminal I_1 as an electromagnetic wave satisfying Maxwell's equations, can never propagate through the undercut region B because in so doing it would loose part of its energy $\hbar\omega$, which in a linear system is prohibitive. On the other hand it is a fact, corroborated by (11.13), that N_{trans} photons of the N photons manage to exit from terminal I_2 with their energy $\hbar\omega$ intact. According to the energy integrity argument they can not come from the input terminal, but where do they come from? In the neoclassical theory this photon quandary, impossible to explain from classical electromagnetics, finds its resolution with environmental photon doublets as the main players. The explanation is a direct application of the general theory of photon routing presented in Chapter 6.

Following the theory of that chapter the regular input $\mathbf{a}$ in the scatter relation (11.10) is supplemented by environmental photon doublets $\boldsymbol{\alpha}$, so that the relation is modified to:

$$\begin{bmatrix} b_1 \\ b_2 \end{bmatrix} = \mathbf{S}(\mathbf{a} + \boldsymbol{\alpha}) \tag{11.14}$$

The doublet vector $\boldsymbol{\alpha}$ is specified by the requirements put on the outputs b_1 and b_2 in relation to total transmission or reflection. The first step is to solve the equation with respect to $\boldsymbol{\alpha}$, making use of (11.7).

$$\boldsymbol{\alpha} = -\mathbf{a} + \begin{bmatrix} 1/\rho & -e^{-\gamma_0 L_0}(1-\rho^2)/\rho \\ -e^{-\gamma_0 L_0}(1-\rho^2)/\rho & 1/\rho \end{bmatrix} \begin{bmatrix} b_1 \\ b_2 \end{bmatrix} \tag{11.15}$$

If the incident photon is reflected, none of its energy is found at the output end, meaning that

$$b_2 = b_{2,refl} = 0 \qquad b_1 = b_{1,refl} = \rho a_1 \tag{11.16}$$

Note that $|\rho| = 1$ so that $|b_{1,refl}| = |\rho a_1| = |a_1|$ verifying full energy reflection. Under these circumstances (11.15) specifies the required doublet admixture $\boldsymbol{\alpha}_{refl}$ by

$$\boldsymbol{\alpha}_{refl} = \begin{bmatrix} 0 \\ -e^{-\gamma_0 L_0}(1-\rho^2)/\rho \end{bmatrix} a_1 \cong \mathbf{0} \tag{11.17}$$

meaning that the required doublet supplement $\boldsymbol{\alpha}_{refl}$ is not exactly zero but very small. On the other hand, if the incident photon is 'transmitted' through the undercut region B, no energy is reflected back, so that

$$b_1 = b_{1,trans} = 0 \qquad b_2 = b_{2,trans} = \rho a_1 \tag{11.18}$$

In this case (11.15) specifies the doublet admixture to

$$\boldsymbol{\alpha}_{trans} = \begin{bmatrix} -1 - e^{-\gamma_0 L_0}(1-\rho^2)/\rho \\ 1 \end{bmatrix} a_1 \cong \begin{bmatrix} -1 \\ 1 \end{bmatrix} a_1 \tag{11.19}$$

With this doublet supplement the photon exits from the output end of the undercut region B. According to the general theory of Chapter 6 the time average of the overall $\boldsymbol{\alpha}$-vector from the N_{refl} reflected photons and the N_{trans} transmitted photons is expected to be zero. That this indeed happens is verified by adding the contributions from $\boldsymbol{\alpha}_{refl}$ and $\boldsymbol{\alpha}_{trans}$ according to

$$\lim_{N\to\infty} \frac{1}{N} \sum \alpha = \frac{N_{trans}}{N} \boldsymbol{\alpha}_{trans} + \frac{N_{refl}}{N} \boldsymbol{\alpha}_{refl} \tag{11.20}$$

With the aid of (11.13) the time average becomes

$$\lim_{N\to\infty} \frac{1}{N} \sum \alpha = \frac{N_{refl}}{N} \begin{bmatrix} -e^{-\gamma_0 L_0}(1-\rho^2)/\rho \\ 0 \end{bmatrix} a_1 \cong \mathbf{0} \tag{11.21}$$

which holds within the accuracy of the approximations.

What is meant by the phrase 'photon being transmitted'? It was already ruled out that the incident photon, as a wave, might propagate through the undercut region B because the photon is forced to retain is energy. So where do the transmitted photons come from? The answer to this question might become a little clearer from Fig. 11.3 showing the photon doublet admixtures for the reflected and transmitted photons. It is recalled, from the general doublet theory in Chapter 6, that the doublets are resonances in bidirectional time with the spatial distributions inside the undercut region B corresponding exactly to the macroscopic specifications as exponentially decaying distributions. These distributions can be thought of as part of the already existing environmental *PDE* pool of doublets. Or, because the doublet resonances have zero net energy they can be excited spontaneously whenever needed and with amplitude distributions determined by the scattering matrix. Armed with this information the 'transmission' or tunneling of an incident photon can be explained as follows: Assume that the momentary situation is characterized by an excess of reflected photons above the equilibrium number, so that there has been a corresponding build up of a surplus doublets α_{refl}. Note from (11.17) that although α_{refl} is small it is not exactly zero, so that a sufficiently large number of reflected photons N_{refl} can build up this surplus. The environmental response to this situation is a spontaneous excitation of the doublet α_{trans} shown in Fig. 11.3b. According to (11.19) the components of α_{trans} at the terminals I_1 and I_2 have the same magnitudes equal to the incident photon a_1. The signs of these components are such as to perform the following: The incident photon is completely *annihilated* at the input terminal, and a photon is *created* and *emitted* from the PDE pool at the output terminal. No transit of photons take place from the input terminal through the undercut region to the output terminal.

The doublet α_{trans} at the terminals can also be interpreted as an *entangled* doublet of a more general form than the ones described in Chapter 2 (see for instance Fig. 2.9). In this interpretation doublet components α_1 and α_2 of equal magnitude are excited spontaneously at I_1 and I_2, both decaying exponentially towards the center of the undercut region B where they annihilate each other. Thus, in the neoclassical theory the tunneling phenomenon is explained as an emission process at the output terminal, where the emitted photons are drawn from the reservoir of photon doublets inside the undercut region serving as a buffer zone. The reservoir consists of photon doublet resonances in bidirectional time. According to the general theory of Chapters 3 and 6 they satisfy Maxwell's equations, which in the present formulation is the scatter relation. The doublet resonances can be visualized as a permanent assembly of zero energy doublet fields existing prior to the arrival of incident photons or, if we like, as being spontaneously excited whenever needed. The spontaneous excitation is explained from the circumstance that no influx of energy is needed in the process. If net power influx were required, the power would necessarily propagate with the velocity of light, so that the excitation process would take some finite amount of time. But the doublets require no influx of power because the net energy is zero. Therefore, the excitation time is expected to be zero. The doublets are spontaneously excited in the required amplitudes so as to satisfy the overall boundary conditions including photon energy integrity.

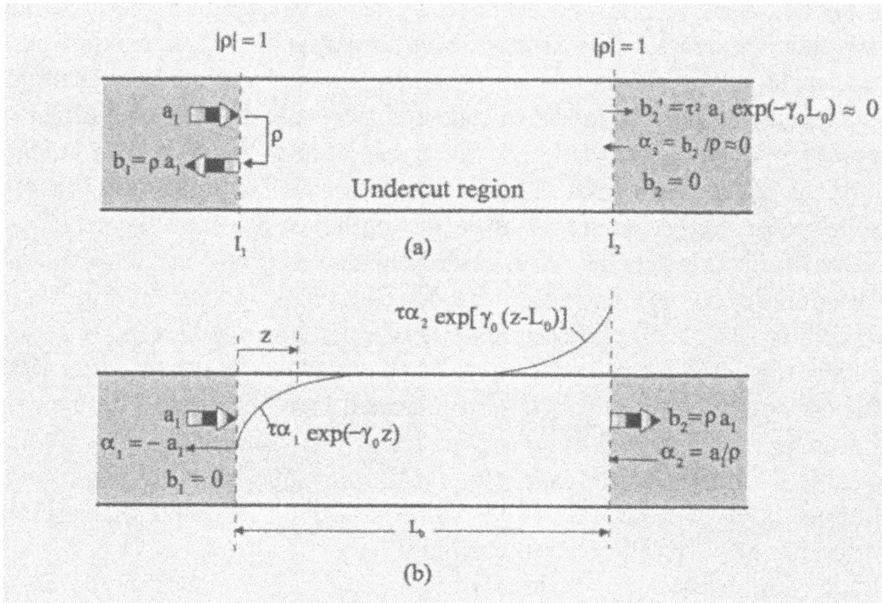

Figure 11.3. **Illustration of photon doublet engagement in the behavior of single photons in the undercut region. (a) Incident photon a_1 reflected: $b_2 = 0$, $b_1 = \rho a_1$. (b) Incident photon 'transmitted': $b_1 = 0$, $b_2 = \rho a_1$.**

11.1.3 Superluminal tunneling

An immediate consequence of the views presented above is the occurrence of superluminal transmission through the undercut region. The tunneling time is expected to be close to zero and the corresponding apparent velocity very large. In this theory the usual concept of photons tunneling through the section is discarded. The photon is never transmitted through; the incident photon is *annihilated* at the input end with a simultaneous *emission* of a photon at the far end. The annihilated photon adds to the internal reservoir of doublets, whereas the emitted photon draws from the reservoir, all in obeyance of the minimization principle of zero doublet energy.

It must be concluded then that the expected 'tunneling' velocity bears scant relation to the velocity of light. The word 'tunneling' is used merely as a reference to the usual description of the phenomenon. The apparent velocity may even be close to infinity. In the present scenario objections to the idea of infinite velocity based on arguments of causality are hardly valid because the photon does not propagate through the undercut region. The described emission process is orchestrated from the internal arsenal of photon doublets which do not adhere to macroscopic causality restrictions. Moreover, if verified by further experimental evidence, recent experimental results suggesting superluminal tunneling times [11] might add support to the alternative tunneling mechanism of photons presented in this chapter.

APPENDIX A TO CHAPTER 11

A.1 SCATTERING MATRIX OF THE TUNNELING SECTION

Referring to Fig. 11.1 in this chapter, the scattering matrix of the tunneling section B is obtained by first determining the fields in the three regions, then matching the tangential fields at the two interfaces I_1 and I_2, followed by evaluation of the elements in the scattering matrix. As indicated in the figure the TE_{10} fields E_y and H_x are adequately represented by their maxima at the center of the cross section. At interface I_1 in region A the fields are given by:

$$\begin{bmatrix} E_1 \\ H_1 \end{bmatrix}_A = \begin{bmatrix} Z_A^{1/2} & Z_A^{1/2} \\ Z_A^{-1/2} & -Z_A^{-1/2} \end{bmatrix} \begin{bmatrix} a_A^f \\ a_A^b \end{bmatrix} \tag{A.1}$$

where the forward and backward components a_A^f and a_A^b are identified by superscripts f and b, respectively, and where $Z_A = Z_C$ is the characteristic impedance given by

$$Z_A = Z_0 \left(\frac{1/\varepsilon_{rA}}{1-(\omega_{cA}/\omega)^2} \right)^{1/2} \tag{A.2}$$

Here, $Z_0 = \sqrt{\mu_0/\varepsilon_0}$ is the impedance of free space, ε_{rA} is the relative dielectric constant, and ω_{cA} is the cut-off frequency in region A. The fields in the undercut region B are given by

$$\begin{bmatrix} E(z) \\ H(z) \end{bmatrix}_B = \begin{bmatrix} Z_B^{1/2} e^{-\gamma_0 z} & Z_B^{1/2} e^{\gamma_0 z} \\ Z_B^{-1/2} e^{-\gamma_0 z} & -Z_B^{-1/2} e^{\gamma_0 z} \end{bmatrix} \begin{bmatrix} a_B^f \\ a_B^b \end{bmatrix} \tag{A.3}$$

where $\gamma_0 = j\beta_0$ is the real attenuation factor specified in (11.5), and where the characteristic impedance Z_B is imaginary.

$$Z_B = \frac{jZ_0}{[(\omega_{cB}/\omega)^2 - 1]^{1/2}} \tag{A.4}$$

From (A.3) the fields inside section B at $z=0$ and at $z=L_0$ are, respectively:

$$\begin{bmatrix} E(0) \\ H(0) \end{bmatrix}_B = \begin{bmatrix} Z_B^{1/2} & Z_B^{1/2} \\ Z_B^{-1/2} & -Z_B^{-1/2} \end{bmatrix} \begin{bmatrix} a_B^f \\ a_B^b \end{bmatrix} \tag{A.5}$$

$$\begin{bmatrix} E(L_0) \\ H(L_0) \end{bmatrix}_B = \begin{bmatrix} Z_B^{1/2} e^{-\gamma_0 L_0} & Z_B^{1/2} e^{\gamma_0 L_0} \\ Z_B^{-1/2} e^{-\gamma_0 L_0} & -Z_B^{-1/2} e^{\gamma_0 L_0} \end{bmatrix} \begin{bmatrix} a_B^f \\ a_B^b \end{bmatrix} \tag{A.6}$$

And finally, the fields in region C at $z=L_0$ are given by

$$\begin{bmatrix} E_2 \\ H_2 \end{bmatrix}_C = \begin{bmatrix} Z_A^{1/2} & Z_A^{1/2} \\ Z_A^{-1/2} & -Z_A^{-1/2} \end{bmatrix} \begin{bmatrix} a_C^f \\ a_C^b \end{bmatrix} \tag{A.7}$$

where the characteristic impedance Z_C is equal to the impedance Z_A in section A. Expressing the boundary conditions in the usual way by continuous tangential fields at the interfaces $z=0$ and $z=L_0$, the conditions are:

$$\begin{bmatrix} E_1 \\ H_1 \end{bmatrix}_A = \begin{bmatrix} E(0) \\ H(0) \end{bmatrix}_B \qquad \begin{bmatrix} E(L_0) \\ H(L_0) \end{bmatrix}_B = \begin{bmatrix} E_2 \\ H_2 \end{bmatrix}_C \tag{A.8}$$

These equations together with (A.1)-(A.7) result in the following scattering matrix $\mathbf{S}$:

$$\mathbf{S} = \frac{\mathrm{Tanh}(\gamma_0 L_0)}{1+\left(\frac{Z_B}{Z_A}+\frac{Z_A}{Z_B}\right)\mathrm{Tanh}(\gamma_0 L_0)} \begin{bmatrix} \frac{1}{2}\left(\frac{Z_B}{Z_A}-\frac{Z_A}{Z_B}\right) & \mathrm{Sinh}^{-1}(\gamma_0 L_0) \\ \mathrm{Sinh}^{-1}(\gamma_0 L_0) & \frac{1}{2}\left(\frac{Z_B}{Z_A}-\frac{Z_A}{Z_B}\right) \end{bmatrix} \tag{A.9}$$

It is assumed that the attenuation in the undercut region B is large, in which case the following approximations hold:

$$\gamma_0 L_0 >> 1 \qquad \mathrm{Tanh}(\gamma_0 L_0) \cong 1 \qquad \mathrm{Sinh}(\gamma_0 L_0) \cong \frac{1}{2} e^{\gamma_0 L_0} \tag{A.10}$$

Under these circumstances the scattering matrix (A.9) is approximated by

$$\mathbf{S} = \frac{1}{1+\frac{1}{2}\left(\frac{Z_B}{Z_A}+\frac{Z_A}{Z_B}\right)} \begin{bmatrix} \frac{1}{2}\left(\frac{Z_B}{Z_A}-\frac{Z_A}{Z_B}\right) & 2e^{-\gamma_0 L_0} \\ 2e^{-\gamma_0 L_0} & \frac{1}{2}\left(\frac{Z_B}{Z_A}-\frac{Z_A}{Z_B}\right) \end{bmatrix} \tag{A.11}$$

It will be convenient to express the matrix elements of $\mathbf{S}$ by the reflection coefficient ρ referred to the interface I_1 between region A and region B.

The following equations apply:

$$\rho = \frac{Z_B - Z_A}{Z_B + Z_A} = \frac{\frac{1}{2}\left(\frac{Z_B}{Z_A} - \frac{Z_A}{Z_B}\right)}{1 + \frac{1}{2}\left(\frac{Z_B}{Z_A} + \frac{Z_A}{Z_B}\right)} \tag{A.12}$$

$$1 - \rho^2 = \frac{2}{1 + \frac{1}{2}\left(\frac{Z_B}{Z_A} + \frac{Z_A}{Z_B}\right)} \tag{A.13}$$

Hereby, the scattering matrix **S** is seen to take the simple form

$$\mathbf{S} = \begin{bmatrix} \rho & (1 - \rho^2)e^{-\gamma_0 L_0} \\ (1 - \rho^2)e^{-\gamma_0 L_0} & \rho \end{bmatrix} \tag{A.14}$$

The formula is applicable for large attenuation $\gamma_0 L_0$ in the undercut region B and serves as the basis for the analysis of microwave tunneling in Chapter 11.

Chapter 12

DELAYED CHOICE INTERFEROMETRIC EXPERIMENTS

As I was walking up the stairs
I met a man who wasn't there.
He wasn't there again today.
I wish, I wish he'd stay away.

Hughes Mearns

12.1 INTRODUCTION

The last few decenniums have seen a flourishing of new sophisticated experimental technique designed to perform experiments on single quantum entities with the aim of gaining new insight in fundamental quantum mechanical phenomena and hopefully resolving some of the paradoxes that regularly pop up in the interpretations [3],[40]. Foremost in these is the problem of dualistic behavior of photons and electrons under *delayed choice* conditions. The reports in the literature on such experiments are firmly anchored in standard quantum theory and the interpretations proposed are restricted by the causality postulate. In this book we are looking upon these experiments with fresh eyes based on the bitemporal neoclassical theory. According to the foundations of this theory the macroscopic causality does not apply in the domain of microcosm where the experiments on single entities take place. It will be demonstrated that rejection of the causal concept leads to interpretations and resolutions of delayed choice experiments drastically different from the standard interpretations. We shall analyze a few of the most famous and best known delayed choice configurations—the Mach-Zehnder interferometer in this chapter—the *EPR* experiment in the next chapter. Because the Mach-Zehnder experiment relies heavily on a particular optical component—the beam splitter—the first section of this chapter is devoted to a description of the expected functioning of the beam splitter, seen from the platform of the neoclassical theory.

12.2 NEOCLASSICAL THEORY OF THE BEAM SPLITTER

The beam splitter, shown schematically in Fig. 12.1, often referred to as a half silvered mirror, is a central component in the quantum experiments to be discussed. Its properties viewed as a circuit can best be appreciated in the context of a neoclassical description, which is the topic of the present section. This optical component represents an electric circuit with well defined macroscopic as well as single photon characteristics. The beam splitter consists of a thin film of highly conducting metal—such as silver—deposited on a glass substrate. The metallic film of silver does not form a continuous layer; it is rather a distribution of tiny silver islands of random size and random spacings formed by sputtering or vacuum deposition. It is a straight forward matter to show that an average of 50 percent coverage splits a macroscopic optical beam into two constituents, one transmitted and one reflected. With a 45 degrees incidence the reflected beam is perpendicular to the transmitted beam. As an optical component it finds numerous applications in regular optics and in several quantum mechanical experiments discussed in this chapter. There are other simpler ways of producing physically different beam splitters consisting of one or several layers of dielectric materials with appropriate dielectric constants producing the same effects as the thin film splitter. The following analysis applies to both kinds.

12.2.1 Macroscopic properties of the beam splitter

The beam splitter is a four terminal circuit whose electromagnetic properties, macroscopic as well as microscopic, are described by its macroscopic 4×4 scattering matrix $\mathbf{S}$, having the form

$$\mathbf{S} = -\frac{1}{\sqrt{2}} \begin{bmatrix} 0 & \mathbf{S}_1 \\ \mathbf{S}_1 & 0 \end{bmatrix} \tag{12.1}$$

where the submatrices $\mathbf{S}_1$ are specified by

$$\mathbf{S}_1 = \begin{bmatrix} e^{j\pi/4} & e^{-j\pi/4} \\ e^{-j\pi/4} & e^{j\pi/4} \end{bmatrix} \tag{12.2}$$

The scattering matrix satisfies the unitary condition of losslessness

$$\tilde{\mathbf{S}}^* \mathbf{S} = \mathbf{1} \tag{12.3}$$

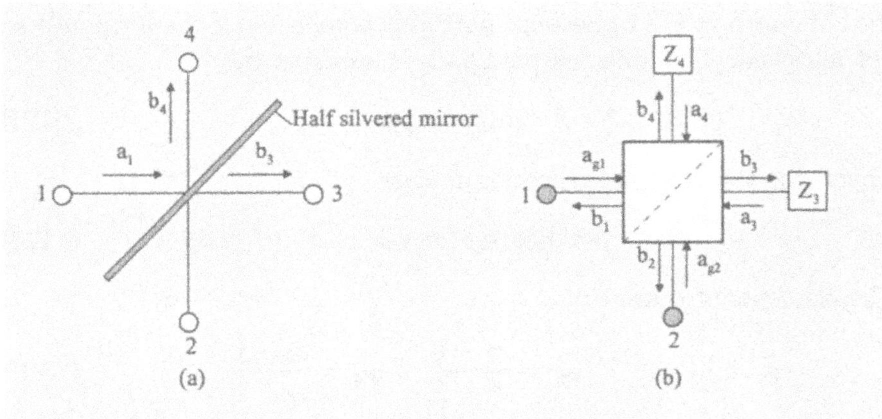

Figure 12.1. **Schematic diagram of the beam splitter. (a) Basic configuration with 4 terminals. (b) Example of a beam splitter set-up with terminals 1 and 2 serving as input terminals and terminals 3 and 4 loaded with impedances Z_3 and Z_4.**

In the scattering formulation the inputs and outputs are related by

$$\mathbf{b} = \mathbf{S}\mathbf{a} \tag{12.4}$$

where power conservation is expressed by

$$\tilde{\mathbf{b}}^*\mathbf{b} = \tilde{\mathbf{a}}^*\mathbf{a} \tag{12.5}$$

With the input $\mathbf{a}$ restricted to terminal 1 the macroscopic response is evaluated to

$$\mathbf{b} = \begin{bmatrix} b_1 \\ b_2 \\ b_3 \\ b_4 \end{bmatrix} = -\frac{1}{\sqrt{2}} \begin{bmatrix} 0 \\ 0 \\ e^{j\frac{\pi}{4}} \\ e^{-j\frac{\pi}{4}} \end{bmatrix} a_1 \tag{12.6}$$

The equation shows that the transmitted and reflected components at terminals 3 and 4 are 90 degrees out of phase and that the macroscopic power is split equally between the two because

$$|b_3|^2 = |b_4|^2 = \frac{1}{2}|a_1|^2 \tag{12.7}$$

These simple results describe the basic splitting function of the device. But a closer look reveals that even the macroscopic behavior is not always adequately represented by (12.4)-(12.7). The details depend on the terminations of terminals 3 and 4. Referring to Fig. 12.1 and the theory of Chapter 6, the power channeled to each terminal is not specified by the column $\mathbf{b}$ in (12.6) because part of it is reflected if the terminals are not matched. The reflected part $\mathbf{a}_{refl}$ is specified by

$$\mathbf{a}_{refl} = \boldsymbol{\rho}\,\mathbf{b} \tag{12.8}$$

where $\boldsymbol{\rho}$ is the diagonal reflection matrix

$$\boldsymbol{\rho} = \mathrm{diag}(\rho_i) \qquad i = 1...4 \tag{12.9}$$

The elements of $\boldsymbol{\rho}$ are

$$\rho_1 = \rho_2 = 0 \qquad \rho_3 = \frac{z_3 - 1}{z_3 + 1} \qquad \rho_4 = \frac{z_4 - 1}{z_4 + 1} \tag{12.10}$$

where the load impedances z_3 and z_4 are normalized with respect to the free space impedance $Z_0 = \sqrt{\mu_0/\varepsilon_0}$. The general theory of Chapter 6 shows that the signals channeled to the four terminations are not the column vector $\mathbf{b}$ but a column vector $\mathbf{c}$ specified by

$$\mathbf{c} = \boldsymbol{\tau}\,\mathbf{b} = \boldsymbol{\tau}\,(\tilde{\mathbf{S}}^* - \boldsymbol{\rho})^{-1}\mathbf{a}_g \tag{12.11}$$

where $\mathbf{a}_g$ is the input column vector, which quite generally will be restricted to terminals 1 and 2, and where the *transmission matrix* $\boldsymbol{\tau}$ is given by

$$\boldsymbol{\tau} = \mathrm{diag}(\tau_i) \qquad i = 1...4 \tag{12.12}$$

The components of $\boldsymbol{\tau}$ are

$$\tau_1 = \tau_2 = 1 \qquad \tau_3 = (1 - \rho_3^*\rho_3)^{1/2} \qquad \tau_4 = (1 - \rho_4^*\rho_4)^{1/2} \tag{12.13}$$

It is readily shown from(12.11) that power is conserved.

$$\tilde{\mathbf{c}}^*\mathbf{c} = \tilde{\mathbf{a}}_g^*\,\mathbf{a}_g \tag{12.14}$$

Equation (12.11) is applicable to circuits in which reflected components of the vector $\mathbf{a}_{refl}$ propagate in well confined waveguides back to the circuits where they serve as extra inputs. But this may not happen in the beam splitter where the optical beams propagate in free space rather than in confined waveguides. Although reflected components from the terminations can not be ruled out, these are more likely to be scattered off to the sides out of the regular signal paths, rather tan finding their way back to the beam splitter. Under these circumstances (12.11) should be modified by dropping the reflections $\boldsymbol{\rho}$ but keeping transmissions $\boldsymbol{\tau}$ so that the equation reduces to the simpler form

$$\mathbf{c} = \boldsymbol{\tau}\,\mathbf{b} = \boldsymbol{\tau}\,\mathbf{S}\,\mathbf{a}_g \tag{12.15}$$

In this formulation the power conservation equation (12.14) does not hold unless $\boldsymbol{\tau}=\mathbf{1}$. Part of the power is scattered and lost. In dealing with optical beams the transmission matrix $\boldsymbol{\tau}$ can also be interpreted as the efficiency of the absorbing device, for instance the trigger efficiency of a counter.

The analysis up to this point is strictly macroscopic electromagnetism. It tells nothing about single photon behavior except that the time average number of photons arriving at terminals 3 and 4 must be the same, so that the macroscopic relation (12.7) is satisfied. The behavior of single photons in the beam splitter is the subject of the next section.

12.2.2 Single photons in the beam splitter

Figure 12.2 shows a configuration typifying the operation of a beam splitter with single photon input at terminal 1. The input a_{g1} is routed to terminals 3 or 4 with nothing channeled to terminal 2. The counters 3 and 4 are registering the arrivals of single photons. In order to avoid possible confusion caused by fairly unessential details let us assume the counters perfectly matched so that the reflection coefficients $\rho_3=\rho_4=0$ and therefore the transmission coefficient $\boldsymbol{\tau}=\mathbf{1}$. The macroscopic scatter relation (12.15) is then

$$\begin{bmatrix} c_1 \\ c_2 \\ c_3 \\ c_4 \end{bmatrix} = -\frac{1}{\sqrt{2}} \begin{bmatrix} 0 \\ 0 \\ e^{j\pi/4} \\ e^{-j\pi/4} \end{bmatrix} a_{g1} \qquad (12.16)$$

Let us visualize the incident power a_{g1} reduced to the point where the photons arrive at the beam splitter one at a time. This is a perfectly valid supposition that can be achieved experimentally as demonstrated for instance in [3]. Briefly, the experimental observations are as follows: The coincidence counter shows that counters 2 and 3 never trigger simultaneously except perhaps for a small coincidence rate attributed to some overlap in the photon stream or imperfect geometry. The accumulated time average photon numbers channeled to the two counters turn out to be the same.

It is clear that the macroscopic behavior described by (12.16) has changed drastically. The input photon is no longer split equally between the two terminals but routed to *either* terminal 3 *or* terminal 4. This behavior is in exact agreement with the general neoclassical theory of Chapter 6 and therefore expected. It is illuminating to compare the essentials of quantum mechanical and neoclassical interpretations of the functioning of the beam splitter, and discuss the differences and disagreements between the two.

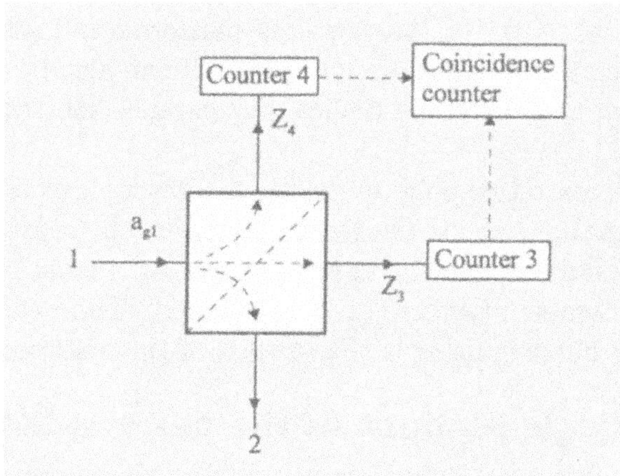

Figure 12.2. **Beam splitter with two identical counters at terminals 3 and 4 and a coincidence counter. Terminal 2 is open.**

12.2.3 Quantum mechanical view

The quantum mechanical explanation is based on the familiar duality principle, according to which the photon is either a particle or a wave depending on the measurement configuration. In the set-up of Fig. 12.2 the quoted experimental observations certainly point to a particlelike behavior, the photon particle being *either* transmitted to terminal 3 *or* reflected to terminal 4. The photon takes one path or the other. It does not seem to behave like a wave which is expected to split its energy evenly between the two terminals. The experimental fact linking the number of trigger events to the time average macroscopic values is ascribed to another quantum concept, the *correspondence* principle. The correspondence principle is a close relative of the concept of wave function *collapse*, a notion that is as puzzling today as it was at its inception some 70 years ago. Some of the problems associated with these concepts are a constant embarrassment in quantum physics, for example the following naive question: Why do not all the photons end up at the same terminal? The question is certainly well justified because a particular photon, well separated from the previous identical ones, sees exactly the same configuration. So why do they not all follow the same path? The reader is referred to the discussion of the same problem in Chapter 7.3.7 in connection with the double slit experiment for photons. Quantum mechanics can point to no memory process that explains the even split between the paths. Instead it resorts to an explanation based on a superposed state of the photon's wave function, with inherent equal probabilities for either path. In this picture the electromagnetics of the photon and the circuit has been discarded and replaced by the nonobjective probability photon.

12.2.4 Neoclassical view

The neoclassical theory of single photons in the beam splitter bears scant resemblance to the quantum mechanical description. Again referring to Fig. 12.2, the all important point is the independent or uncoupled loads of terminals 3 and 4 represented by the two separate counters. A single photon can not spread its energy between two independent loads because it would violate its energy integrity. In the parlance of circuit theory the two paths represent equivalent separate 'waveguides' leading the photon waves to their loads. The photon has to take one of the two paths, and this would be true in *any* theory of photon behavior, be it quantum theory or neoclassical theory, or for that matter some different theory, provided photon energy integrity is accepted. If the two loads were interconnected, the situation would be entirely different, as we shall see shortly in connection with a more complex configuration, the Mach-Zehnder interferometer.

The neoclassical theory describes the single photon in the beam splitter as a truly electromagnetic wave entering at terminal 1 and routed to terminals 3 or 4, through the intermediary of supplemental inputs from environmental doublet fields. For details of the basic theory of photon routing the reader is referred to Chapter 6. With N photons per second entering the beam splitter at terminal 1 let the composite input of one photon be given by

$$\mathbf{A}_g = \frac{1}{N}\mathbf{a}_g + \boldsymbol{\alpha}(\omega, -\omega) = \bar{\mathbf{a}}_g + \boldsymbol{\alpha}(\omega, -\omega) \tag{12.17}$$

where $\bar{\mathbf{a}}_g$ is the input associated with a single photon, and where the supplemental doublet inputs $\boldsymbol{\alpha}(\omega,-\omega)$ presumably are restricted to terminal 2. With this composite input the scattered components at terminals 3 and 4 due to the single photon $\bar{\mathbf{a}}_g$ are specified by

$$\bar{\mathbf{c}} = \begin{bmatrix} \bar{c}_3 \\ \bar{c}_4 \end{bmatrix} = -\frac{1}{\sqrt{2}} \begin{bmatrix} e^{j\pi/4} & e^{-j\pi/4} \\ e^{-j\pi/4} & e^{j\pi/4} \end{bmatrix} [\bar{\mathbf{a}}_g + \boldsymbol{\alpha}(\omega, -\omega)] \tag{12.18}$$

where $\bar{\mathbf{c}}$ is the single photon response. The doublet supplement $\boldsymbol{\alpha}(\omega,-\omega)$ is obtained through inversion of the equation.

$$\boldsymbol{\alpha}(\omega, -\omega) = -\frac{1}{\sqrt{2}} \begin{bmatrix} e^{-j\pi/4} & e^{j\pi/4} \\ e^{j\pi/4} & e^{-j\pi/4} \end{bmatrix} \begin{bmatrix} \bar{c}_3 \\ \bar{c}_4 \end{bmatrix} - \begin{bmatrix} \bar{a}_{g1} \\ 0 \end{bmatrix} \tag{12.19}$$

Now consider two photons q and p routed to terminals 3 and 4, respectively. The corresponding $\bar{\mathbf{c}}$ vectors are then specified by:

$$\bar{\mathbf{c}}_q^{(3)} = -\begin{bmatrix} 1 \\ 0 \end{bmatrix} \sqrt{2} e^{j\frac{\pi}{4}} \bar{a}_{g1} \qquad \bar{\mathbf{c}}_p^{(4)} = -\begin{bmatrix} 0 \\ 1 \end{bmatrix} \sqrt{2} e^{-j\frac{\pi}{4}} \bar{a}_{g1} \tag{12.20}$$

By substitutions into (12.19) the accompanying doublet admixtures achieving these routing tasks are evaluated to:

$$\boldsymbol{\alpha}_q^{(3)} = \begin{bmatrix} 0 \\ j \end{bmatrix} \bar{a}_{g1} \qquad \boldsymbol{\alpha}_p^{(4)} = \begin{bmatrix} 0 \\ -j \end{bmatrix} \bar{a}_{g1} \qquad q, p = 1, 2, ..N/2 \tag{12.21}$$

where the frequency dependence in the doublet term $\boldsymbol{\alpha}(\omega, -\omega)$ has been left out. Going back to (12.19), adding the contributions from all the N incident photons, the total becomes

$$\boldsymbol{\alpha}_{tot} = \frac{1}{\sqrt{2}} \begin{bmatrix} e^{-j\pi/4} & e^{j\pi/4} \\ e^{j\pi/4} & e^{-j\pi/4} \end{bmatrix} \begin{bmatrix} e^{j\frac{\pi}{4}} \\ e^{-j\frac{\pi}{4}} \end{bmatrix} \sqrt{2} \frac{N}{2} \bar{a}_{g1} - \begin{bmatrix} 1 \\ 0 \end{bmatrix} \bar{a}_{g1} \tag{12.22}$$

The right hand side of the equation is just the macroscopic scatter relation, readily verified by comparison with (12.19). Accordingly, the right hand side is zero, so that the following result applies:

$$\boldsymbol{\alpha}_{tot} = \mathbf{0} \tag{12.23}$$

This result is also confirmed by direct inspection of (12.21), which shows that the supplementary photon doublet inputs at terminal 2 have opposite phases for photons q and p, thus maintaining zero time average $\boldsymbol{\alpha}_{tot}$. Therefore, the beam splitter configuration is in agreement with the general theory of Chapter 6.

The routing process is governed by a kind of memory bank keeping track of the previous history of photon arrivals and routings. The routing of the next photon to terminal 3 or terminal 4 is decided entirely from the principle of minimization of accumulated doublet expenditure $\boldsymbol{\alpha}$. Apparently, the total $\boldsymbol{\alpha}$ is minimized with a periodic reversal of doublet excitation phase at terminal 2, meaning that each of the counters 3 and 4 receives a photon every second time. The routing appears to be governed by a simple periodic engagement of positive and negative doublet components. Figure 12.3 illustrates the details of the routing process in the two cases.

In the neoclassical model of the beam splitter the photons triggering the counters are considered to be truly electromagnetic waves. There is no need to introduce the particle aspect of quantum theory. The details of how the counters themselves are triggered by the photons were shown in Chapter 5, where the wavelike to particlelike transition in the counter regions was analyzed. But this part of the process is quite separate from the two-way routing operation itself and need not be included in the basic beam splitter theory of the present section.

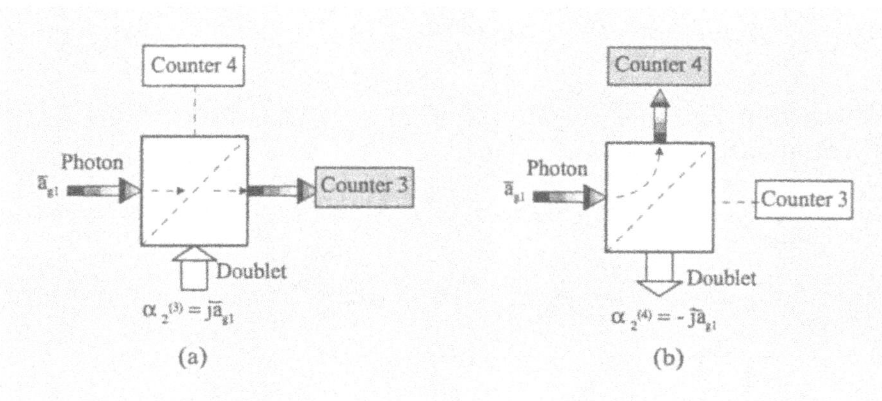

Figure 12.3. **Sketch showing the two photon routes of the beam splitter. (a) Incident photon $\bar{a}_{g1}$ at terminal 1 transmitted to terminal 3 with the aid of photon doublet $\alpha_2^{(3)} = j\bar{a}_{g1}$ at terminal 2. (b) Incident photon reflected to terminal 4 with the aid of doublet $\alpha_2^{(4)} = -j\bar{a}_{g1}$ at terminal 2.**

At the conclusion of this section it should again be emphasized that the reason for the photons to travel *one* path *or* the other is the separate and independent loads at terminals 3 and 4. The photon energy integrity leaves them no choice. The accounting is kept by the environmental *PDE* fields, maintaining zero time average energy. Further insight into these processes is gained from studying more complex configurations involving several beam splitters such as the Mach-Zehnder interferometer discussed in the following sections.

12.3 THE MACH-ZEHNDER INTERFEROMETER

The building blocs of the Mach-Zehnder interferometer are two *beam splitters* A and B separated by lengths L_1 and L_2 as shown in Fig. 12.4. The interferometer is a four terminal electromagnetic circuit and, as such, specified completely by its scattering matrix $\mathbf{S}$, readily determined from the data (12.1) and (12.2) of the beam splitter.

$$\mathbf{S} = e^{-jk(L_1+L_2)/2} \begin{bmatrix} \mathbf{0} & \mathbf{S}_\Delta \\ \tilde{\mathbf{S}}_\Delta & \mathbf{0} \end{bmatrix} \tag{12.24}$$

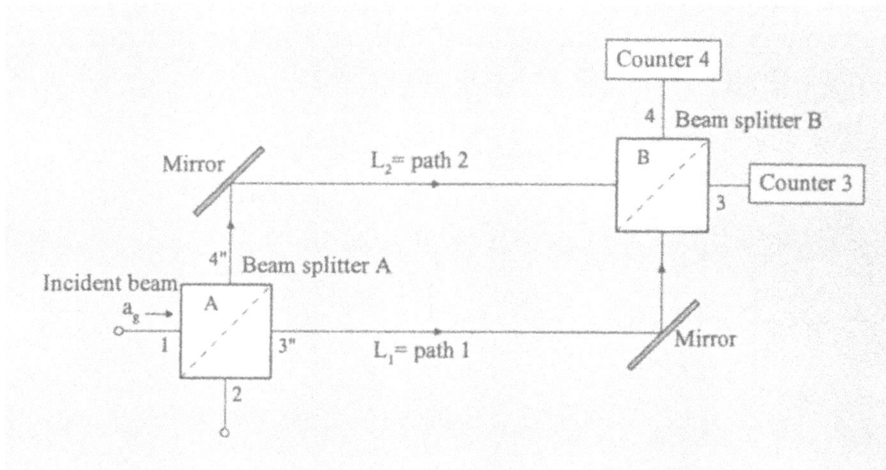

Figure 12.4. **Sketch of the basic configuration of the Mach-Zehnder interferometer.**

where the submatrix $\mathbf{S}_\Delta$ is specified by

$$\mathbf{S}_\Delta = \begin{bmatrix} \cos(\Delta) & \sin(\Delta) \\ -\sin(\Delta) & \cos(\Delta) \end{bmatrix} \tag{12.25}$$

The angle Δ appearing in the equation is one half the normalized path difference.

$$\Delta = k\frac{L_1 - L_2}{2} \tag{12.26}$$

The relation between inputs **a** and outputs **b** of the four terminal network is specified by the usual form of the scatter formula:

$$\begin{bmatrix} b_1 \\ b_2 \\ b_3 \\ b_4 \end{bmatrix} = \mathbf{S} \begin{bmatrix} a_1 \\ a_2 \\ a_3 \\ a_4 \end{bmatrix} = e^{-jk(L_1+L_2)/2} \begin{bmatrix} \mathbf{0} & \mathbf{S}_\Delta \\ \tilde{\mathbf{S}}_\Delta & \mathbf{0} \end{bmatrix} \begin{bmatrix} a_1 \\ a_2 \\ a_3 \\ a_4 \end{bmatrix} \tag{12.27}$$

where we have retained a completely general input vector **a**. According to the general theory of Chapter 6 the scattering matrix tells us everything there is to know about the macroscopic as well as single photon behavior. In the analysis to follow unessential complications arising from possible reflections from counters are avoided by the assumption that these are matched, i.e., the counter efficiencies are supposedly one hundred percent.

12.3.1 Macroscopic response of the interferometer

In the normal operation of the interferometer the incident beam excites terminal 1, with $a_1 = a_g$, in which case the macroscopic response is specified by

$$\begin{bmatrix} b_1 \\ b_2 \\ b_3 \\ b_4 \end{bmatrix} = e^{-jk(L_1+L_2)/2} \begin{bmatrix} \mathbf{0} & \mathbf{S}_\Delta \\ \tilde{\mathbf{S}}_\Delta & \mathbf{0} \end{bmatrix} \begin{bmatrix} a_g \\ 0 \\ 0 \\ 0 \end{bmatrix} \tag{12.28}$$

which evaluates to

$$\begin{bmatrix} b_1 \\ b_2 \\ b_3 \\ b_4 \end{bmatrix} = e^{-jk(L_1+L_2)/2} \begin{bmatrix} 0 \\ 0 \\ \cos(\Delta) \\ \sin(\Delta) \end{bmatrix} a_g \tag{12.29}$$

The output components $b_1 ... b_4$ are then:

$$b_1 = b_2 = 0$$

$$b_3 = e^{-jk(L_1+L_2)/2} \cos(\Delta)\, a_g \tag{12.30}$$

$$b_4 = e^{-jk(L_1+L_2)/2} \sin(\Delta)\, a_g$$

The macroscopic powers at terminal 3 and 4 are given by

$$P_3 = P_{tot} \cos^2(\Delta) \qquad\qquad P_4 = P_{tot} \sin^2(\Delta) \tag{12.31}$$

so that the power ratio becomes

$$\frac{P_4}{P_3} = \frac{|b_4|^2}{|b_3|^2} = \frac{\sin^2(\Delta)}{\cos^2(\Delta)} = \tan^2(\Delta) = \frac{N_4}{N_3} \tag{12.32}$$

The equations show that the incident beam a_g is channeled to both terminals 3 and 4 of beamsplitter B, unless $\Delta = n\pi$ or $\Delta = \pi(n+\frac{1}{2})$, where n is an integer. If the path difference Δ is varied, the relative powers at terminal 3 and 4 show the familiar cosine and sine interference pattern of the Mach-Zehnder device.

12.3.2 Single photon response of the interferometer

Let us now turn the attention to the behavior of single photons, with photon counts registered by means of the counters placed at terminal 3 and 4 of the second beam splitter. The counters are expected to register numbers proportional to $\cos^2(\Delta)$ and $\sin^2(\Delta)$, respectively. The details of the analysis is quite similar to that of the beam splitter alone discussed in the first part of the chapter. With the assumption that the supplementary environmental doublet inputs are restricted to beamsplitter A the scatter relation becomes

$$b_1 = b_2 = 0$$

$$\begin{bmatrix} b_3 \\ b_4 \end{bmatrix} = e^{-jk(L_1+L_2)/2} \begin{bmatrix} \cos(\Delta) & -\sin(\Delta) \\ \sin(\Delta) & \cos(\Delta) \end{bmatrix} [\bar{\mathbf{a}}_g + \boldsymbol{\alpha}(\omega, -\omega)] \tag{12.33}$$

where $\bar{\mathbf{a}}_g$ is the single photon input. The required photon doublet admixture vector $\boldsymbol{\alpha}(\omega,-\omega)$ is determined by inversion of the equation.

$$\boldsymbol{\alpha}(\omega, -\omega) = e^{jk(L_1+L_2)/2} \begin{bmatrix} \cos(\Delta) & \sin(\Delta) \\ -\sin(\Delta) & \cos(\Delta) \end{bmatrix} \begin{bmatrix} b_3 \\ b_4 \end{bmatrix} - \begin{bmatrix} 1 \\ 0 \end{bmatrix} \bar{a}_g \tag{12.34}$$

A particular photon $\bar{a}_g$ incident at terminal 1 has the option of being routed to either terminal 3 or terminal 4. If the qth photon is routed to terminal 3, it is clear that the component $b_q^{(4)}$ must vanish. The component $b_q^{(3)}$ is selected so that the doublet supplement at beamsplitter A is the simplest possible, which is $\alpha_1 = 0$ and $\alpha_2 \neq 0$. With these additional conditions on $\boldsymbol{\alpha}$ the appropriate components are

$$b_q^{(4)} = 0 \qquad b_q^{(3)} = e^{-jk(L_1+L_2)/2} \frac{1}{\cos(\Delta)} \bar{a}_{g,q} \tag{12.35}$$

These conditions inserted into (12.34) result in the following doublet admixture:

$$\boldsymbol{\alpha}_q^{(3)} = \begin{bmatrix} 0 \\ -\tan(\Delta) \end{bmatrix} \bar{a}_{g,q} \tag{12.36}$$

Similarly, with the routing of photon p to terminal 4 the appropriate $\mathbf{b}_p$-components are

$$b_p^{(3)} = 0 \qquad b_p^{(4)} = e^{-jk(L_1+L_2)/2} \frac{1}{\sin(\Delta)} \bar{a}_{g,p} \tag{12.37}$$

yielding the following doublet admixture:

$$\boldsymbol{\alpha}_p^{(4)} = \begin{bmatrix} 0 \\ \cot(\Delta) \end{bmatrix} \bar{a}_{g,p} \tag{12.38}$$

In both cases the doublet components of $\boldsymbol{\alpha}_q^{(3)}$ and $\boldsymbol{\alpha}_p^{(4)}$ at the *input terminal 1* are zero. Thus, the photon doublet vector $\boldsymbol{\alpha}$ is always orthogonal to the actual photon input vector $\bar{\mathbf{a}}_g$ so that

$$\tilde{\boldsymbol{\alpha}}^* \bar{\mathbf{a}}_g = 0 \tag{12.39}$$

a relation that will be needed shortly. Let us first demonstrate that the routing of a large number of incident photons is consistent with zero time average of the overall doublet admixture. Of the N photons per second, let N_3 and N_4 be routed to terminals 3 and 4. With the ratio N_4/N_3 specified by (12.32) the contributions from $\boldsymbol{\alpha}_q^{(3)}$ and $\boldsymbol{\alpha}_p^{(4)}$ add up to a total $\boldsymbol{\alpha}$ of

$$\boldsymbol{\alpha}_{tot} \sim \begin{bmatrix} 0 \\ -\tan(\Delta) \end{bmatrix} \cos^2(\Delta) + \begin{bmatrix} 0 \\ \cot(\Delta) \end{bmatrix} \sin^2(\Delta) = \mathbf{0} \tag{12.40}$$

showing that the overall time average doublet admixture vanishes.

12.3.3 Photon signal flow in the interferometer paths

Let us look at the detailed signal flow of the single photon along the two paths L_1 and L_2 from beam splitter A to beam splitter B. The intermediary output $\bar{\mathbf{b}}$ at terminals 3" and 4" of beam splitter A is readily evaluated, with the following result: If the incident photon q is routed to terminal 3 in beamsplitter B, the output $\bar{\mathbf{b}}_q^{(3)}$ from beam splitter A is given by:

$$\bar{\mathbf{b}}_q^{(3)} = -\frac{1}{\cos(\Delta)} \begin{bmatrix} e^{j(\pi/4-\Delta)} \\ e^{-j(\pi/4-\Delta)} \end{bmatrix} \frac{\bar{a}_g}{\sqrt{2}} \tag{12.41}$$

If the incident photon p is routed to terminal 4 in beamsplitter B the output $\bar{\mathbf{b}}_p^{(4)}$ from beam splitter A is then:

$$\bar{\mathbf{b}}_p^{(4)} = -\frac{1}{\sin(\Delta)} \begin{bmatrix} e^{-j(\pi/4+\Delta)} \\ e^{j(\pi/4+\Delta)} \end{bmatrix} \frac{\bar{a}_g}{\sqrt{2}} \tag{12.42}$$

In both cases the *amplitudes* of the two branch components are observed to be the same. It is quite remarkable that the photon energy, in both cases, is equally divided between the two signal paths regardless of the difference Δ in path length. Thus, the neoclassical theory predicts that the photon always takes *both* paths after the *first* beam splitter. The explanation for this is tied in with the fact that the two paths do not eventually end up in two separate loads for the photon. Therefore, there is no reason to take only one path. It is first after the second beam splitter the photon must take a single path, because of the two *independent* loads of terminals 3 and 4 of beam splitter B. There is nothing mysterious about a photon splitting up into two different paths. It merely expresses the more general property of the photon that its distribution in k-space is quite arbitrary; it is decided only by whatever boundary conditions prevail.

The reader may perhaps have noticed from (12.41) and (12.42) that the two photon constituents traveling along paths L_1 and L_2 appear to exceed the expected amplitude $\bar{a}_g/\sqrt{2}$ for half the incident photon. It must remembered that the components are composite entities, containing the incident actual photon plus the supplemental photon doublet. The latter has nonzero amplitude but carry no power. This point is perhaps made clearer by considering the output $\mathbf{b}$ from the *second* beam splitter B. From (12.33), forming the product $|\mathbf{b}|^2$:

$$|\mathbf{b}|^2 = |b_3|^2 + |b_4|^2 = |\bar{\mathbf{a}}_g + \boldsymbol{\alpha}(\omega, -\omega)|^2 = |\bar{a}_g|^2 + \tilde{\boldsymbol{\alpha}}^*\boldsymbol{\alpha} + 2\,\mathrm{Re}(\tilde{\boldsymbol{\alpha}}^*\bar{\mathbf{a}}_g) \qquad (12.43)$$

It follows from the orthogonality condition (12.39) that $\tilde{\boldsymbol{\alpha}}^*\bar{\mathbf{a}}_g = 0$. Furthermore, the doublet power $\tilde{\boldsymbol{\alpha}}^*\boldsymbol{\alpha} = 0$ always. Therefore

$$|\mathbf{b}|^2 = |\bar{a}_g|^2 \qquad (12.44)$$

verifying the proper power balance.

In summary, the experimental results quoted in [40] are fully substantiated by the present neoclassical theory of the beam splitter and the Mach-Zehnder interferometer. But the explanations offered by the author of [40] and others, being heavily embedded in the quantum mechanical duality concept, are not supported. In their interpretation the experimental observation of photons routed to terminal 3 *or* terminal 4, but never to both simultaneously, are explained as meaning that photons travel as particles along path L_1 *or* path L_2, which is in disagreement with (12.41) and (12.42). The observation of the build up over time of an interference pattern at terminals 3 and 4 is interpreted as meaning that photons travel both paths. The inevitable end result is the appearance of the 'which way' or 'both ways' paradox so familiar from the double slit experiments discussed in Chapters 7 and 8. Paradoxes of this nature are bound to appear in both actual and 'Gedanken' experiments with single photons, if one sticks to the duality concept.

In the present neoclassical theory the photon is an honest and goodness electromagnetic wave, never a particle. Its energy integrity forbids fragmentation of its energy between independent loads. The routing of the incident photons to the various loads is orchestrated by the environmental photon doublets, adhering strictly to the principle of minimum time average doublet energy. The energy minimization process appears as a fundamental principle in explaining single photon behavior in any electromagnetic circuit. The principle implies that the accumulated trigger counts at each terminal corresponds exactly to the relative *macroscopic* values. In this respect the principle replaces the *correspondence* principle in quantum mechanics. For a more detailed account the reader is referred to the general theory of Chapters 2-6.

12.4 MACH-ZEHNDER DELAYED CHOICE EXPERIMENT

In quantum physics so-called 'delayed choice experiments' on single photons, aimed at resolving basic questions in quantum mechanics, tend to confirm the photons' seemingly paradoxical behavior. Originally these were 'Gedanken experiments' which were analyzed in light of the duality concept of photons. More recently, improved experimental proficiency has allowed some of these to be carried out as practical experiments. One of these is a delayed choice experiment with the Mach-Zehnder interferometer. Let us give a brief description of the experiment and demonstrate how it is fully explained by the neoclassical theory without incurring any paradox or weirdness.

The experiment was proposed in 1978 by John A. Wheeler and later carried out by several groups of investigators. Figure 12.5 shows a sketch of the essentials of the experimental set-up. The configuration of the regular Mach-Zehnder interferometer of Fig. 12.4 is supplemented with a switch, a so-called Pockels cell inserted into path L_2. The switch can be actuated in a few nanoseconds, which is less than the transit time between beam splitter A and the switch. The crystal in the Pockels cell becomes birefringent when a switching voltage is applied across it. The configuration is arranged so that light entering the cell is either transmitted right through—this is the *voltage off* position—or the light is diverted into the side path L_3 with a counter C at the end—this is the *voltage on* position. The experimental situation corresponds to equal path lengths $L_1 = L_2$ so that the path difference $\Delta = 0$. In the original configuration shown in Fig.12.4 the incident single photon then always goes to terminal 3, with counter A triggered. According to (12.41) the incident photon at the first beam splitter separates into two equal constituents of amplitudes $\bar{b} = \bar{a}_g/\sqrt{2}$ propagating along the two paths L_1 and L_2. The Pockels cell is actuated *after* the photon has left beam splitter A but *before* it arrives at the cell, an operation made possible by sophisticated experimental technique. How does the photon react? The experiments show that the photon reacts as if the Pockels cell had been on *prior* to the photon's departure from the beam splitter, with

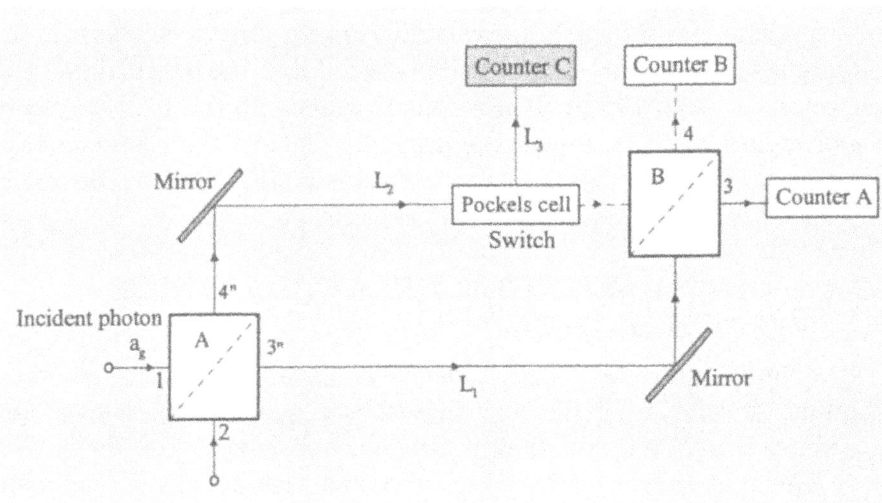

Figure 12.5. **Delayed choice experiment with a Mach-Zehnder interferometer supplemented with a Pockels cell in path L_2.**

observed trigger rates at counters C, B, and A as expected from that particular experimental situation. The photon takes *either* path L_1 *or* path L_2 in spite of the fact that it had already started out by splitting itself into two equal components to travel *both* paths. It appears as if the photon somewhat mysteriously knew in advance that the Pockels cell would be switched on and acted accordingly. This clever trick of magic is extremely difficult, essentially impossible, to explain by any *objective causal* theory but, as we shall see shortly, is readily resolved in the bitemporal domain of the neoclassical theory.

Actuation of the Pockels cell represents an abrupt change of the electromagnetic circuit from one configuration to a different one. Talking in general terms, the analysis of time varying circuits is notably quite difficult, even if limited to the macroscopic domain. Most often the treatment is restricted to special time functions such as simple periodic variations of some circuit parameter, giving rise to the familiar parametric interactions. Abrupt changes from one state to another, as in the present experiment, can be analyzed in time frame by considering pulse propagation and reflection along transmission paths. These kind of analyses, described in many textbooks on classical transmission theory, and often supported by computer packages for demonstrational purposes, provide quite vivid visual impressions of pulse behavior. But these classical theories are not capable of describing the corresponding *single photon* behavior. As expected from the general treatments in Chapters 2-6 the neoclassical theory, with the causal restriction lifted, provides the additional boundary and initial conditions that are needed. But first, let us see how quantum physics interprets the experimental results of the delayed choice experiment.

Quantum physics claims that the paradoxical situation is explained by the principle of duality. The photon behaves as a particle when particlelike properties are measured, corresponding to the Pockels cell on. In this experimental situation the path L_2 is blocked off behind the cell so that the incident photon is routed to one of the counters A, B, or C with just one of the optional paths taken. With the Pockels cell switched off the wavelike properties are measured; the photon is taking both paths L_1 and L_2, interfering in beamsplitter B in the manner explained earlier. To uninhibited observers the explanations merely confirm the weirdness of quantum theory. The causal restriction appears to be the underlying reason for its inability to offer a more objective interpretation.

The neoclassical theory is bitemporal and not limited by the causal restriction. In the context of this theory the interpretations of the experimental observations are straight forward and quite trivial, although the details are a little bit involved. The predictions of the neoclassical theory about photon behavior in the delayed choice experiment are sketched in Fig. 12.6. The two photon constituents, each of energy $\hbar\omega/2$, are supplemented with photon doublets α_1 and α_2. The doublet α_2 restores the photon at counter C to full energy status $\hbar\omega$, thereby allowing the photon to trigger the counter. The doublet α_1 at beam splitter B cancels the photon energy, thereby annihilating this component. The two leftover doublet components $-\hbar\omega/2$ and $\hbar\omega/2$ travel backward in time along L_2 and L_1, meeting up at beamsplitter A where they annihilate each other.

The following is an attempt to clarify the details in a somewhat unconventional way by presenting a scenario of an imaginary transcript of a dreamed-up discussion between a materialized photon and a duly appointed comptroller of the *PDE* environmental energy balance sheet. The scenario is set to the scene in which the Pockels cell has been switched on *after* the photon left beam splitter A in its travel along path L_2 but *before* it arrives at the Pockels cell. The photon, having been caught off balance with only half its normal energy and arriving at the counter C, is in a bind because of its inability to trigger the counter with its reduced energy. In its predicament it turns to the Comptroller and asks for an energy loan. To allow the two participants time for their conversation we imagine the passing of time temporarily suspended.

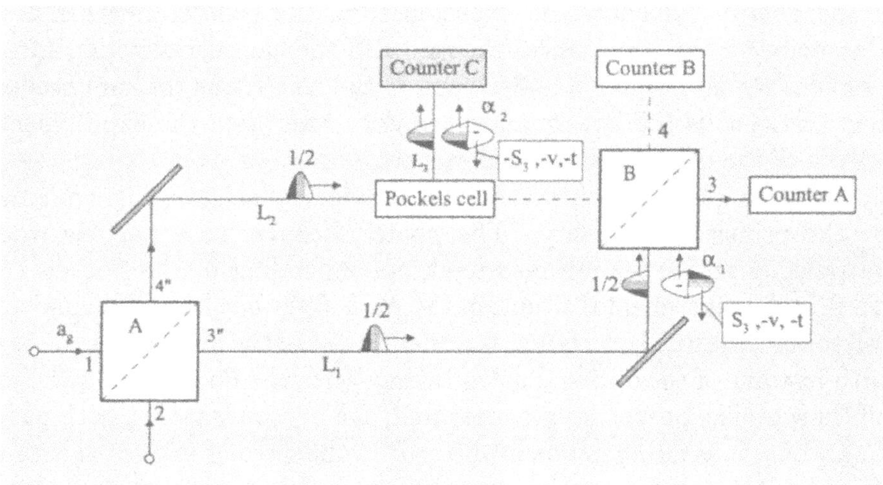

Figure 12.6. Photon in Mach-Zehnder delayed choice experiment. (a) The two constituents in branch L_1 and L_2, each of energy $\hbar\omega/2$, are supplemented with photon doublets α_1 and α_2. The doublet α_2 at counter C restores the photon to full energy status $\hbar\omega$, triggering the counter and leaving a surplus of *negative* energy $-\hbar\omega/2$. The doublet α_1 cancels the photon constituent $\hbar\omega/2$ at beam splitter B leaving a *positive* excess energy $\hbar\omega/2$.

Comptroller:—Let me see if I can meet your request. Our balance sheet seems to be a tiny bit off into the positive. You can have the loan. Be aware though that the loan is payable in photon doublet currency.—That should allow you to handle the new situation.

Photon:—I am a little bit lost. Can you explain what I have to do to get out of my predicament?

C:—I shall be glad to do just that. You realize of course that you really are in a bind because of the dirty trick played by the Pockels cell, blocking your normal travel to beam splitter B and preventing the lineup with your sibling from path L_1. In your crippled state of half normal energy, dumping the energy directly at counter C is out of the question. I am sure you are well aware of that point. Neither can you travel back along path L_2 as a regular reflected photon with your energy intact. That would require a reflection coefficient of unity which you can not muster. Remember your wave impedance is that of free space. You really are in a bind, not being able to dump your energy and not reflect it back!

Photon:—I know all that too well. Get to the point and tell me how to get out of the quandary.

C:—Sorry for lingering over trivialities. I did not mean to upset you. This is how your dilemma is resolved: The positive energy component of the doublet loan is simply added to your own energy, restoring you to full photon status and enabling you to dump your full photon energy $\hbar\omega$ to counter C which is then triggered. That takes care of you. Enjoy your new status in the counter environment.

Photon:—I sure will. But what to do with the leftover doublet component of negative energy? It looks like I have absolutely no use for that part of the loan.

C:—That is true, but don't be so selfish. You should be a little bit concerned about the fate of your sibling along path L_1 who is in exactly the same quandary as yourself because of your failure to meet up with him at the beamsplitter B. He was supplied with a similar loan, but restraints on my balance sheet permit use of the negative energy component only. Unfortunately for him, rather than being restored to full photon status, he is simply annihilated, the only trace of his earlier existence being the positive energy part of his doublet loan.

Photon:—I forgot all about my sibling in rejoicing over my own good fortune—sorry he didn't make it. What puzzles me now is what is going to happen to the two leftovers, one positive and one negative energy doublet component?

C:—I am glad you asked. The two leftover components represent an entangled doublet traveling backwards in time towards beamsplitter A in the negative time direction, arriving at exactly the same time you and your sibling left on your first journey.

Photon:—Sound like magic to me but I believe you, although it is hard to get physicists to accept negative time. But tell me, what happens next to the two leftovers?

C:—Your remark about magic is quite inappropriate—there is not a shred of magic in it. Our bylaws are spelled out in all details forming a consistent set of rules and formulae obeying late Maxwell to his full satisfaction, and also satisfying all initial and boundary conditions in the bitemporal frame. Your question about the fate of the leftovers has a ready answer. The two constituents represent an entangled doublet and simply annihilate each other at beamsplitter A, leaving no traces at terminals 1 and 2.

Photon:—I am pleased I was not asked to violate any basic law of nature in the tricky process of being restituted to full photon status and allowing me to trigger counter C. There are enough of those around pushing us poor photons to do exactly that, coming up with all sort of preposterous propositions. Some even claim we have a dual, or split personality.—Thank you so much for helping me out. Hope I did not ruin your balance sheet. But before taking off, tell me, what would have happened to me if your original balance sheet had been off into the red. Would you still have provided me with a positive energy loan?

C:—Absolutely not. The function bestowed on me by the Great Creator is to maintain the PDE sheet balance without any bookkeeping swindle. In the case you are referring to I would be obliged to supply you with a negative energy loan, which would immediately annihilate you. Sorry about that my friend; if it gives you any comfort, your sibling in path L_1 would then be the survivor.

Photon:—It certainly is comforting that one of us is always surviving. What you imply is that photons along path L_2 can expect to survive roughly every second time and their siblings along path L_1 the other half time.

C:—Correct.— You have grasped the gist of it. It turns out that acting to maintain our sheet balance guarantees the routing of exactly the right number of you photons to the destinations, all to the full satisfaction of auditors at the three counters A, B, and C.—Best of luck to you!

Photon:—Wait a minute! It looks like I need another loan to convert my usual comfortable wave costume into a concentrated particlelike attire in order to hit one of those damned small electrons or microcrystals at the counter surface!—Can you accommodate me?

C:—That is outside my district. Talk to the branch manager at the counter terminal. He'll be sure to help you out and explain what you have to do. If it is hard to grasp read Chapter 5!

And then the photon—aided by the environmental doublets to gradually convert from a plane electromagnetic wave to a particlelike electromagnetic pulse—hurried along path L_3 towards the counter C, where it finally plunged into a counter surface point minimizing the offset of the balance sheet of the local branch manager. End of conversation.

A few commentaries are in place. Referring to Fig. 12.7, the drawing shows the Pockels cell switched to the *on* position, blocking the truncated photon from getting to its beamsplitter B destination and leaving the photon in a paradoxical situation from the point of view of physical theories limited by causality restrictions. These can simply not supply a full set of field constituents for satisfying overall boundary and initial conditions. There are no solutions of this problem in classical electromagnetism. And there are no *objective* solutions in quantum mechanics, although quantum theory make claims to the opposite based on the nonobjective duality principle. In both these theories strict adherence to causality with a truncated time variable is taken as an absolute imperative, a condition that in itself prohibits resolution of the delayed choice experiment in objective terms. Various paradoxical situations and weird behavior are unavailable adjuncts of this restrictive concept. It appears that a much simpler way out is to take the experimental results from the delayed choice experiment as strong indications, if not actual proofs, of the noncausal nature of the microcosm. In the neoclassical theory the added field constituents represented by the photon doublets provide exactly what is needed to satisfy all conceivable boundary and initial conditions in bitemporal space.

For these reasons cleverly contrived delayed choice experiments can never fool the photon into unexpected or paradoxical behavior. They are doomed to failure, a fact that might be presented as a general consequence of the neoclassical theory. And this claim applies to all such experiments, not only the Mach-Zehnder experiment, but to a fairly large number of Gedanken and real experiments in quantum physics. Of these experiments perhaps the most celebrated is the Einstein-Podolsky-Rosen experiment, most often cited as the *EPR* paradox. In the neoclassical theory this particular experiment reduces to a mere triviality. For a detailed discussion of the *EPR* problem the interested reader is referred to Chapter 13.

But let us get back to the Mach-Zehnder delayed choice experiment. The last step in the process, where the two leftover backward traveling doublet components cancel each other at beamsplitter A, warrants a more detailed explanation. With reference to beamsplitter A in Fig. 12.7 let us consider the inputs at terminals 3" and 4" from the entangled doublet originating at the end stations an traveling towards beamsplitter A. We expect the doublet to be annihilated at A so that the corresponding responses at terminals 1 and 2 vanish. The positive energy doublet component α_3, serving as input to terminal 3", gives rise to the following

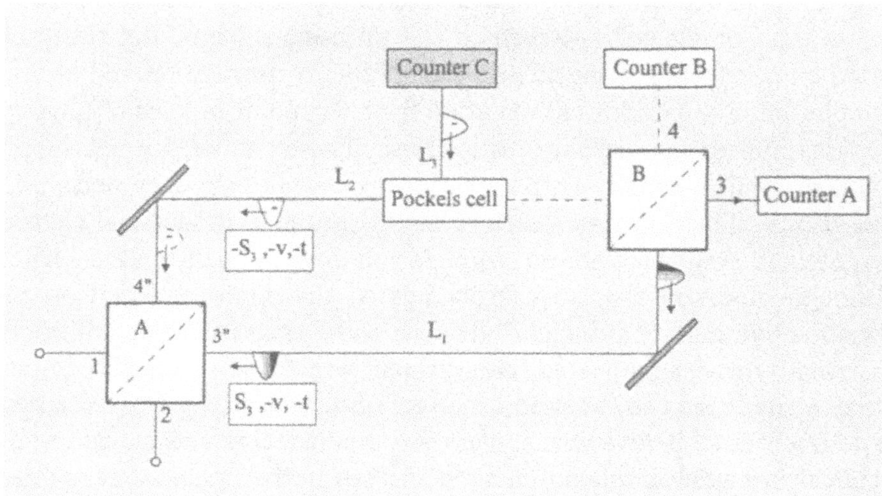

Figure 12.7. **The two surplus doublet pulses represent an entangled doublet One pulse carries positive energy and the other negative energy, both traveling backwards in branches L_1 and L_2 in the *negative* time direction, arriving simultaneously at terminals 3" and 4" of beam splitter A, where they annihilate each other.**

response at terminals 1 and 2, obtained from the scattering equation.

$$\begin{bmatrix} b_1 \\ b_2 \end{bmatrix}^{(3)} = -\frac{1}{\sqrt{2}}\mathbf{S}_1 \begin{bmatrix} \alpha_3 \\ 0 \end{bmatrix} = -\frac{1}{\sqrt{2}} \begin{bmatrix} e^{j\pi/4} \\ e^{-j\pi/4} \end{bmatrix} \alpha_3 \tag{12.45}$$

The response to the negative energy doublet input α_4 at terminal 4" is equivalent to a negative frequency input which is specified by the appropriate scattering relation for negative frequencies:

$$\begin{bmatrix} b_1 \\ b_2 \end{bmatrix}^{(4)} = -\frac{1}{\sqrt{2}}\tilde{\mathbf{S}}_1^* \begin{bmatrix} 0 \\ \alpha_4 \end{bmatrix} = -\frac{1}{\sqrt{2}} \begin{bmatrix} e^{j\pi/4} \\ e^{-j\pi/4} \end{bmatrix} \alpha_4 \tag{12.46}$$

The total response at terminals 1 and 2 is then:

$$\begin{bmatrix} b_1 \\ b_2 \end{bmatrix} = \begin{bmatrix} b_1 \\ b_2 \end{bmatrix}^{(3)} + \begin{bmatrix} b_1 \\ b_2 \end{bmatrix}^{(4)} = -\frac{1}{\sqrt{2}} \begin{bmatrix} e^{j\pi/4} \\ e^{-j\pi/4} \end{bmatrix} (\alpha_3 + \alpha_4) = \mathbf{0} \tag{12.47}$$

because $\alpha_4 = -\alpha_3$. Thus, the absence of any response at terminals 1 and 2 verifies the complete annihilation of the doublet leftovers. The interpretation offered in Fig. 12.8 is an alternative but equivalent description of the process. It was established, in Chapter 6, that pulses in the negative

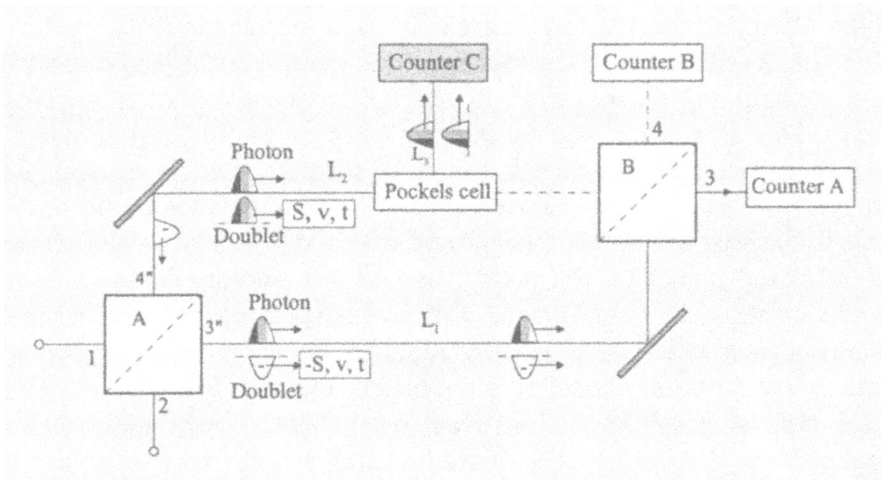

Figure 12.8. Alternative description of the photon doublet involvement. Observed in the *positive* time frame the two doublet pulses in Fig. 12.7 are everywhere synchronous with the original forward photon pulses in this figure. In branch L_2 the two pulses add to form a full photon, triggering counter C. In branch L_1 they cancel each other.

time direction are equivalent to pulses in the positive time direction but with negative frequency. At any given moment in time the two doublet pulses of Fig. 12.8 along L_1 and L_2 are everywhere synchronous with the *original* pulses of Fig. 12.6. They appear to travel together from beamsplitter A towards the respective end stations. In path L_1 they add up to a full photon, in path L_2 they cancel each other. Hence, in this description it looks as if the photon at the original departure at time t_0 had some mysterious advance knowledge that the Pockels cell would be switched on before it arrives at the cell. Causal theories, such as quantum theory and also classical electromagnetic theory label this phenomenon as paradoxical behavior. In the neoclassical theory it is a perfectly consistent and objective phenomenon satisfying all boundary conditions in space and bidirectional time.

12.4.1 Cosmic delayed choice experiments

The following experiment was apparently proposed by John A. Wheeler in an effort to underscore the weirdness of delayed choice experiments carried to the extreme. As illustrated in Fig. 12.9 let us visualize separate photons coming from quasars, these extremely bright mysterious objects of controversial origin near the edge of the universe. The light from a quasar could conceivably arrive along transmission paths involving galaxies acting as gravitational lenses serving to focus the light to some observational platform on earth. The neoclassical theory holds that the contribution to the focusing from gravitational lenses must be minor compared to the effect of the environmental photon doublets. The theory of routing and focusing of photons in this cosmic domain is a special case of the general theory of wave to particle transition in Chapter 5. Without the intermediacy of the doublets we would not be able to observe the quasars, nor would we be able to see the closer stars as twinkling points on the night sky. All the photons in the heavens would be wavelike distributions over the celestial sphere with none of them focused to pointlike entities required for observation by the human eye or by some recording equipment. The sky would be eternally black with no pointlike photons reaching the earth and, there would be no sunshine. Fortunately, the photon routing processes administered by the environmental sea of doublets save us from such a dismal scenario.

Experimental observations indicate that the light from the quasars is coherent over some width w of the order of several meters, and the spatial widths of the photons constituting the light beam must be the same, equal to w. If counters are placed across the width w they trigger at random times but never simultaneously. The coincidence rate is essentially zero. A particular photon goes to just one counter at a time.

How are these experimental results explained in quantum mechanics? Quantum theory insists that the photon triggering a particular counter must have particle rather than wave costume. Therefore, the photon triggering counter A must have traveled a path L_1 different from the path L_2 of a photon triggering counter B. How could the photons originating billions of years ago at the quasar have a premonition of how to chose the right paths in order to satisfy the experimental set-up of unborn astronomers on a still nonexistent planet? Whereas, in the absence of counters, the photons are happy to spread their energy over the entire image of width w, producing interference fringes showing that each has indeed gone both ways through the gravitational lens, perhaps being spread out over millions of light years. The photon quandary is exactly the same as in a number of other quantum delayed choice experiments, for instance the Mach-Zehnder interferometer. The main difference is the time scale involved, making it an extreme example of a delayed choice experiment. But the essence is the same. The Copenhagen Interpretation of quantum physics is quite inept at providing an objective explanation. It resorts to stating that quantum entities are neither wave nor particles but rather undefined until they are measured, existing only as possibilities. In the measurement process they collapse to whatever the experimenter set up to measure.

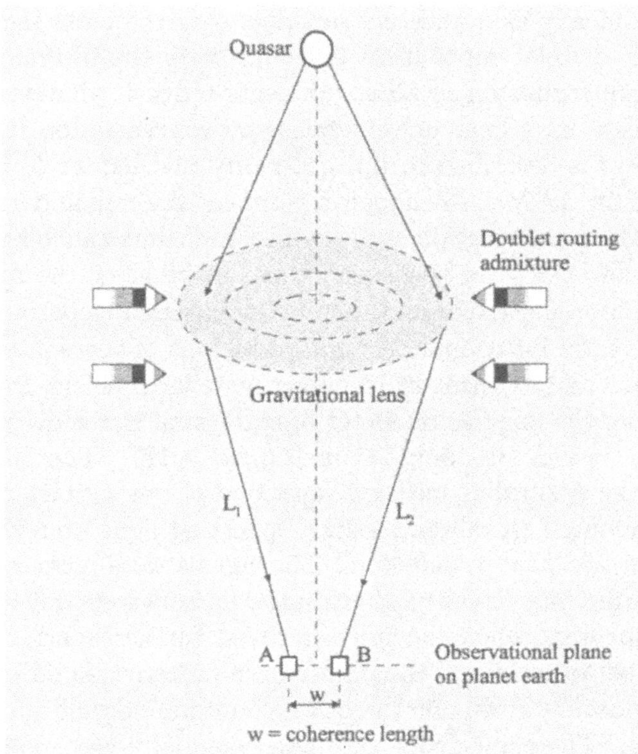

Figure 12.9. **Cosmic delayed choice experiment. Photons from a quasar are deflected in a gravitational lens and by environmental doublets, and subsequently arriving on earth.**

The neoclassical theory explains the observations essentially in the same way as in the delayed choice Mach-Zehnder experiment. On its way from the quasar, before the signal path is terminated by absorbers in the form of counters, the photon spreads its energy over any cross section, propagating as a regular wave. In the figure this means that part of its energy travel along path L_1 and part of it along L_2, in addition to many other paths not shown. There are no immediate restrictions on its spatial distribution, i.e., on its distribution in k-space, only in the frequency or energy domain where the total photon energy is constant and equal to $\hbar\omega$. The situation is quite analogous to the Mach-Zehnder set-up in Fig.12.5. Placing a counter at some spot A represents a delayed choice action, changing the overall earth-quasar configuration. If the photon is registered by counter A, the photon is forced to dump its full energy at the counter, a process achieved through the now familiar engagement of photon doublets excited over the entire travel route from the quasar, and more locally as evanescent doublets over the cross section w. The evanescent doublet supplement combines with the incoming photon so that the superposition of all components reduces whatever spread the photon might have to an extremely narrow concentration at the counter. The process is described in full generality in Chapter 5. In the same manner as in the Mach-Zehnder experiment the remainder propagating doublets do not engage in the superposition but can be visualized as traveling backwards in space and time, arriving at the quasar at the time of photon departure and thereby modifying its spatial distribution so as to exactly hit counter A perhaps billions of years later!

A similar real experiment involving a stellar intensity interferometer determining the angular diameter of stars from the width of their photon 'proxy' waves is described in [23], on p.127. The interferometer, set up in the Australian outback, consisted of two six meter search light mirrors mounted on railroad tracks, collecting light from the same star focused on two photon detectors. The signals were processed electronically, showing interference pattern if the mirrors were not too far apart, but disappearing when the mirrors rolled further apart. As we have seen in other experiments the interference pattern was built up of single spots originating at one or the other counters, never from both at the same time. The coincidence rate was essentially zero, indicating that the incident photons were dumping their energy selectively to one or the other counter.

The scenario emerging from these cosmic delayed choice experiments sounds a bit preposterous. However, if the bitemporal concept in microcosm is accepted, the actual time scale must be irrelevant. If the concept holds for short time spans, it must hold for any time scale. The visualization of pulses traveling forward and backward in time is perhaps more difficult to reconcile with our ingrained causal way of thinking than the more abstract equivalent representation by photons having positive as well as negative frequencies traveling in the positive time direction and satisfying a set of overall *bitemporal* and spatial boundary conditions. Except for the unessential difference in the transmission paths these cosmic experiments are nothing more than large scale demonstrations of the delayed choice phenomena described in detail in the Mach-Zehnder interferometer.

Chapter 13

THE FAMOUS EPR PARADOX

'The opposite of a correct statement is a false statement. But the opposite of a profound truth may well be another profound truth.'

Niels Bohr

13.1 INTRODUCTION

The famous *EPR* paradox is named after Einstein, Podolsky, and Rosen, who presented, in their paper published in 1935 [15], what originally was just a 'Gedanken experiment' formulating for the first time the dichotomy of locality versus non-locality. Since then very few problems in physics have received more attention than the *EPR* paradox, challenging the basic physical as well as philosophical notions of the world we live in. It more than matches the paradoxical situations attributed to the double slit experiments on photons and electrons.

More recently, it was realized by John S. Bell [6] that the nonlocal proposition of the *EPR* paradox could be tested in real experiments. With the launching of his famous 'Bell's Theorem' the *EPR* paradox turned from a purely academic endeavour to an experimentally verifiable proposition. The most celebrated of many such experiments appears to be the one performed by A. Aspect, P. Grangier, and G. Roger [3] employing sophisticated experimental technique for testing the *EPR* propositions.

The primary goal of the present chapter is to analyze the *EPR* proposition and the experimental results from the neoclassical platform. In the process we shall occasionally dwell on some of the basic differences between this theory and the quantum mechanical interpretation. In light of the earlier chapters' success of explaining other quantum experiments, it comes as no great surprise that the neoclassical theory predicts exactly the experimental observations, with full objectivity retained throughout, and no need of the nonlocal assumptions characterizing the orthodox quantum explanations.

13.2 THE BASIC EPR CONFIGURATION

The original *EPR* 'Gedanken experiment' was concerned with two momentum-correlated electrons, but the experiments carried out recently use the simpler entities of two polarization-correlated photons shown in the basic configuration of Fig. 13.1. Pairs of polarization-correlated photons are emitted from a calcium atom source located at the midpoint $z = 0$ and triggered by a laser beam not shown in the diagram. The two polarizers measure the polarizations of the constituent photons under various experimental conditions involving different relative orientations of the polarizers. The polarizer is a Calcite birefringent crystal, splitting light into two beams, one polarized along the optical axis—the ordinary ray—and one polarized perpendicular to the first one—the extraordinary ray. The ordinary ray goes straight, but the extraordinary ray bends, thus permitting registration of the two rays through appropriately positioned photon counters V and H.

13.2.1 Macroscopic properties of the polarizer

The *EPR* configuration represents a special case of an electromagnetic circuit and must necessarily comply with the general circuit analysis of Chapters 2-6 with respect to macroscopic as well as single photon behavior. We saw that the electromagnetic properties of any circuit are fully described by its scattering matrix. In the *EPR* case we shall need only the scattering matrix of the polarizers. With reference to Fig. 13.2, showing polarizer 1, it is legitimate to disregard reflections from the front face. A somewhat simplified but for our purpose adequate scattering relation is then

$$\begin{bmatrix} b_v \\ b_h \end{bmatrix} = \mathbf{S} \begin{bmatrix} a_v \\ a_h \end{bmatrix} = \begin{bmatrix} 1 & 0 \\ 0 & 1 \end{bmatrix} \begin{bmatrix} a_v \\ a_h \end{bmatrix} \tag{13.1}$$

which shows that the scattering matrix **S** is equal to the unity matrix. The componentsb_v and b_h are the vertically and horizontally polarized components beyond the polarizer, whereas a_v and a_h are the correspond-

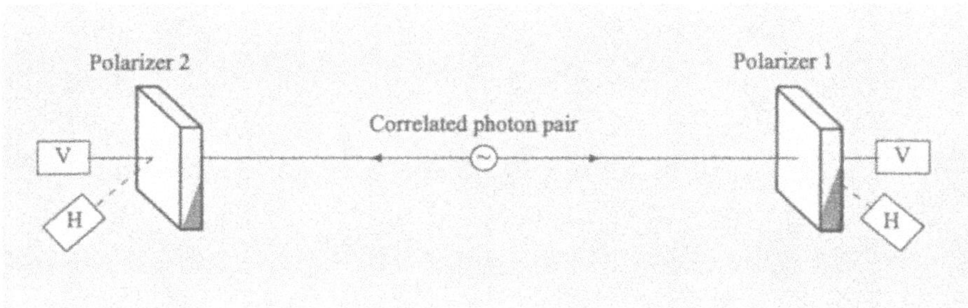

Figure 13.1. **Basic EPR configuration. Photon source at the center emits two oppositely directed polarization correlated photons towards two identical polarization analyzers.**

ing components at the front side. In the neoclassical theory an incoming ray a has a definite polarization θ with respect to the vertical axis, its components specified by

$$\begin{bmatrix} a_v \\ a_h \end{bmatrix} = \begin{bmatrix} \cos(\theta) \\ \sin(\theta) \end{bmatrix} a \tag{13.2}$$

From (13.1) the macroscopic response is then

$$b_v = a\cos(\theta) \qquad\qquad b_h = a\sin(\theta) \tag{13.3}$$

With N_v and N_h photons out of a total of N photons per second channeled to the two counters the relative macroscopic powers in the two channels are

$$\lim_{N\to\infty}\frac{P_v}{P_h} = \frac{|b_v|^2}{|b_h|^2} = \frac{\cos^2(\theta)}{\sin^2(\theta)} = \frac{N_v}{N_h} \tag{13.4}$$

The equation is clearly expressing the macroscopic behavior, i.e., the average over a large number of photons. According to these equations the probabilities of vertical and horizontal trigger counts are specified by

$$\begin{aligned} P(\theta, V) &= \cos^2(\theta) \\ P(\theta, H) &= \sin^2(\theta) \end{aligned} \tag{13.5}$$

These are all macroscopic responses. Let us now proceed to look at the expected behavior of a single photon.

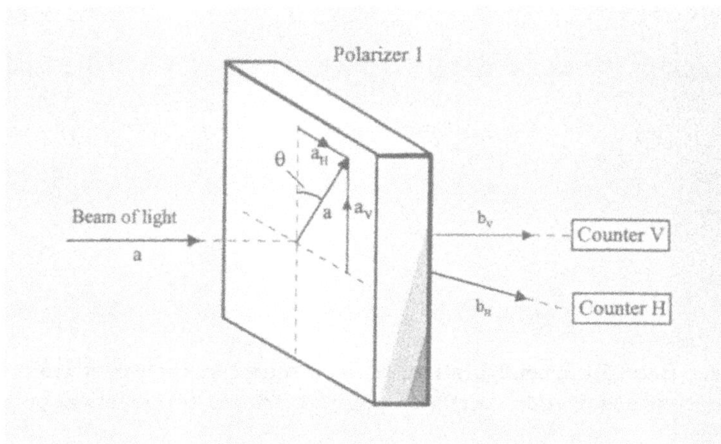

Figure 13.2. **Calcite polarizer with orthogonal inputs a_V and a_H, and corresponding outputs b_H and b_V.**

13.2.2 Single photons in the polarizer

Let us assume that the photon source strength is reduced to the point where only a single photon $\vec{a}$ of polarization θ arrives at the polarizer at a given time. According to the neoclassical theory the counters V and H represent independent terminals, and photon energy integrity prevents a single photon from splitting its energy between them. The photon goes either to counter V or to counter H, a well known fact substantiated by experimental evidence. At this stage of the game the photon finds itself in a similar quandary as in the Mach-Zehnder interferometric experiment described in Chapter 12. The vertical and horizontal input components are $a\cos(\theta)$ and $a\sin(\theta)$, respectively, so that none of the two outputs specified by (13.3) matches the input photon energy unless $\theta = 0$ or $\theta = \pi/2$. From a classical point of view the photon is really in a bind, having absolutely no options compatible with classical macroscopic electromagnetism and photon energy integrity. Neither the vertical nor the horizontal component is large enough for the photon to dump its full energy at one of the counters. Nor can the photon be fully reflected because the polarizer medium does not posses the necessary electromagnetic properties to reflect the photon, i.e., the reflection coefficient is not unity. Hence, the photon is faced with an impossible situation; it can not be transmitted through the polarizer and it can not be reflected back.

13.2.3 Quantum theory of the polarizer

How is this quandary resolved? Macroscopic electromagnetism is not concerned with single photon behavior and does not address the problem. Quantum theory gives only statistical predictions for a large number of photons with random polarization, which is a 50 percent probability for vertical and the same for horizontal polarization. These numbers correspond exactly to the results of macroscopic electromagnetic theory. But quantum theory dares to do more than electromagnetics; it pushes the probability predictions all the way to the behavior of single photons. The theory takes the point of view that the photon does not by itself have a definite polarization until it is measured by the polarizer as being vertical or horizontal, the only two possible outcomes. Quantum physics rejects the idea that it makes sense to talk about a photon with a given polarization attribute specified by the angle θ. In quantum theory the polarization is a nonobjective elusive concept represented only as possibilities expressed by its wave function. It is first through the measurement that the photon materializes itself as vertically or horizontally polarized, through the 'collapse' of the wave function to its macroscopic representation.

This bold step is paid for by heavy penalties. First of all it requires a nonobjective probabilistic interpretation of specific physical attributes of the single photon, such as polarization, which exists only as a set of possibilities. The reader familiar with regular electromagnetics or optics is likely to view this concept with great suspicion, because he is used to consider the polarization of laser light and electromagnetic waves as a very real physical attribute represented by the direction of the wave's electric field component. After all, even single photons are electromagnetic waves and must adhere to the verifiable reality of such waves. But the heaviest penalty is the actual measurement problem, the 'collapse' of the elusive probabilistic polarization vector to one of the two possible outcomes, vertical or horizontal polarization. The concept of wave function collapse is a recurring problem in quantum physics in general, being subject to unresolved and controversial viewpoints very much in the forefront of attention by physicists and philosophers alike. Viewed from the platform of the neoclassical theory the collapse problem is subject to further discussion in Chapter 14.

13.2.4 Neoclassical theory of the polarizer

Before proceeding to discuss the *EPR* problem let us describe how the photon's dilemma is resolved in the neoclassical theory. The explanation follows the general theory of Chapters 2-6, and reference might be made directly to that text. But the general interest and standing of the *EPR* problem warrants a detailed discussion from the platform of the neoclassical theory. In this theory the photon arriving at the polarizer is not in a superposed state of possibilities but has a well defined polarization θ relative to the vertical axis. Unless θ is exactly zero or exactly $\pi/2$ the photon can not by its own propagate through to either counter V or counter H. Nor can it reflect back where it came from. In its predicament it engages the help of a photon doublet vector $\vec{\boldsymbol{\alpha}}(\omega)$ adding to the input vector $\vec{a}$ as shown in Fig. 13.3 for photons routed to counter V and counter H, respectively.

Referring to Fig. 13.2 the regular photon a in the scatter relation (13.2) is supplemented by the photon doublet $\boldsymbol{\alpha}$ with components α_1 and α_2 so that the equation is modified to

$$\begin{bmatrix} a_v \\ a_h \end{bmatrix} = \begin{bmatrix} \cos(\theta) \\ \sin(\theta) \end{bmatrix} a + \begin{bmatrix} \alpha_1 \\ \alpha_2 \end{bmatrix} \tag{13.6}$$

where the frequency variation in the components of $\vec{\alpha}$ are not shown. According to the general theory of Chapter 6 the doublet vector $\vec{\alpha}$ should be orthogonal to the photon vector $\vec{a}$, yielding the following condition:

$$\alpha_1 \cos(\theta) + \alpha_2 \sin(\theta) = 0 \tag{13.7}$$

It is observed that the orthogonality condition is satisfied in the drawings of Fig. 13.3. Let us consider the two cases separately, with counter V or counter H triggered, specifying vertical or horizontal polarization of the *composite* entity of regular photon and doublet additive.

If the composite photon is vertically polarized, with $a_h = 0$ and routed to counter V, the doublet component $\alpha_2^{(v)}$ is equal to $-a\ \sin(\theta)$. Furthermore, the orthogonality condition (13.7) requires $\alpha_1^{(v)} = a\ \sin^2(\theta)/\cos(\theta)$. Accordingly, the doublet column $\boldsymbol{\alpha}^{(v)}$ is given by

$$\boldsymbol{\alpha}^{(v)} = \begin{bmatrix} \sin^2(\theta)/\cos(\theta) \\ -\sin(\theta) \end{bmatrix} a \tag{13.8}$$

which is the case shown in Fig. 13.3a. If the photon is routed to the alternative horizontal counter H, then $a_v = 0$ so that $\alpha_1^{(h)}$ equals $a\cos(\theta)$, in which case the doublet column $\boldsymbol{\alpha}^{(h)}$ becomes:

$$\boldsymbol{\alpha}^{(h)} = \begin{bmatrix} -\cos(\theta) \\ \cos^2(\theta)/\sin(\theta) \end{bmatrix} a \tag{13.9}$$

This situation corresponds to Fig. 13.3b. Before looking at the implications of these routing processes let us prove that the time average of

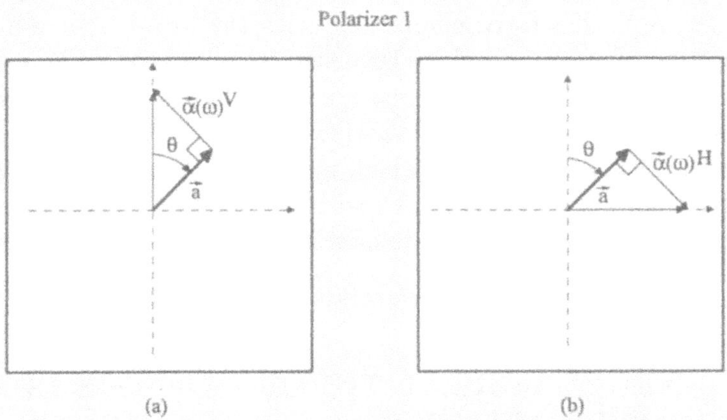

Figure 13.3. **Schematic diagram of polarizer 1 showing the orthogonal photon doublet supplements of the input photon vector. (a) Incident photon $\vec{a}$ routed to the vertical counter V by means of doublet $\alpha(\omega).^V$ (b) Photon routed to the horizontal counter H by means of doublet $\alpha(\omega).^H$**

the total doublet admixture vanishes. This property was proved quite generally in Chapter 6 and we would be much surprised if it did not apply in the present case. Assuming all incident photons to be polarized in the same θ-direction, the overall doublet admixture is given by

$$\sum \boldsymbol{\alpha} = \boldsymbol{\alpha}^{(v)} N_v + \boldsymbol{\alpha}^{(h)} N_h \qquad (13.10)$$

Inserting from (13.4), (13.8), and (13.9) the overall doublet admixture is evaluated to

$$\sum \boldsymbol{\alpha} = \begin{bmatrix} \sin^2(\theta)/\cos(\theta) \\ -\sin(\theta) \end{bmatrix} \cos^2(\theta) + \begin{bmatrix} -\cos(\theta) \\ \cos^2(\theta)/\sin(\theta) \end{bmatrix} \sin^2(\theta) = \mathbf{0} \qquad (13.11)$$

showing that the sum vanishes. In accordance with the general discussions in Chapter 6 the result is interpreted as a routing constraint imposed by the requirement of vanishing time average environmental photon doublet energy. This condition serves to subdivide the incoming photons of polarization θ to vertical or horizontal counters so that the macroscopic populations N_v and N_h are exactly matched, a condition valid for photons of arbitrary polarization θ.

Quantum mechanics has no explanations for such matching except Bohr's *correspondence* principle which is a statement of facts bearing no relation to any objective physical principles or phenomena. It simply expresses an unknown or unspecified ad hoc link between the microcosm and the macrocosm—a link that necessarily must somehow exist in order for quantum physics to be compatible with the world we live in.

It is clear that the neoclassical description of single photon behavior in the polarizer bears little resemblance to the quantum description. The nonobjective probabilities associated with superposed photon states are replaced by a principle in which the environmental *PDE* fields are routing the now objective photons with specified polarization attributes to the respective counters. In the next sections we are investigating how this new theory of operation of the polarizers is affecting the interpretation of *EPR*.

13.3 NEOCLASSICAL THEORY OF THE EPR CONFIGURATION

With reference to Chapter 6 the doublets $\boldsymbol{\alpha}^{(v)}$and $\boldsymbol{\alpha}^{(h)}$are composite entities of positive and negative energy components. At the polarizer 1 the supplemental doublet input appearing in (13.8)-(13.11) can be expressed in the form

$$\boldsymbol{\alpha} = \boldsymbol{\alpha}(\omega) + \boldsymbol{\beta}(-\omega) \tag{13.12}$$

where $\boldsymbol{\alpha}(\omega)$ is the positive frequency component and $\boldsymbol{\beta}(-\omega)$ the additive negative frequency component. The complete scattering relation for the full doublet components was derived in Chapter 6.

$$\boldsymbol{\beta}(\omega) + \boldsymbol{\alpha}(-\omega) = \mathbf{S}\left[\boldsymbol{\alpha}(\omega) + \boldsymbol{\beta}(-\omega)\right] \tag{13.13}$$

where the left hand side is the 'output' doublet at the polarizer exit. The doublet *resonance* condition is specified by zero input and output components

$$\begin{aligned} \boldsymbol{\alpha}(\omega) + \boldsymbol{\beta}(-\omega) &= \mathbf{0} \\ \boldsymbol{\beta}(\omega) + \boldsymbol{\alpha}(-\omega) &= \mathbf{0} \end{aligned} \tag{13.14}$$

Let us concentrate the attention on (13.14) which tells us that the doublet supplements $\boldsymbol{\alpha}^{(v)} = \boldsymbol{\alpha}(\omega)^{(v)}$ and $\boldsymbol{\alpha}^{(h)} = \boldsymbol{\alpha}(\omega)^{(h)}$ in (13.8) and (13.9) are accompanied by the negative frequency components:

$$\boldsymbol{\beta}(-\omega)^{(v)} = -\boldsymbol{\alpha}(\omega)^{(v)} \qquad \boldsymbol{\beta}(-\omega)^{(h)} = -\boldsymbol{\alpha}(\omega)^{(h)} \tag{13.15}$$

By definition, the right hand side of the scattering formulation (13.13) represents waves traveling *into* the polarizer. In Fig. 13.4a these are doublets traveling to the right towards polarizer 1. The doublet components behave in the manner described earlier in Fig. 6.2 of Chapter 6.

The negative frequency component $\beta(-\omega)$ traveling *forward* in positive time is fully equivalent to a backward traveling component in negative time. With these preliminaries let us look at the *EPR* configuration in Fig. 13.4.

13.3.1 EPR with aligned polarizers

In the basic configuration of Fig. 13.4a both polarizers are aligned in the *vertical direction*. The two correlated photons $\vec{a}(\theta)$ and $-\vec{a}(\theta)$, with arbitrary polarization θ, travel in opposite directions to the polarizers 1 and 2. Provided θ is different from 0 or $\pi/2$, the photons are supplemented with doublets $\vec{\alpha}$ and $-\vec{\alpha}$ according to (13.8) or (13.9), so that the two vertical or the two horizontal counters trigger simultaneously. The requirement of zero time average of the *overall* doublet supplement forces the two additions $\vec{\alpha}$ and $-\vec{\alpha}$ to be equal and of opposite polarization.If one of the horizontal counters trigger, then the other horizontal counter trigger. The experimental fact that the coincidence ratio is unity regardless of the random photon polarization θ is explained in Fig. 13.4a. The two polarizers are linked together by the doublet components $\vec{\beta}(-\omega)$ at polarizer 2 and $-\vec{\beta}(-\omega)$ at polarizer 1, as specified in (13.15). These components, forming an entangled doublet of the type shown earlier in Fig. 2.10b of Chapter 2, travel backwards in time towards the photon source at $z=0$. The component arrives at $z=0$ at time t_0, i.e., at the exact time of excitation of the correlated photons from the source.

As discussed more generally in Chapter 2 there exists considerable flexibility in interpretation of bitemporal phenomena, illustrated for instance by Fig 2.10 of that chapter. The diagram in Fig. 13.4b is an example of this flexibility, picturing the components of the entangled doublet as being created at $z=0$ simultaneously with the correlated photons $\mathbf{\vec{a}}(\theta)$ and $-\mathbf{\vec{a}}(\theta)$, and then traveling in synchronism with these to both polarizers. In this interpretation it appears that the photons have a *precognition* of the aligned state of the two polarizers. They seem to 'know' in advance not only that the two are aligned, but also at which angle. But of course the photons are unable to know anything. In the bitemporal neoclassical theory, with absence of stringent causal relations, statements about *knowledge* are foreign concept. Although the *apparent* precognition is bound to arise vehement opposition from physicists conditioned to think in terms of a causal microworld, the neoclassical theory simply claims that the photon boundary conditions can not be satisfied by classical electromagnetic theory, and therefore requires bidirectional time components. In this theoretical framework concepts such as 'knowledge about future' or lack of it, and 'delayed

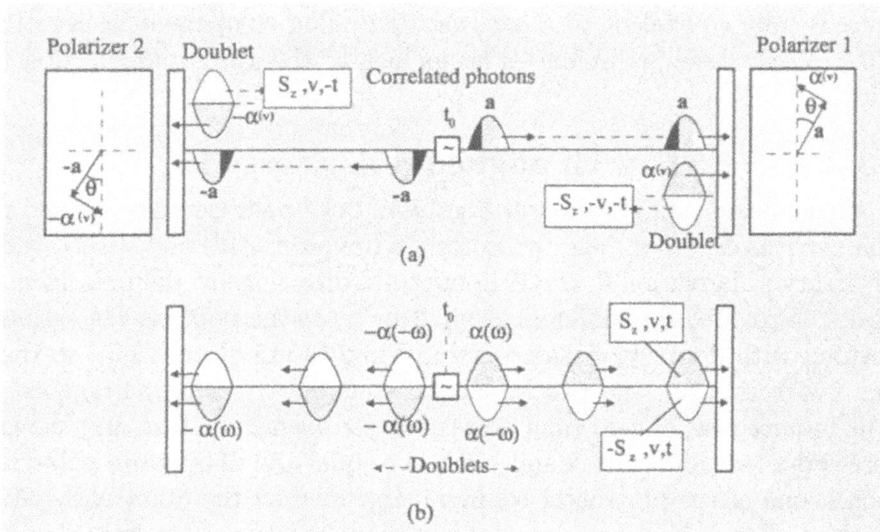

Figure 13.4. **Basic EPR configuration. The polarizers are aligned in the same direction. The photon polarization θ is arbitrary. (a) Illustration of the two polarization correlated photons a and $-$a and the corresponding doublet admixture $\alpha^{(v)}$ and $-\alpha^{(v)}$ causing counters V to trigger (vertical polarization in the figure). The two polarizers are entangled by the two surplus doublet components $(-S_z, -v, -t)$ at polarizer 1 and $(S_z, v, -t)$ at polarizer 2, representing an entangled doublet traveling backwards in time according to Fig. 2.9b in Chapter 2. (b) The diagram expresses the equivalent entanglement in *positive* time frame. An entangled doublet with positive and negative doublet frequency components $\alpha(\omega)$ and $\alpha(-\omega)$, of type shown in Fig. 2.9a, is excited simultaneously with the primary correlated photon pair (not shown). The components travel in opposite directions in synchronism with the primary correlated photons.**

choice arguments' have no place. In the microcosm the future and the past are symmetric concepts.

13.3.2 EPR with nonaligned polarizers

The theory of the last section is readily generalized to the case of arbitrary polarizer tilts δ_1 and δ_2 relative to the global vertical, as shown in Fig. 13.5. Assume that photon $\vec{a}(\theta)$, traveling to the right, trigger counter V_1 of polarizer 1 through a doublet supplement $\vec{\alpha} = \vec{\alpha}(\omega)^{(v)}$ as shown in the figure. Referring to the discussion in the last section, the doublet supplement at polarizer 2 has to be equal to $-\vec{\alpha}(\omega)^{(v)}$. Otherwise, the two components would not form an entangled doublet, which would violate the requirement of zero time average energy. Hence, the *composite* photon input at polarizer 2 is not $-\vec{\mathbf{a}}$ but rather $-\vec{\mathbf{a}}$"

specified by

$$-\vec{a}" = -[\vec{a} + \vec{\alpha}(\omega)^{(v)}] \tag{13.16}$$

As shown in Fig. 13.5 the polarization of this composite vector is at the angle δ_1 with respect to the *global* vertical, and not θ which would be expected in the absence of any link between the two polarizers. In polarizer 2 the tilt $-\delta_2$ adds to the input vector δ_1 so that the overall angle δ between the composite input vector $\vec{a}"$ and the local vertical of polarizer 2 is given by

$$\delta = \delta_1 + \delta_2 \tag{13.17}$$

which is the relative angle between the polarizers. The demonstration shows that this angle is constant and independent of the *original* photon polarization θ. The fact that the composite input vector $-\vec{\mathbf{a}}"$ at polarizer 2 points in the direction of the local axis of polarizer 1 (the V_1 axis) is interpreted as an *instantaneous* coupling between the two polarizers. Rotating polarizer 1 by some angle causes an immediate and corresponding change in the polarization of the composite input at polarizer 2. This effect has the flair of apparent nonlocality, or alternatively, an apparent superluminal connection between the two. The effect is discussed more thoroughly in the subsequent section on Bell's theorem and the Aspect experiment.

The composite input vector $-\vec{\mathbf{a}}"$ forms the angle δ with the *local* vertical of polarizer 2, which calls for additional doublet supplement for triggering either the vertical or the horizontal counters. This process is entirely independent of the original supplements $\vec{\alpha}$ and $-\vec{\alpha}$ which assure that counter V_1 of polarizer 1 is triggered. Because of this independence, the original discussion of the basic functioning of the polarizers applies. Accordingly, the probabilities of trigger counts V_2 and H_2 are specified by (13.2)-(13.4):

$$\begin{aligned} P(V_2) &= \cos^2(\delta) \\ P(H_2) &= \sin^2(\delta) \end{aligned} \tag{13.18}$$

The routing processes at polarizer 2, described in this set of equations, result in zero time average doublet energy expenditure due to a *sequential* compensation of the triggering action of alternate counters at rates according to (13.4). In this respect the sequential compensation differs from the *spatial* compensation of the entangled doublets in the original trigger process of counter V_1.

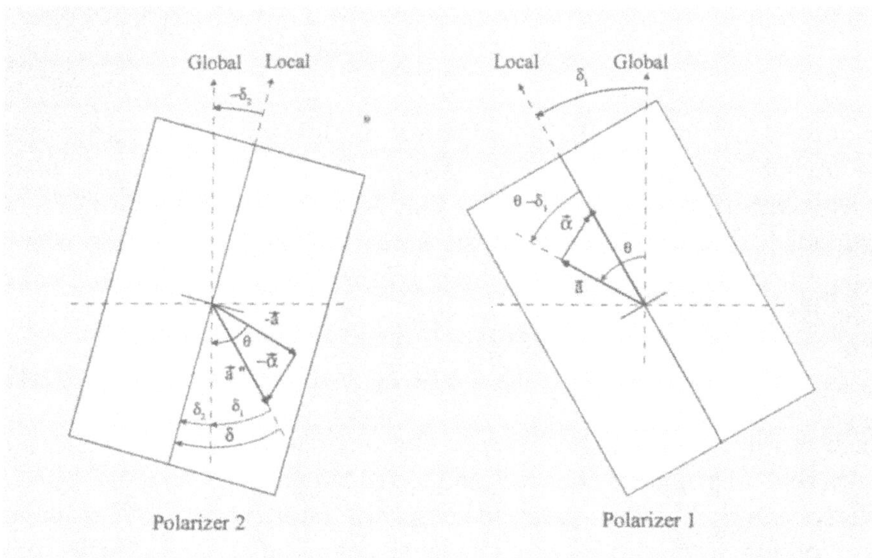

Figure 13.5. The two polarizers are tilted with angles δ_1 and $-\delta_2$ relative to the *global* reference line. The correlated photon pair $\vec{a}$ and $-\vec{a}$ are polarized at the angle θ with respect to the global line. The right traveling photon $\vec{a}$ triggers the vertical counter V_1 with supplemental aid from the photon doublet $\vec{\alpha}(\omega)$. Similarly, the left traveling photon $-\vec{a}$ is augmented with the other half $-\vec{\alpha}(\omega)$ of the entangled doublet, the sum $-(\vec{a}+\vec{\alpha}(\omega))$ forming the angle $\delta=\delta_1+\delta_2$ with the *local* vertical line of polarizer 2. The entangled doublet pair is identical to the pair shown in Fig. 13.4.

It was originally *assumed* that photon $\vec{a}$ was triggering counter V_1 in polarizer 1. Accordingly, (13.18) also specifies the polarization correlations:

$$\begin{aligned} P(V_1, V_2) &= \cos^2(\delta) \\ P(V_1, H_2) &= \sin^2(\delta) \end{aligned} \tag{13.19}$$

The certainty that either counter V_2 *or* counter H_2 trigger implies that

$$P(V_1, V_2) + P(V_1, H_2) = 1 \tag{13.20}$$

which is verified by (13.19). If counter H_1 had been triggered instead of counter V_1 as the primary event, the results would be exactly the same, with obvious permutations of V_1 and H_1 yielding:

$$\begin{aligned} P(H_1, H_2) &= \cos^2(\delta) \\ P(H_1, V_2) &= \sin^2(\delta) \end{aligned} \tag{13.21}$$

which also shows that

$$P(V_1, V_2) = P(H_1, H_2)$$
$$P(V_1, H_2) = P(H_1, V_2) \tag{13.22}$$

The theoretical predictions of polarization correlations specified by the listed formulae are quite remarkable in several respects. They are in perfect agreement with the experimental results [3] to be discussed later, and supply the basis of an alternative theory for the *EPR* problem. Already at this stage it appears that the neoclassical theory provides a full resolution of the *EPR* paradox, with the photon doublet excitations serving as the essential ingredient for the sought after 'instantaneous' communication link between the polarizers. The following sections discuss more fully how the neoclassical theory conceives and resolves the *EPR* phenomenon, with emphasis on the conceptual differences with orthodox quantum theory. But first a brief presentation of the *EPR* paradox.

13.4 THE EPR PARADOX IN NEOCLASSICAL INTERPRETATION

It is well documented that Albert Einstein never accepted quantum mechanics as a *complete* description of nature, and he and Niels Bohr debated the quantum reality question throughout their lives. In the famous publication [15] Einstein and his coauthors presented the *EPR* paradox in an attempt to show the incompleteness of quantum theory. They presented a 'Gedanken' experiment involving two momentum-correlated electrons, which is conceptually equivalent to the polarization-correlated photon experiment of Fig. 13.1. The idea was to demonstrate that some kind of 'hidden variable' had to be added in to explain the results.

Quantum mechanics, represented by Bohr, held the view that the photon polarization does not exist *before* it is measured. Polarization is a relational attribute that does not come into existence until some apparatus is deployed to measure it. The probabilistic polarization wave function *collapses* to either one of the two possible values permitted by the polarizer, vertical or horizontal. Einstein held the realistic view, in line with his profound notion of nature, that the photons have a *definite*, but unknown polarization prior to measurement, reflecting a *classical* ignorance rather than a *quantum* ignorance. His argument on hidden variables goes as follows. It is an experimental fact that the two photons, with specified polarization θ and $-\theta$ with regard to the presumably aligned polarizers trigger either both the vertical or both the horizontal counters. If the quantum concept of complete nonexistence of photon polarization before measurement were correct, the vertical or horizontal triggering of the two would be completely *uncorrelated*. But this is contrary to experiments which show that they are indeed completely *correlated*. And then comes Einstein's crucial argument. If we assume that there is no information flowing between the two polarizers, and this information would certainly have to be *instantaneous*, the correlations can only be explained by a *polarization attribute*, implying a functional relationship between the photon polarization θ and the selection by the photon of vertical or horizontal counters. This 'hidden variable' is required to explain the experimental results of complete coincidence between right and left counters. This, in brief, is the *EPR* paradox.

Bohr explains the problem away fairly vaguely in terms of correlation entanglement between the photon pairs, involving in some ways the experimental arrangement. He did not offer a convincing solution to the *EPR* paradox which has remained an ongoing debate until this day. Einstein's *EPR* argument did not convince supporters of the orthodox quantum interpretation, but this seems to be due more to the obscurity of Bohr's response rather than any good arguments it brought forward. The reader interested in the detailed history of this decisive and turning point in physics is referred to the book by Mara Beller [7] which presents an exceedingly well documented and comprehensive discussion of the circumstances surrounding the dialogue between the two, as well as the subsequent development of the Copenhagen Interpretation into its widespread acceptance by a large fraction of the physics community. The account leaves the definite impression that its promotion was more due to the persistent and at times aggressive campaigning persuaded by Bohr and followers, rather than its substance.

We shall shortly return to the later developments of the *EPR* problem, including the work of John Bell and others, in addition to the interpretation offered by the neoclassical theory. But let us first discuss briefly some of the issues in the original debate between Einstein and Bohr. The neoclassical theory supports Einstein's objective view of the polarization as a *real attribute* of the photon. But Einstein's views differed from the bitemporal proposition, being firmly based on the concept of a *causal* microworld and on the so-called *locality assumption,* permitting no communication whatsoever between the two polarizers. In retrospect, it must have been an impossible thought for the father of relativity to even consider the possibility of any kind of nonlocal superluminal connection between the polarizers, not to speak of a bitemporal microworld. As it turned out much later his assumption of locality was doomed to failure. The fact remains that Einstein has been proven wrong by the fairly recent correlation experiments [3], which are in disagreement with predictions based on locality assumptions. Hence, the locality argument does not stand up to experimental scrutiny. In the neoclassical theory the concept of nonlocality is intimately entwined with a noncausal bitemporal microcosm. In fact, nonlocal phenomena are bound to be part of a bitemporal description of microcosm, essentially a consequence of photons propagating in the positive as well as negative time direction. Quantum physics has been forced to accept nonlocality as a physical reality, but the concept has created more problems than it solved, through the ineptness of quantum physics of providing a physical theory for the experimental fact of nonlocality. Largely due to this failure the literature abounds with philosophical and quasi-religious speculations on the true meanings of nonlocality in relation to our perception of the universe as a whole, and the world around us.

We saw that Bohr, in his debate with Einstein, held the orthodox view that the polarization-correlated photons do not posses any definite direction of polarization before being measured, and this so-called 'Copenhagen Interpretation' appears to be the view of the majority of quantum physicists. To scientists working with coherent electromagnetic waves or laser beams which definitely consist of photons polarized in a verifiable direction, it is difficult to accept the proposition that single photons have no definite polarizations. They tend to look at such propositions as far fetched ideas with no firm basis in the real world, bordering on philosophical speculations rather than physics. Bohr's proposition is certainly not supported by the neoclassical theory, which sides with Einstein on this particular part of the *EPR* question. But in the next stage of the process—the collapse of the photon polarization to either vertical or horizontal—explained by Bohr's concept of phase entanglement or correlation entanglement of the photon pair, his view is surprisingly similar to the neoclassical theory involving excitation of entangled photon doublets. The main differences are that Bohr's explanations border on the metaphysical, providing no objective theory of neither the collapse nor the entanglement. In the neoclassical theory the nonlocality is a triviality arising from bitemporal photon doublets serving to complement the regular correlated photon pair in satisfying overall boundary conditions of the *EPR* configuration.

The two combatants never agreed on the *EPR* problem, and numerous other physicists and philosophers alike have subsequently struggled with the *EPR* paradox. The theoretical work of John Bell [6] in 1964 threw new light on the locality problem. His famous 'Bell's Theorem' provides an experimentally verifiable criterion deciding on the question of local or nonlocal processes.

13.4.1 Bell's theorem in light of the neoclassical theory

In his pioneering theoretical work, Bell focused attention on the locality assumption which was foremost in Einstein's argument. In his famous theorem Bell presents a criterion expressed as an inequality relating the polarization correlation $P(H_1, H_2)$ of the photon pair to the locality assumption. Expressed briefly, if the correlation $P(H_1, H_2)$ exceeds certain values—meaning that Bell's inequality is violated—then the locality assumption is also false, meaning that the two polarizers are somehow interconnected by nonlocal effects. Subsequent experiments on nonaligned polarizers [3] give conclusive evidence of violation of Bell's inequality, at least for certain relative tilt angles. Hence, the locality argument that went into Bell's theorem must also be false. It leads to a prediction that disagrees with experimental results and must therefore be wrong. Locality must be denied.

Bell's theorem, paired with experimental verification of nonlocality, has a deep impact on quantum theory, which is now being forced to accept the concept of nonlocality as a real issue. In the neoclassical theory the theorem has a minor impact because the concept of nonlocality follows directly from the theory, as indicated briefly in the discussion following (13.17). Thus, the neoclassical theory does not need Bell's theorem in order to show nonlocal phenomena. Taken together with the experimental verification of nonlocality it merely serves to support the neoclassical theory of *EPR* with its base of a bitemporal microcosm, reducing the *EPR* paradox to a mere triviality. Because of its minor relevance for the neoclassical theory we shall abstain from digging into Bell's reasoning arriving at his famous 'Bell's Theorem'. The interested reader will find the derivation in most textbooks on the subject. Instead we shall discuss in some details the scenarios emerging from the neoclassical theory and the quantum theory.

The theoretical predictions expressed by (13.17)-(13.22) of the neoclassical theory are quite remarkable in several respects. First of all the derivations are based on a well defined theoretical platform specifying the photon as an *objective* entity in bitemporal microcosm possessing a *real* attribute of polarization. Second, the theoretical predictions based on the theory are in perfect agreement with experimental results. Equations (13.17)-(13.22) predict not only that the polarization correlation $P(H_1, H_2)$ depends on just the relative angle $\delta = \delta_1 + \delta_2$ between the polarizers, but also specifies its value to $\cos^2(\delta)$. These predictions are in perfect agreement with experimental results to be discussed later [3]. Thus, we have a well defined theory which, applied to the extremely controversial *EPR* paradox, makes theoretical predictions that exactly match the experimental facts.

How does this scenario compare to the handling of the *EPR* problem by quantum theory? In the orthodox quantum theory the photon does not posses a polarization attribute before it is registered by the polarizers. Therefore, the polarization correlation must simply depend on the difference of angular position δ between the polarizers, and quantum theory predicts the same variation (13.21) as the neoclassical theory.

$$f(\delta) = P(V_1, V_2) = P(H_1, H_2) = \cos^2(\delta) \qquad (13.23)$$

This functional variation is confirmed by the measurements in [3] to within experimental error. The detailed variation is the central issue in Bell's theorem, stating: 'If $f(\delta)$ exceeds certain values specified in the theorem, then locality must be denied'. The experimental results of [3] provide conclusive evidence of violation of Bell's inequality. Thus, the quantum mechanical conclusion that locality is untenable. And these experimentally verified results imply that quantum mechanics, in order to save the theory, is forced to accept the concept of nonlocality and the associated ideas of superluminal connections as part of the quantum scenario.

It is important to realize what Bell's theorem and the experiments do say and what they don't say. They definitely kill Einstein's notion of locality through the indirect demonstration of the necessity of nonlocal connections. Neither Bell's theorem nor the experimental results provide any theory whatsoever for specific mechanisms of nonlocality. Specifically, they do not 'prove' that quantum mechanics is right although claims to this effect are commonplace. What is proved is that quantum mechanics, in order to survive as a physical theory, must accept the concept of nonlocality.

In sharp contrast to this elusive scenario the neoclassical theory provides an objective theory of the *EPR* problem, with theoretical predictions in full agreement with available experimental results. Not only does the theory provide a detailed physical mechanism for the effect; it also predicts correctly the probabilities of the various trigger counts in the overall *EPR* configuration. But it does more than this; it explains equally well the *delayed choice EPR* experiment described shortly. But first, let us look at the physics behind nonlocal effects.

13.4.2 The physics of nonlocal connections

The discussion following (13.16)-(13.22) already touched upon the problem of nonlocality, and this section is just an elaboration of the subject. With reference to Fig. 13.5 the concept of nonlocality follows in a straight forward manner from the composite input vector $\vec{a}''$ at polarizer 2, forming the angle δ with respect to the local vertical specified by

$$\delta = \delta_1 + \delta_2 \tag{13.24}$$

where the angles δ_1 and δ_2 are the tilts of polarizer 1 and 2, respectively. The physical interpretation is tied in with the circumstance that the composite input photon $\vec{a}''$ engaging polarizer 2 carries information on the tilt angle δ_1 of polarizer 1. And this information is instantaneous because $\vec{\alpha}$ is an entangled doublet of the general type shown in Fig. 2.10 of Chapter 2. Under the assumption that polarizer 1 is triggering the vertical counter V_1, the probability of triggering V_2 at polarizer 2 is given by (13.18).

$$P(V_2) = \cos^2(\delta_1 + \delta_2) \tag{13.25}$$

with the polarization correlation $P(V_1, V_2)$ specified by the same formula. The nonlocal connection, observed directly from Fig. 13.5, is further elaborated on in Fig. 13.6. If the tilt angle δ_1 of polarizer 1 is moved an angle $\Delta\delta_1$ within the shaded sector, the polarization of the composite input vector $\vec{a}''$ at polarizer 2 changes *instantaneously* by the same angle within the shaded sector. The instantaneous effect is independent of the polarization of the primary photon pair, and of the distance between the polarizers. The physics behind the instantaneous effect is due to the bitemporal phenomena of entangled photon doublets discussed more fully in Chapter 2. We can think of the entangled photon doublet pair $\vec{\alpha}$ and $-\vec{\alpha}$ as being excited at the two polarizers at the time of arrivals of the two primary photons. The doublets then propagate with the velocity of light towards each other in the negative time direction towards the source, arriving at exactly the departure time of the primary photons, and then *annihilating* each other. According to the description in Fig. 2.10 of Chapter 2 this is entirely equivalent to a *creation* of an entangled doublet pair at the source simultaneously with the primary photon pair, propagating together in positive time towards the polarizers. Because the velocities never exceed the velocity of light no superluminal effects are involved. The phenomenon is due to bitemporal effects just creating the *appearance* of nonlocality and superluminality.

13.4.3 The Aspect delayed choice EPR experiment

In their publication [3] A. Aspect and his coauthors described an experiment on the *EPR* problem employing very sophisticated experimental set-ups, as shown schematically in Fig. 13.7. The experiment was

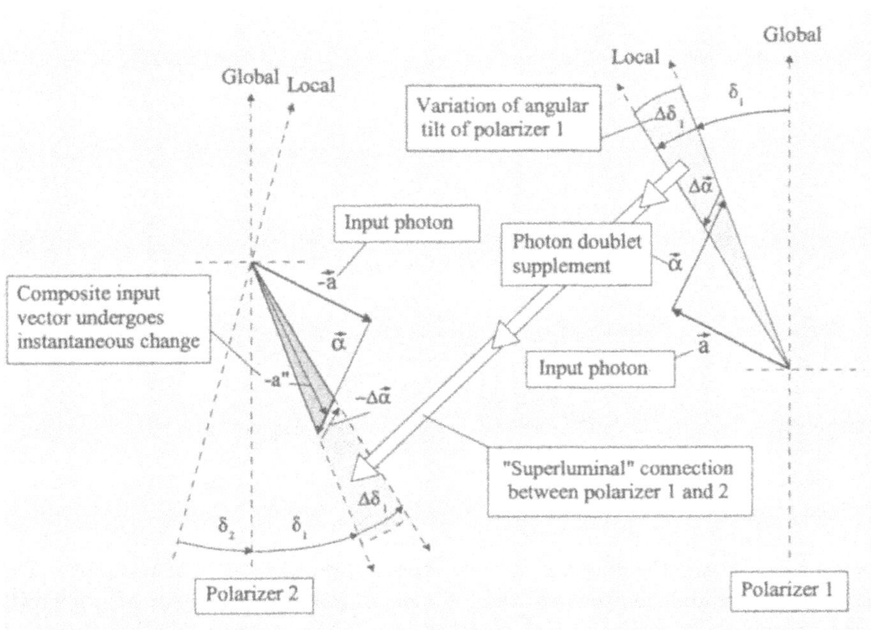

Figure 13.6. **Illustration of the instantaneous connection between polarizer 1 and 2. If the tilt angle δ_1 of polarizer 1 is changed within the shaded sector, the polarization of the composite input vector $-\vec{a}$ " at polarizer 2 changes instantaneous within the shaded sector by the same angle. The instantaneous effect is independent of the polarization $\vec{a}$ of the primary photon pair. The phenomenon can be interpreted as a non-local effect or a 'superluminal' connection between the polarizers, but the effect is only apparent, being due to bitemporal phenomena rather than non-locality.**

basically performed in an effort to check Bell's inequality under conditions more general than the basic *EPR* configuration of Fig. 13.1. The additions involved extremely fast acousto-optical switches inserted into the transmission paths between the photon source and the composite set of two linear polarizers in each branch. The switching time was considerably less than the transit time between the source and the switch, allowing 'delayed choice' experiments of a sophisticated nature. Hence, the choice of routing the photon $\vec{a}$ to the differently oriented polarizer 1 or polarizer 1' could be made *after* the photon left the source. Any beforehand knowledge on the part of the photon $-\vec{a}$ of the final destination of photon $\vec{a}$ could then be ruled out. The experimental set-up had an elaborate set of detectors in the form of counters and a fourfold coincidence counter monitoring the coincidence rates of the four channels. In the experimental results presented in [3] it made no difference

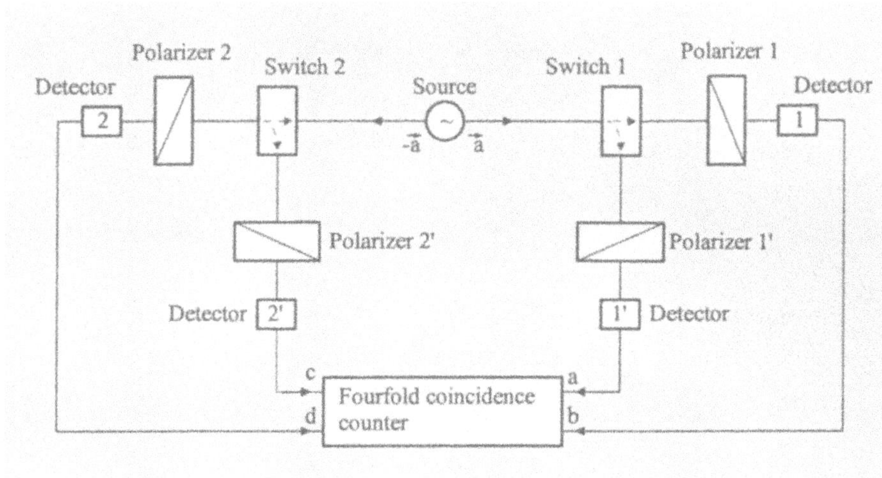

Figure 13.7. **Schematic diagram of the Aspect experimental configuration. The polarization-correlated photons $\vec{\alpha}$ and $-\vec{\alpha}$ can be routed to a choice of differently aligned polarizers by means of fast switches inserted in the transmission paths. A fourfold coincidence counter is monitoring the four channels a, b, c, and d.**

in the results whether the system were set to the delayed choice mode of operation, or were stationary. This experimental fact is in complete agreement with the neoclassical explanation in terms of photon doublet supplements added to the primary photon pair in order to match electromagnetic boundary conditions required in the photon state. With reference to Fig. 13.6 the delayed settings of the polarizers make no difference to the entangled doublets $\vec{\alpha}$ and $-\vec{\alpha}$ because the appropriate doublet excitation takes place exactly at the times of arrival of the correlated photon pair at the two polarizers. Thus, the process is *local* to the polarizers; a delayed setting has absolutely no effect if it just takes place sometime *before* the arrival of the photons. Thus, the details of this process are identical to the earlier experiment with nonaligned polarizers but without the delayed choice of polarizer settings. To save the reader for another detailed description the analysis is not repeated for this latest experiment.

The results from the Aspect experiment serve as further confirmations of earlier statements on the impossibility of dreaming up any delayed choice experiment designed to fool the photon, provided we accept that the phenomena take place in bidirectional time. Earlier statements on bitemporal effects were formulated in more general terms involving photon doublet components required to satisfy extended boundary conditions. If these deliberations are applied to the Aspect configuration, they simply mean that the electromagnetic boundary conditions for photons can not be satisfied by causal components alone, but require bitemporal supplements added to the regular photons. The overall amplitudes of the composite field entities adjust to satisfy all boundary conditions in regular space and bidirectional time, including photon energy integrity. We may look upon the Aspect experiment as an exercise in solving a neoclassical electromagnetic problem in a configuration with well defined boundary conditions in space and bidirectional time, a problem that is guaranteed to have a unique solution. And this solution is exactly the results observed by Aspect and coworkers.

It is important to realize what the Aspect experiment proves. Seen from the quantum theoretical platform it shows that Bell's inequality is violated at least for certain polarizer alignments, supporting a view that nonlocal phenomena must in fact be accepted as part of the nonobjective theory of quantum mechanics. The concept of nonlocality should not be viewed as a real physical phenomenon, but rather as another aberration added to the already existing set of paradoxes of quantum theory. In the same sense the commonly voiced interpretation by physicists and philosophers that the results show a superluminal connection between counters has no real basis. Faced with the facts of the Aspect experiment they seem to accept, perhaps reluctantly, some unknown phenomenon propagating *faster* than light, although they are painfully aware that neither Einstein nor Maxwell would have liked it. Any superluminal phenomenon that can be dreamt up—if it involves electromagnetic phenomena—can not be made to satisfy Maxwell's equations.

Seen from the neoclassical theory the interpretation of the Aspect experiment is drastically different. Bell's theorem and the locality question are not part of the issues. The experimental results simply verify the predictions of the theory (13.19)-(13.22). In particular, they verify the detailed variation of the polarization correlation $P(H_1, H_2)$ with respect to the angular difference δ of the two polarizers' orientation. From the neoclassical viewpoint the results of the Aspect experiment would, in principle, either vindicate or reject the theoretical propositions based on a bitemporal microcosm. As it turned out they were fully confirmed. In contrast, Bell's theorem and quantum mechanics did not propose any mechanism for the concept of nonlocality. Therefore, the theorem is limited to a check on the veracity of the initial proposition of nonlocality expressed in the famous inequality of Bell. As it turned out locality must be denied and, as a result, is creating more problems with the *interpretations* of quantum theory than it solves. Because quantum mechanics has no theory for nonlocal effects, the paradox of an experimentally verified concept of nonlocality is added to the already overburdened assortment of paradoxes and weirdos in the quantum family. This is a striking contrast to the neoclassical theory, which came through the process void of paradoxes, explaining the Aspect experiment as a trivial example of the theory.

Chapter 14

QUANTUM BASES–NEOCLASSICAL VIEW

'I think it is safe to say that no one
understands quantum mechanics.
Do not keep saying to yourself,
if you can possibly avoid it,
"but how can it be like that?"
because you will go 'down the drain'
into a blind alley from which nobody
has yet escaped. Nobody knows how
it can be like that.'

Richard Feynman

14.1 INTRODUCTION

The theory of quantum mechanics emerging shortly after the turn of the century did quite rapidly develop into a dominant section of physics. The theoretical predictions have been confirmed by empirical measurements to unprecedented levels of precision. Despite its predictive successes quantum mechanics has been plagued with conceptual problems that have not been resolved with the passing of time. The interpretational problems have their roots in the circumstance that quantum mechanics is a statistical theory of quantum events. The experimental verification hinges on the presumption of a large sample of statistically similar events taking place in systems of identical preparation. This so-called ensemble interpretation associating the quantum state function with a sampled set of similar systems is a platform that is inept for describing individual events in an unambiguous way.

Thus, the empirical success of quantum theory has very little, if anything to do with interpretation of individual events, except that it tends to promote an attitude of acceptance. Success breeds success. 'Since the theory works so well statistically it must also be true at the event level'. The logic behind such statements is at best questionable. All the same, quantum theory does predict single event behavior but in a very peculiar framework encumbered with interpretational dilemmas. The problems are centered around the dichotomy between the theoretical definitions of the quantum entities and the physical attributes of those entities. This conceptual problem has evoked an abundance of bizarre answers and promoted a continued climate of questioning the basis of quantum theory. The theoretical description of experimental situations as well as 'Gedanken' experiments invariably lead to logical inconsistencies and strange correlations, none of which are quite satisfactory to the classical way of thinking. To compensate the uneasiness created by this wretched state of affairs, many problems of this category are labeled as weird and paradoxical, only to be understood in the context of quantum philosophy.

The discussion in this chapter does not by any means profess to be an exhaustive account of the foundations and various interpretations of quantum mechanics. It merely serves to point out some of the fundamental differences between orthodox quantum theory and the neoclassical theory. This purpose is hopefully achieved by selecting areas in quantum theory that are typically plagued with interpretational problems, and comparing these with the alternative neoclassical description. In this sense the discussion serves as a supplement to the more occasional arguments put forward in the descriptions of the various quantum experiments in the main sections of the book.

14.2 QUANTUM PRINCIPLES AND 'FOOLISH QUESTIONS'

In the period following the inception of statistical quantum theory a number of theorems and principles were put forward or invented in attempts to legitimate and expand the theory to individual events as well. These principles, forming the basis of the Copenhagen Interpretation, go way beyond what can be expected from a statistical theory. They are generally expressed in vague terms and, moreover, formulated in 'Catch 22' type jargon[1]. These characteristics are probably the main explanation for the surprising robustness of quantum theory against attacks on

[1] Joseph Heller's novel: Catch-22—expression for an 'insoluble dilemma, a double bind'.

its foundations. This arsenal of defences are not formulated in terms of propositions on realistic physical attributes. They do not satisfy the minimum requirement which any set of principles should pass, namely that the mathematical formulations relating to the principles should be derivable from them. And most if not all of the principles listed below do not comply with this criterion.

- Heisenberg's uncertainty principle.
- Principle of complementarity, and wave-particle duality.
- Principle of quantum-classical correspondence.
- Principle of wave function collapse.
- Principle of nonlocality.
- Principle of the nonobjective quantum world.

It is an interesting observation that none of the listed principles are ever used in actual quantum computations, which adhere strictly to ensemble interpretations rather than description of single events. Without exceptions, one or several of the listed items play a role in the explanations of the paradoxes that regularly pop up in quantum theory's attempts to describe single event behavior. A typical example is the familiar paradox associated with the double slit experiment for photons and electrons, characterized by the 'which way' quantum mechanical enigma discussed in Chapters 7 and 8. This example is just one out of an endless series of attempts to apply quantum theory to individual events exhibiting similar interpretational problems.

If we stick to the belief of a *causal* microcosm, held by the founding fathers of quantum phenomena and by the majority of the physics community, the available set of boundary conditions is simply not sufficient for objective descriptions. They circumvented the problem of missing boundary conditions by launching the duality or complementarity concept, with pretty much disastrous consequences for the overall interpretations of single quantum events. First, there is the heavy penalty arising from the numerous bizarre and paradoxical explanations. Second, the duality proposition of nonobjective entities represents nothing more than a postponement of the basic boundary value problem to the end process—the registration of a quantum event by some measurement process. At this stage the quantum entity has to face the realities of an objective world with a complete set of boundary conditions. It is not surprising that the interfacing of the nonobjective microcosm with the objective macroworld, regularly referred to as 'the measurement problem', is a recurring and largely unsolved problem in quantum physics.

This glum state of affairs is a constant discomfort to physicists concerned with the interpretation of quantum mechanics. They find themselves in a quagmire of paradoxes and logical inconsistencies that are difficult to accept. Since the inception of quantum theory by the founding fathers there has been little progress in the interpretations, although such topics have not been laid dead with the passing of time. It is unfortunate that the predominant trend during this period appears to be temptations by the physics community to convert the prevailing theory into prevailing dogma.

A large group of physicists are not concerned with interpretations, but stick to the statistical theory, with ensemble models satisfying their computational needs. Central in the ensemble model is the state function Ψ, with Ψ^2 specifying event probability, although Ψ itself develops in a perfectly deterministic manner specified by solutions of the wave equation. An attitude that by no means is foreign among physicists is that questions beyond ensemble interpretations are considered inappropriate, or foolish, and should not be asked. Herman Feshbach and Victor F. Weisskopf, in a reference frame column in Physics Today [16], suggested that questions whose answers are left out of the statistical quantum mechanical description are foolish. This certainly is an attitude towards interpretation of quantum mechanics shared by many in the physics community—that the wave function Ψ contains 'all there is to know about the system'. Bohr and Heisenberg were firm proponents of this attitude, carrying it to the extreme, by holding that quantum entities are void of any physical attributes until they are measured. They denied the existence of an observer-independent physical reality, at least insofar microscopic, or quantum, phenomena are concerned.

Judging from the vehement responses by a number of scientists, the attitude portrayed in the reference column [16] is far from being universally accepted. Many physicists insist on asking 'foolish' questions on the nonstatistical nature of individual quantum events. The response to the quoted column [16] is unison on the importance of asking questions, these being 'foolish' or not.

In spite of the somewhat unfortunate rhetorics, Feshbach and Weisskopf might possibly be correct in their assessment of orthodox quantum theory. It is foolish, or inappropriate, to ask questions on individual events *in the context of quantum theory,* simply because such questions transcend the validity of the theory. Quantum theory is just incapable of providing *objective* answers to such questions—they should not be asked. But it should be equally emphasized that the questions are foolish only in the very restricted context of orthodox quantum theory. In a more general context, the commentators referred to in [16] were certainly well justified in their insistence on the importance of asking 'foolish' questions on individual quantum events. Unfortunately, if directed at orthodox quantum theory, such questions tend to fall on barren soil. Quantum theory is simply unable to answer them, they fall outside the domain of its validity. Moreover, they are easily fended off by the arsenal of defences listed as quantum principles in the first part of the chapter. Perhaps a more fruitful approach would be to direct questions towards the validity of the principles, i.e., towards quantum foundations, not interpretations of the established orthodox theory.

Einstein, one of the more famous proponents of the alleged incompleteness of quantum theory, in his well known interludes with Bohr, was looking for a 'hidden variable' that might somehow make quantum mechanics complete. Firmly rooted in his belief of an objective world, he considered quantum mechanics a mathematical recipe for predicting statistical outcomes of a large number of experiments, the ensemble interpretation. This led to the famous debate between Einstein and Bohr, which was to continue throughout their lives, putting deep marks on both participants. It is a myth that has gained wide acceptance in the physics community that Einstein came out the looser in their dialogue, but this appears to be more due to the obscurity of Bohr's arguments and a persistent campaigning for his views rather than real substance [7]. In retrospect it appears that Einstein fell prey to accepting off-hand some of the ground rules of Bohr, including theorems such as the uncertainty relation and the duality concept, although he appeared to question both as a fundamental fact of nature. Bohr was firmly rooted in his belief in the standard quantum interpretation, stubbornly avoiding questions of the reality of quantum entities—'they do not exist before being measured'—a phrase typical of Bohr. In retrospect it can be said that Einstein was asking 'foolish' questions, but only because they were directed at the wrong targets. He failed in his attack on quantum theory, not because quantum theory is necessarily a complete theory, but because he was deeply rooted in his belief in the concept of *locality* as well as a *causal* microcosm. Seen from the neoclassical platform these concepts represent preconditions essentially blocking any objective resolution of the locality issue. On the other hand it can certainly be questioned whether Bohr was right in his assessment of quantum theory as 'telling everything there is to know about nature'. His statement was only right in the strict context of quantum theory.

With regard to his basic views on quantum objectivity, Einstein may perhaps have his day in court, with the jury still undecided. An increasing number of physicists and philosophers alike are not willing to off-hand accept the conceptual foundations of orthodox quantum theory with the arsenal of traditional dogma handed down to them by the founding fathers. The renewed interest in interpretation of quantum phenomena is witnessed by the large number of publications on the issue, of which [10],[20],[21],[23],[36],[37],[38],[48] are typical, but by no means exhaustive. Much of the late interest has been triggered by the relatively recent developments in quantum measurement relating to the *locality* problem [3],[6], an issue that was foremost also in the mind of Einstein. Unfortunately, the experimental results on locality has added to the quantum baggage of paradoxes, rather than resolving the issue. It is unavoidable that 'foolish' questions are being asked about the significance of quantum nonlocality phenomena in general, and more specifically their repercussions on the conceptual interpretation of the macroworld surrounding us. Such 'foolish questions' of the kind referred to earlier, can not be fended off as inappropriate, by statements that the weirdness of the quantum world must be accepted at face value.

The bitemporal neoclassical theory of electromagnetics presented in the present book is a modest attempt to ask 'foolish' questions from a different platform, rejecting part of quantum theory dealing with individual quantum events, replacing it with the objective and realistic neoclassical theory introduced in Chapter 1 and exposed further in the following chapters. The theory, being mainly concerned with photons and, to a lesser degree, with electrons, does not by any means purport to be a full quantum theory. Nevertheless, introduction of a bitemporal platform for single quantum events of photons and electrons represents a major deviation from the regular doctrines, creating serious repercussions on the general interpretations. In the following we shall discuss the impact of this platform on some of the quantum principles listed in the introductory passage.

14.3 THE UNCERTAINTY RELATION—HOW UNCERTAIN?

Heisenberg's uncertainty relation, easily one of the most celebrated theorems in physics since its inception in 1927, is perhaps the very cornerstone of quantum physics. It shares this standing with Bohr's principle of complementarity. The two are quantum twins, one can not exist without the other. The uncertainty equation also displays a certain semantic uncertainty in its interpretation. If you ask physicists of their opinion on the subject you get different proposals. If you pry further, the majority admit, perhaps reluctantly, that they do not really understand it. Those with an analytical mind do not understand the consistency or the logic behind its *derivation*, and those with a more philosophical attitude might have problems with *interpretation.*

When I was a young student of physics at the Norwegian Institute of Technology[2] around 1950, the Physics Department was visited by Niels Bohr who gave a set of lectures and seminars. I had the privilege of being able to listen to a few of his lectures on the new physics, and have a vivid remembrance of one of the seminars where he elaborated on what is known as the 'Copenhagen Interpretation' of quantum theory. Because Scandinavians by definition are supposed to understand each other's languages, Bohr was talking in his native tongue, a somewhat unfamiliar Danish dialect. Bohr was also a notorious mumbler and all this, together with the elusiveness of the subject, did not contribute to much understanding on my part of the subject nature. My foremost reaction was an impression of listening to a lecture on philosophy rather than physics, where the lecturer was attempting to mold physics to fit his philosophical mind. After the seminar I still recall my foremost thoughts: How is it that a great physicist have such difficulties in conveying his theories to others? It was much later I gradually came to realize that his theories are not easily interpreted in a rational way. But the experience made a profound impact on the attitude towards quantum theory in general, promoting a fairly critical view of the orthodox Copenhagen Interpretation.

The various interpretations of the uncertainty relation can be studied in almost any book on quantum physics. The account presented in the following does not by any means purport to be a comprehensive discussion of the topic, but rather an attempt to relate the uncertainty equation to some of the concepts arising from the neoclassical theory, in particular its possible connection with the universal photon doublet energy fluctuations (PDE).

The quantum mechanical concept of zero point energy (ZPE) was discussed in Chapter 3. Not only is quantum theory predicting infinite ZPE in the universe, but infinite energy *density*. The quantum mechanical infinite ZPE concept derives from the uncertainty principle applied to the ground state of quantum entities. At absolute zero, where the quantum entities are supposedly at rest, i.e., their positions known exactly, the momentum must necessarily be nonzero, that is, if the uncertainty relation holds. Arguments of this nature links the infinite quantum mechanical zero point energy directly to the uncertainty relation. If the first is discarded, the other one must be viewed with a great deal of suspicion. In the neoclassical theory the utterly hopeless concept of infinite zero point energy density does not appear. As

[2]Now: The Norwegian University of Science and Technology

shown in the earlier chapters it is replaced by the photon doublet energy (*PDE*). This concept is a consequence of the basic hypothesis of a bitemporal microcosm, is characterized by zero time average energy everywhere, and does not in any sense depend on the uncertainty relation.

14.3.1 Heuristic derivation of the uncertainty relation

In classical physics the concept of space-time is considered a fundamental postulate. Physical phenomena are described in terms of time (the asymmetric macroscopic time) and the three spatial coordinates. These form the four-dimensional reference frame to which the human brain is so well adapted in its ability to interpret and orient itself in the world we live in. The space-time coordinates must be accepted at face value as the basic reference system relating to human consciousness. Philosophical discussions on the actual meaning and significance of space-time tend only to muddle the picture and should be held apart from classical physics.

In the *classical* space-time reference frame there is no problem visualizing and defining a point particle of mass m located at point x along the x-axis and moving with velocity v_x specified by the usual formula

$$v_x = \frac{dx}{dt} \tag{14.1}$$

The *classical* particle momentum p_x is given by

$$p_x = mv_x = m\frac{dx}{dt} \tag{14.2}$$

The momentum might be determined by two seauential measurements of position and time

$$p_x = m\frac{x_2 - x_1}{t_2 - t_1} \tag{14.3}$$

The quantum mechanical uncertainty relation asserts that the uncertainty of position and momentum are related by

$$\Delta p_x \Delta x \geq \hbar \tag{14.4}$$

Let us tentatively accept this condition and apply it to (14.3). Let us assume that x_1 and x_2 can be measured accurately so that Δx is zero. Then, according to (14.4) the momentum uncertainty Δp is completely undetermined. But according to (14.3), if x_2 and x_1 are accurate and p_x completely undetermined, the denominator $t_2 - t_1$ must be completely undetermined. This is only possible if time itself is completely undetermined. With classical time put in such an unfortunate situation, how can any classical theory, in particular the theory of relativity, survive under such circumstances?

Quantum theorists will probably be quick to object to this presentation, telling us that the quantum domain and the classical domain are conceptually different so that the comparison of (14.3) and (14.4) is not valid. And they may very well be right in this claim. The basic split or dichotomy between the classical and the quantum world prohibits such a comparison. But it is a fact that most presentations in textbooks of the uncertainty relation are heuristic approaches based on the classical transformation from configuration space to k-space, or momentum space, of which the following represents a typical abstract.

A particle of finite width $\Delta\ell$ moving with the velocity v in classical configuration space is conveniently described by the function $f(x-vt)$ where the function f is arbitrary except for the finite width. The particle can be equally well defined in classical configuration space and in classical momentum space, i.e., in Fourier space, through the Fourier transform

$$f(x-vt) = \int_{k=-\infty}^{k=\infty} F(k)e^{jk(x-vt)}dk = \int_{k=-\infty}^{k=\infty} F(k)\, e^{jkx}e^{-jkvt}\, dk \qquad (14.5)$$

The left hand representation in configuration space is a function $f(x)$ of constant shape moving in the positive x-direction with constant velocity v. The right hand representation in momentum space is an infinite sum of plane wave components of the form

$$F(k)\, e^{j(kx-kvt)} \qquad (14.6)$$

where the wave number is k, the frequency $\omega = kv$, and the amplitude $F(k)$. Their phase velocities $v_{phase} = \omega/k = v$ are all the same and equal to the particle velocity in configuration space. The amplitude $F(k)$ is related to $f(x)$ through the inverse Fourier transform

$$F(k) = \frac{1}{2\pi}\int_x f(x)\, e^{-jkx}dx \qquad (14.7)$$

The two are further related through the unitary condition, or Parceval's theorem

$$\int_x |f(x)|^2 dx = \int_k |F(k)|^2 dk \qquad (14.8)$$

which expresses particle conservation in the transformation process. The classical spectral widths, defined in the usual manner, satisfy the inequality

$$\Delta\ell\,\Delta k \geq 1 \tag{14.9}$$

where the equal sign applies to the special case of a Gaussian distribution $f(x)$, in which case the transformed amplitude $F(k)$ is also a Gaussian.

There is not a shred of physics in these transformations. They represent purely mathematical transformations specifying the kinematics of the classical particle. As such, the representation of the particle as the pulse $f(x{-}vt)$ in configuration space is entirely equivalent with the Fourier representation as a superposition of wave components (14.6) of different wave numbers and frequencies, all having the same phase velocities. If you add them all up through the integration procedure (14.5), you get exactly the pulse $f(x{-}vt)$. Whether it is preferable to represent a particle by $f(x{-}vt)$ or by its spectrum of infinitely many components, is a matter of convenience, often dictated by details of interaction with the environment.

The mathematicians tell us that there is an infinite number of different unitary transformations (14.8), meaning that the pulse $f(x{-}vt)$ can be represented equally well by an infinite number of different superpositions. The Fourier transform is just one of these. What makes the transform so special? The Fourier transform is unique on account of the circumstance that the exponential function is the eigenfunction of the differential operator.

$$\frac{df(x)}{dx} = jk\,f(x) \qquad\qquad f(x) = e^{jkx} \tag{14.10}$$

It is an empirical fact that most physical phenomena in our everyday life satisfy differential equations, which might be of very high order in coupled systems, but nevertheless are reducible to sets of first order differential equations such as (14.10), at least in the linear domain. The variety of wave phenomena of all kinds pervading mother nature are just solutions of such sets of differential equations. Through relation (14.10) the Fourier transform is perfectly adapted to solutions of differential equations and therefore to physical wave phenomena. For this reason it holds a special ranking amongst all possible transforms.

Up to this point the discussion is purely classical and could be taken out of any textbook in a variety of technical and physical fields, for instance spectral analysis or signal theory. Basically, the representation is not concerned with physics as such, but is rather describing different but equivalent geometric or kinematic ways of describing a classical particle

$f(x{-}vt)$ in regular space-time and in Fourier space. If the particle spectrum is known, the space-time function is readily obtained from (14.5). Such function pairs are commonplace in many branches of physics. They are routinely displayed on spectral analyzers in electronic laboratories and radar installations. Electronic technicians are experts in interpreting their shapes and significance, in particular the mutual dependence of spectral widths expressed by (14.9). In spite of its formal likeness with the quantum uncertainty relation (14.4) the two are conceptually eons apart.

Nevertheless, the uncertainty relation, as it is presented in most textbooks, is just based on the spectral relation (14.9) supplemented with some quantum postulates. In fact, Heisenberg himself is using essentially this approach in [22]. The quantum argument goes as follows: The well defined spectra $f(x)$ and $F(k)$ in (14.7) are replaced by *single* components $f(x_0)$ and $F(k_0)$. But since the single components obviously do not satisfy the classical kinematic transform pairs (14.5) and (14.7), quantum theory substitutes the uncertainty concept, stating that x_0 and k_0 are really not known. They can take on any values within the actual classical spectra $\Delta\ell$ and Δk. In essence, quantum theory interprets (14.9) as an uncertainty relation, which it is not. What is gained by the bold step of representing the spectrum Δk by a *single* wave component is to introduce the well known relation between quantum *momentum p* and the now *single* wavenumber k_0 of a plane wave

$$p = \hbar\, k_0 \tag{14.11}$$

so that (14.9) transforms to

$$\Delta p\, \Delta\ell \geq \hbar \tag{14.12}$$

which is Heisenberg's uncertainty relation, relating the uncertainty of momentum Δp and position $\Delta\ell$, where now the particle is a delta function $\delta(x_0)$ in configuration space and a delta function $\delta(k_0)$ in momentum space. The relation seems to support the empirical but mysterious observations of photons and electrons behaving in a dualistic manner, sometimes as particles and sometimes as waves. The logical somersault represented by the step from (14.9) to (14.12) essentially means that

$$\delta(x - vt) \leftrightarrow F(k_0)e^{jk_0(x_0 - vt)} \tag{14.13}$$

The expression is a strictly formal relation that must be interpreted as a duality equivalence between the point particle $\delta(x{-}vt)$ and the plane wave $F(k)\exp[jk(x{-}vt)]$. The equation is only symbolic, expressing one particular interpretation of the uncertainty principle.

The mathematics of replacing the actual classical spectra with localized 'uncertainties' is obviously questionable and can not be justified on logical grounds. There exists a deep conceptual discontinuity between the classical and the quantum domain, and the uncertainty relation must be interpreted with this point in mind. In essence, the discussion retraced here was followed by the founders of quantum physics. In spite of the commonplace appearance in most textbooks, it is an unfortunate derivation because it leaves the impression that the uncertainty relation can be derived in a logical manner from simple classical Fourier transforms, which it can not. The mingling of the concepts can even be found in texts on signal processing, a purely classical subject, making reference to the spectral width relation (14.9) as 'Heisenberg's uncertainty relation'. In more abstract formulations of quantum theory the uncertainty relation is expressed in a more general form which follows as a consequence of the basic quantum postulates, with no need of the Fourier transform argument. Its general form is a relation between noncommuting observables. The validity of the uncertainty relation then rests on the soundness of the basic postulates, which perhaps should be subject to scrutiny rather than the uncertainty relation itself.

The quantum particle should be interpreted as some kind of a pseudoparticle, which is different from the classical particle. The uncertainty principle, being intimately entwined with the duality concept, relates strictly to the quantum pseudoparticle. In particular, the momentum uncertainty Δp in (14.12) is not a *classical* momentum uncertainty because the classical momentum mv is not equal to $\hbar k_0$ in (14.11). It does not make sense to talk about momentum uncertainty of a classical particle; the concept should be viewed as strictly belonging to quantum theory. Therefore, the problems raised earlier in connection with classical velocities and momenta expressed by (14.1)-(14.3) are not affected by the uncertainty relation, although it is not uncommon in quantum theory and quantum texts to make claims to this effect.

There exists in the literature dozens of interpretations of Heisenberg's uncertainty relation, as discussed for instance in [48]. During the time following the inception of the theorem the key to understanding was thought to lie in the *measuring apparatus.* If we first measure the momentum p with a certain accuracy Δp followed by a second measurement of position x on the *same* particle, we will find a statistical spread Δx specified by the uncertainty relation. This interpretation describes an uncontrollable perturbation Δx caused by the act of measurement of Δp. The blame is on the measuring apparatus. On the other hand, in the Copenhagen Interpretation, consistent with the views of Bohr and Heisenberg, the uncertainty is not basically associated with the measurement procedure. It is a fundamental uncertainty between noncommuting variables. According to this view it is impossible to get useful knowledge violating the uncertainty equation. In this interpretation the uncertainty does not rest with the measurement process itself but is rather an inherent uncertainty associated with the quantum entities.

Most theoreticians, who never use the uncertainty relation in their quantum calculations, shed away from what they consider futile discussions on interpretations. The main effect of the uncertainty principle, along with complementarity and the arsenal of principles listed in the introduction, is perhaps psychological. They introduce semi-philosophical doctrines having little basis in reality. Einstein, as one of the most prominent critics of Bohr's interpretation of quantum theory, once compared the 'Bohr-Heisenberg tranquillizing philosophy' to a soft pillow on which to rest one's head.

The complementarity principle reflects the fact that quantum mechanics is a less detailed description of a physical system than classical physics. The principle was launched by Bohr in 1928, shortly after Heisenberg set forth his uncertainty relation. The complementary principle states that wave and particle aspects of physical systems are complementary—both are needed for a complete description of nature. Duality and uncertainty are quantum twins; one can not exist without the other. In the abundance of examples in quantum texts, supposed to verify the uncertainty relation, the duality proposition is either directly or implicitly involved. The following few examples serve to illustrate this point.

14.3.2 Heisenberg Microscope—uncertainty and duality

A typical example is the familiar 'Gedanken' experiment first discussed by Heisenberg [22], in textbooks often referred to as the Heisenberg Microscope. With reference to Fig. 14.1 the experiment is concerned with measurement of the position of a particle, such as an electron, by observing the image through a microscope. If λ is the wavelength of the illuminating radiation, the x-component of the particle can be determined to within a precision Δx given by the resolving power of the microscope:

$$\Delta x = \frac{\lambda}{\sin(\theta/2)} \tag{14.14}$$

where θ is the lens aperture. In principle, the resolving power Δx can be made very small by reducing the wavelength λ. Because of the electron's extremely small size this would require gamma rays, which certainly

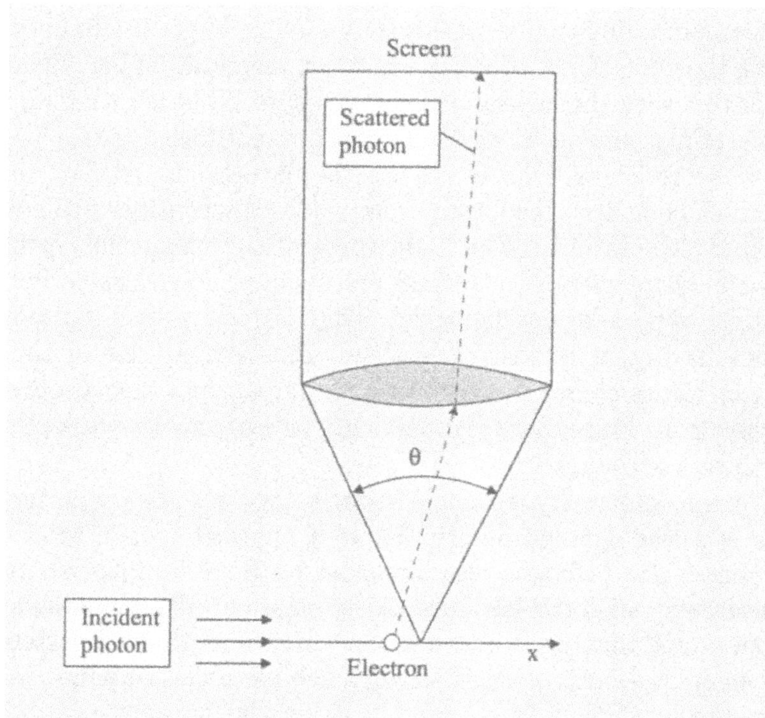

Figure 14.1. **The Heisenberg Microscope. Incident photons from the left are scattered into the microscope from an electron located underneath the lens.**

would present experimental challenges bordering on the impossible, but remember this is a 'Gedanken' experiment. The Heisenberg argument goes as follows: For any measurement to be possible at least one photon must be scattered from the electron and enter the microscope. The so-called Compton scattering of the photon imposes to the electron a certain recoil momentum which is not exactly known because the direction of the scattered photon is undetermined to within the angle θ. Thus, there is an uncertainty in the recoil in the x-direction by the amount

$$\Delta p_x = \frac{h}{\lambda}\sin(\theta/2) \tag{14.15}$$

It follows, upon combination (14.14) and (14.15) that

$$\Delta x \Delta p_x \approx h \tag{14.16}$$

which is consistent with Heisenberg's uncertainty relation. The reasoning sounds so simple, so what is wrong with it? The crucial argument leading to (14.16) is the direct association of the recoil Δp_x with the possible spread of angular deflections θ of the scattered photon which presumably enter the microscope as a *particle,* not as a wave. On the other hand, the scattered photon, supposedly as a particle, must somehow transform to a wave just in time to pass through the lens system of the microscope. It is clear that the description contains the implicit assumption of photon duality.

Let us next see how the same experiment is described in the neoclassical theory. This theory describes the scattering process quite differently. The direct relation between the imparted recoil Δp_x and the angular width θ is nonexistent, meaning that (14.15) does not apply. The reason for this is the direct intermediary of the environmental PDE fluctuations in the scattering process itself. The simple two-element scattering in the quantum account—photon and electron—is replaced by a multi-element scattering process—photon, electron, and environmental doublets. The naive duality concept of the scattered photon behaving as a particle is rejected. The neoclassical theory retains the wave nature of the input photon as well as the scattered photon. The process is viewed as a perturbation of the input photon wave by the electron, with the scattered photon visualized as some kind of asymmetric spherical wave originating at the electron position. The scattered photon is also subject to the usual energy integrity restriction, forcing it to dump its entire energy $\hbar\omega$ at a *localized* load somewhere in the universe. The photosensitive spots of the microscope screen represent just one set of such localized loads. It is unlikely that the wavelike scattered photon by itself can make its way to any localized load, wherever it is located in the universe. It must engage supplements of photon doublets from the PDE environment, exactly in the same manner as described in the quantum experiments of earlier chapters. The routing of the photon with doublet admixture to the various possible 'energy sinks' is governed by the principle of minimization of the time average PDE fluctuations. In one of these possible route selections the composite photon is scattered into the microscope, not as a particle as assumed in the quantum description, but rather as a wave with its wavefront distribution necessarily confined within the angle θ. The angular confinement, achieved with an appropriate doublet supplement, is the first step in the overall routing of the photon to the particular spot on the screen that minimizes the PDE fluctuations. The number of possible wavefront distributions is determined by the wavelength λ and the aperture θ, yielding an overall resolution determined from (14.14).

These deliberations show that there is absolutely no relation between the spot position on the screen and the momentum recoil because the doublet supplements represent *unknown* momentum components contributing to the overall momentum balance. Accordingly, the claims (14.15) and (14.16) made in the quantum description of the Heisenberg Microscope are not supported by the neoclassical theory. The uncertainty relation does not apply in this case.

14.3.3 Electron diffraction in a slit

A second example discussed by Heisenberg in his book [22] concerns electron diffraction in a slit. This particular case was discussed in Chapter 8.4 but with no reference to uncertainty. Let us now look at the diffraction problem in light of the uncertainty relation. As shown in Fig. 14.2 the incident electron with presumably zero transverse momentum traverses the slit of width d. In Heisenberg's description—the quantum mechanical view—the electron is also a De Broglie wave and is subject to diffraction in the slit. By the simplest law of optics, or directly from (8.65) in Chapter 8, the deflection angle is contained within

$$\tan(\theta) \approx \frac{\lambda_e}{d} = \frac{\hbar}{mv_0 d} \tag{14.17}$$

The transverse momentum is then confined within the limit

$$\Delta p_x = mv_0 \tan(\theta) = \frac{\hbar}{d} \tag{14.18}$$

The lateral uncertainty Δx in the electron's position is equal to the slit width d so that

$$\Delta p_x \Delta x = \Delta p_x d = \hbar \tag{14.19}$$

which is the uncertainty equation expressing the uncertainty between the electrons position Δx and its momentum Δp_x. In this description the electron enters the slit as a wave and exits as a particle. Duality is implicitly part of the explanation. The neoclassical theory holds that the electron is always a particle, never a wave. The electron enters the slit as a particle at point x and is deflected by an unknown angle within the range $\pm\theta$ due to the slit field set up through interaction with the doublet evanescent mode of the slit plate (see Chapter 8.4). The uncertainty in the deflection angle does not arise from the various possibilities of the electrons superposed wave function, but is rather specified by the state of the environmental *PDE* fluctuations. In this theory the position of the electron is not due to quantum mechanical uncertainty. In fact, faced with the problem of measuring the electron position with the transverse velocity presumably zero (as also assumed in the Heisenberg model) we could just as well throw away the slit plate, replacing it with a phosphorescent screen registering the electron impact at position x. In such a scenario there would be no transverse deflection, with Δx as well as Δp_x being zero. Or, we could let the electron hit the phosphorescent screen behind the slit at the definite position x'. The momentum $p(x')$ of the particle would be known to arbitrary accuracy.

$$p(x') = mv_0 \frac{x'}{L} \tag{14.20}$$

because the slit width d can be made arbitrarily small and the spacing L arbitrarily large.

This argument can be carried a bit further with the modified experiment shown in Fig. 14.3. The single slit configuration in Fig. 14.2 is supplemented with an identical second slit plate located a distance L from the first one. A phosphorescent screen located immediately behind the second slit would presumably register the arrival of an electron. In order to hit the second slit the maximum lateral momentum of the electron is given by

$$\Delta p = mv_0 \frac{d}{L} \tag{14.21}$$

where v_0 is the electron velocity. The maximum lateral uncertainty Δx is equal to the slit width d, so that

$$\Delta p \, \Delta x = mv_0 \frac{d^2}{L} \tag{14.22}$$

In principle, there is no lower limit to the product which can be made as small as we like through appropriate choice of the geometrical dimensions d and L. Thus, provided the electron makes it through the second slit, the uncertainties Δx and Δp *immediately* before the slit, at point 1 in Fig. 14.3, are arbitrarily small.The fact that the electron might be subject to some arbitrary deflection during the subsequent traverse of the second slit is of no consequence for the state of the electron at point 1. How is it to understand that the uncertainty relation is not upheld in these simple examples? The quantum mechanical explanation to this question is that knowledge of position and momentum to arbitrary accuracy is useless for subsequent experiments because the electron has been absorbed by the phosphorescent screen in Fig. 14.2 and deflected randomly in Fig. 14.3. However, it looks that we are justified in taking the experiments as a demonstration of the failure of the special *Copenhagen Interpretation* of the uncertainty principle referred to in the discussion under 14.3.1. This interpretation asserts that the uncertainty is an inherent property quite independent of measurement procedure. This

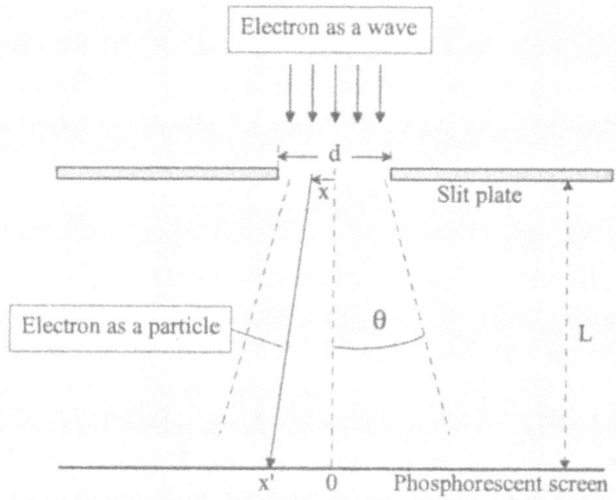

Figure 14.2. **Quantum mechanical view of electron diffraction in a single slit.**

particular claim is not upheld because the momentum *and* position of the electron can be specified to arbitrary accuracy. Quantum theory gets away with problems of this nature by denying that electrons have definite trajectories. Instead, the theory holds that the electron propagates through the configurations as a wave suffering diffraction phenomena in the slits. The registration of an electron on the phosphorescent screen is considered to be an entirely different phenomenon, the familiar quantum mechanical *collapse* of the electron wave function to a point. We shall shortly return to the concept of wave function collapse viewed from the neoclassical theory. But first, let us take a look at uncertainty from the same theory.

14.3.4 Neoclassical uncertainty—a different kind

The neoclassical theory of single events presented in the introductory chapters and the set of experiments analyzed in the ensuing chapters do not support the standard quantum interpretations of single photon events. In particular, Heisenberg's uncertainty relation is not unconditionally upheld. In the neoclassical theory duality and uncertainty are foreign elements. These notions are, in a sense, replaced by an entirely different concept—the interaction of photons and electrons with the environmental sea of photon doublets (PDE) serving to restrict and govern the behavior of single quantum events. In the presented exam-

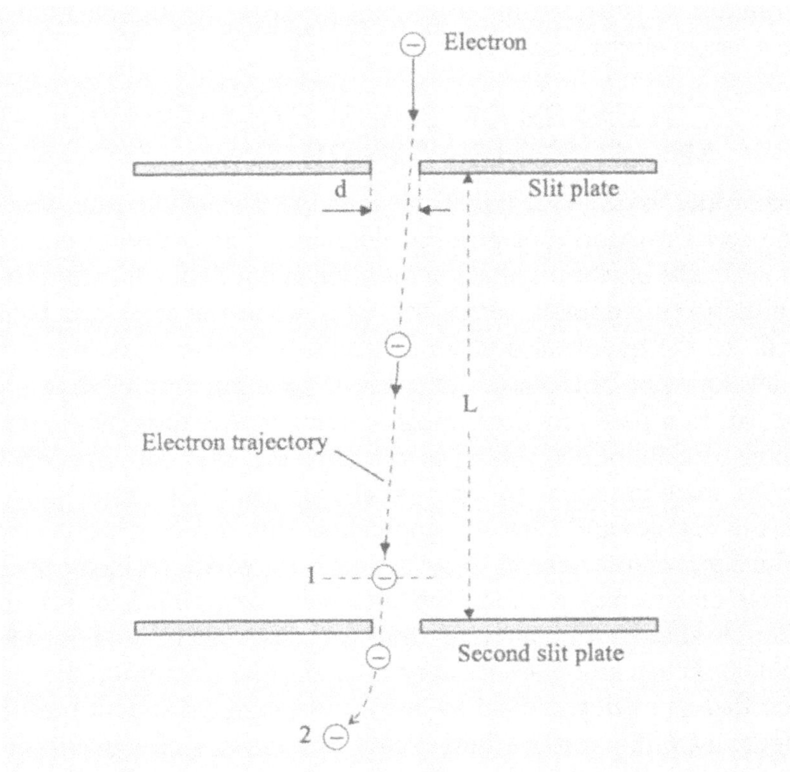

Figure 14.3. **Sketch illustrating the trajectory of an electron traversing two slits of small size *d* and separated a large distance *L* apart.**

ples throughout the book we have seen that photons and electrons are still subject to a kind of uncertainty, reflecting a *classical* ignorance rather than a *quantum* ignorance. The uncertainty in the development of a quantum incident as well as its 'collapse' to the macroscopic equivalent is due to our lack of knowledge of the state of the environmental *PDE* fluctuations which fully determine the outcome in strict adherence to the principle of minimization of the time average of the energy fluctuations. Rejection of duality and engagement of the environmental body of *PDE* fluctuations in the overall quantum process imply that the uncertainty of an outcome can not unconditionally be expressed by Heisenberg's uncertainty relation. The uncertainty in the outcome is not due to a *conceptional* uncertainty but to our *ignorance* about the state of the environmental *PDE* fluctuations. Thus, Heisenberg's uncertainties are replaced by these less stringent uncertainties which are not random in the quantum sense, but governed by minimization of the time average environmental doublet energy.

14.4 COLLAPSE OF THE WAVE FUNCTION–THE MEASUREMENT PROBLEM

According to orthodox quantum views the laws of quantum mechanics refer to two categories of physical domains. The *quantum* domain applies when the physical system in question is not being directly observed; the *macroscopic* domain comes into play when the system is being observed. In the quantum domain Schrødinger's wave equation expresses the development of the state function Ψ as a function of time and position, all in a perfectly deterministic manner. Microscopic systems are the constituents of everything in the universe; therefore one would expect the wave functions to tell 'everything there is to know' about the quantum entities and their developments. But this conclusion seems to fail during *measurement* of quantum entities, where an element of pure chance seems to play a role in the outcomes, exemplified for instance by the double slit experiment on photons and electrons treated elsewhere in the book. When the measurement is made, the photon or the electron is recorded as having arrived at just one seemingly random position.

Figure 14.4 illustrates schematically the essence of a quantum measurement, showing the separation into the quantum domain and the macroscopic domain. Let us visualize a particular wave function Ψ representing mathematically the behavior of an unobserved quantum entity. The function is expressed in a superposed state $\Sigma\Psi_i$ containing all the possible outcomes that can happen to the entity at the time it is observed through an act of measurement. When the measurement is made, the quantum entity is recorded as having arrived at just one given position, in the figure represented by Ψ_q. In this process the wave function, originally containing all possible outcomes, is reduced to one particular value recorded in the measurement. All the others are lost in the process. The transformation from the overabundance of possibilities $\Psi_1 ... \Psi_n$ to the measured single state Ψ_q is described in quantum mechanics as 'the collapse' of the wave function. No mechanism for the collapse has been generally accepted, except that the process is irreversible and random and, moreover, intimately associated with the superposed state of the wave function.

The founders of quantum mechanics had very little to offer in the way of providing acceptable distinctions between an ordinary physical quantum process and the process of measurement. It borders on the impossible to distinguish between the conditions under which the two kinds of behavior applies. The separation into two profoundly different

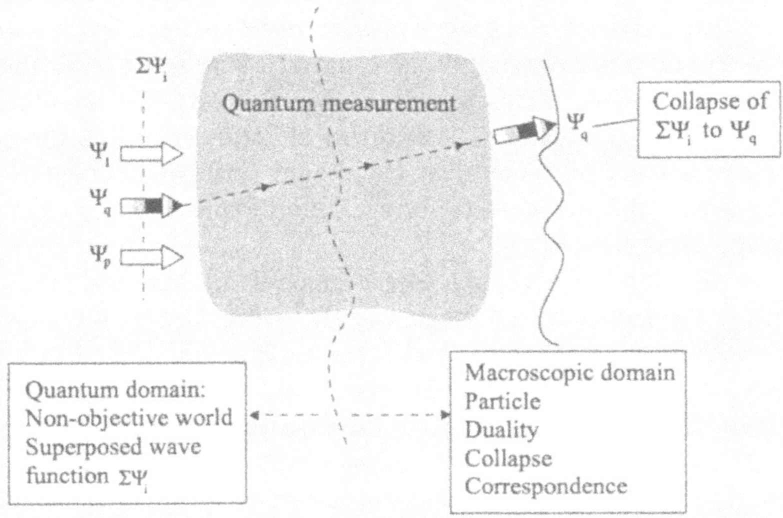

Figure 14.4. **Essentials of a quantum measurement in the orthodox quantum theory. In the left *quantum domain* the wave function exists in a superposed state $\Psi_1 ... \Psi_p$ containing all possible outcomes of the measurement. In the right *macroscopic domain* the wave function collapses to the measured single value Ψ_q.**

domains is highly artificial, putting undue significance in the process of observing. The measurement procedure needs an observer, but who qualifies as a player in this process? Is it limited to the conscious mind of a human observer, or can your personal computer qualify? What makes the human brain and human consciousness so very special; are they not part of the physical universe, with adherence to wave function development?

Many concerned physicists and philosophers have viewed this state of affairs as deeply unsatisfactory. It seems unreasonable that a basic law of nature should be subject to such elusive distinctions. The problem goes under a whole set of different names: The Schrødinger's cat problem, quantum state reduction, and collapse of the wave function. Due to the many interpretational problems arising in connection with the collapse, it is common in the physics community to refer to the process as the 'measurement problem'. The collapse can not be described by Schrødinger's wave equation itself. It is an irreversible process quite independent of the deterministic wave equation. Quantum theory has no good explanation for the randomness of the collapse process. In the plethora of possibilities the outcome of a particular measurement can not be predicted in advance. The fact that a sequence of measurements on identical particles in the same measurement environment always gives rise to the macroscopically observed pattern is referred to as the *correspondence* principle. This principle, together with the quantum principles listed in the introduction to this chapter, including the collapse principle, are not by themselves stand-alone principles; they all are interconnected in some ways to form a kind of consistency in the overall quantum theory.

It is illuminating to illustrate the general measurement problem of quantum mechanics shown schematically in Fig. 14.4 by a typical quantum experiment, and compare with the explanation offered by the neoclassical theory. The double slit experiment on photons serves well to illustrate the basic incompatibility between the two viewpoints. For details of this experiment the reader is referred to Chapter 7. Let us assume a situation in which the incident photons arrive at the double slit experimental set-up one at a time, each photon passing through the slits and finally dumping its energy at some localized spot on the photosensitive screen behind the slit plate.

14.4.1 The measurement problem in quantum theory

In the terminology of quantum theory the incident photon is specified by its wave function Ψ which supposedly exists in a superposed state containing all the possible outcomes, with probabilities for a specific outcome equal to Ψ^2. In the Copenhagen Interpretation the photon exists only as a pattern of probabilities. In the introduction this view was listed as the *principle of the nonobjective* quantum world. When the photon, after its flight through the experimental set-up, hits a point on the photosensitive screen representing one particular state, all the remaining states collapse, i.e., they disappear in the irreversible process. This is the *collapse principle* reducing the photon state to its macroscopic equivalent. The mere concept of collapse is implicitly tied to the superposed states of the wave function. Thus, it is a fair statement that the measurement problem owes its existence to the acceptance of the nonobjective superposed state of the wave function. If the state of wave function Ψ of the photon were a *real* attribute, rather than being nonobjective, the measurement problem would be nonexistent. This is exactly the situation mirrored in the neoclassical view discussed shortly.

The photon, during its transfer through the experimental set-up, is also transforming itself from its input wave costume to a particlelike appearance at the screen. This somewhat mysterious transformation is explained by the *duality principle*. As more and more photons are arriving at the screen the subsequent build-up of the typical diffraction pattern is explained by the heuristic *correspondence principle*—the idea that the quantum interaction with the macroworld must necessarily be compatible with the macroscopically observed result.

Thus, it is clear that the quantum explanation of the overall measurement process engages a fair number of the principles listed in the introduction: the *nonobjective* quantum world, the *duality*-, the *uncertainty*-, the *collapse*-, and the *correspondence* principles. The double slit experiment discussed here is merely a typical example out of many other experiments illustrating the ineptness of quantum theory to put real physical substance into the description of single events. The overall impression left by the quantum explanations borders on conjuring tricks. But these tricks are so cleverly contrived that quantum theory gives the right final answer if we abstain from posing 'silly questions' on interpretations.

14.4.2 Neoclassical view—the measurement problem gone

In the neoclassical description of the same double slit experiment the picture looks entirely different. For details the reader is referred to Chapter 7. Instead of the schematic diagram in Fig. 14.4 the neoclassical view of the measurement problem is illustrated in Fig. 14.5. The incident photon is a real entity specified by its electromagnetic fields, but in this discussion we should have no qualms associating it with the wave function Ψ. However, in this theory Ψ does not exist as a pattern of probabilities but is rather a physically well defined objective entity. The various outcomes of the experiment do not at all depend on any probability attached to Ψ but is determined by the interaction of Ψ with the extraneous medium, represented by the environmental doublet field α supplementing the photon field and routing the photon to the particular spot on the screen that minimizes doublet energy expenditure, with strict adherence to the particular geometry of the double slit configuration. The blackening of the silver atoms on the screen appears to be a random process, each photon hitting quite arbitrary locations on the plate. But the randomness is not real. The apparent randomness stems from our *ignorance,* not from a *conceptual* randomness. The photon's point of arrival $\Psi + \alpha_q$ is not arbitrarily selected by the photon itself, i.e., by its wave function Ψ but by the particular doublet supplement α_q minimizing the expenditure of accumulated environmental energy supplement. The next incident photon, arriving at $\Psi + \alpha_p$, is subject to the same minimization principle, which is orchestrating the entire sequential process. It has been amply demonstrated throughout the book that the

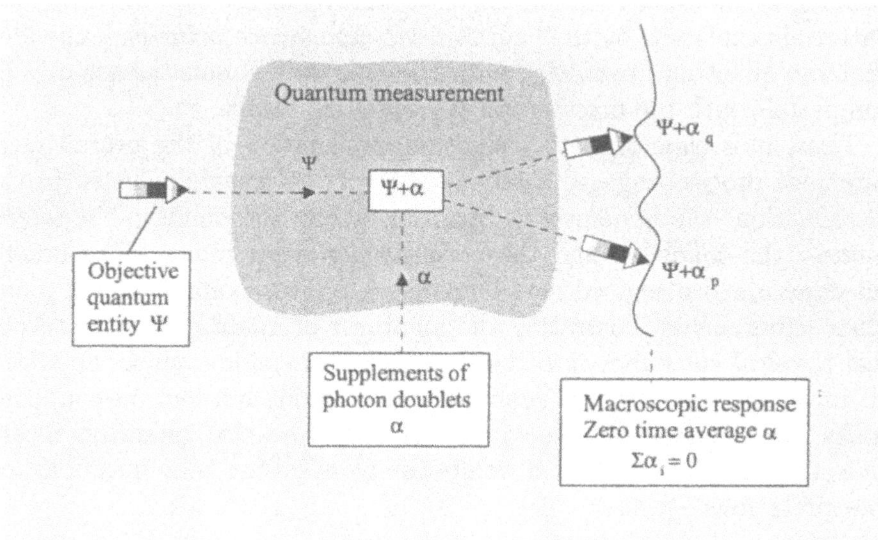

Figure 14.5. **Quantum measurement in the framework of neoclassical theory. In the left *quantum domain* the photon wave function Ψ is an objective entity. During the measurement the photon is supplemented by a set of photon doublets α routing the photon to the particular spot $\Psi + \alpha_q$ in the *macroscopic domain* minimizing the time average doublet energy. The overall macroscopic response, the 'diffraction pattern', is determined by zero time average doublet expenditure $\sum \alpha_i = 0$.**

time average build-up of the blackened spots to the macroscopic 'diffraction pattern' is fully governed by the same principle—the vanishing of the time average PDE. The processes describing single events as well as the time average final result are governed by the same minimization principle. Macroscopic electromagnetism is equivalent to the time average of the individual routing sequences. In this description there is no mystery with the photon 'collapsing' to a particular spot on the screen.

The collapsing process is interpreted as a bitemporal electromagnetic phenomenon in which the photon, not as a true particle but in its particlelike wave transition described in Chapter 5, is dumping its electromagnetic energy $\hbar\omega$ to the energy absorbing entity on the screen. In a photosensitive screen the absorbers are single molecules. A particular molecule hit by the photon's concentrated electromagnetic flux is transformed to a higher energy state, being blackened in the process. Even if the operation itself may be a fairly complex multi-level nonlinear process, there is no mystery associated with it. If the screen is conducting rather than photosensitive, the absorbers are charge carriers interacting with the photon's electromagnetic fields enhancing their kinetic energy. The energy is eventually converted to heat through lattice collisions. Again, there is no mystery associated with the energy conversion, and no processes that can be labeled as collapse of photon states take place. The artificial distinction between the 'observer' and the 'observed' is absent. Thus, in the neoclassical theory the quantum mechanical 'measurement problem' and the associated paradoxes do not exist. The environmental doublet fields adding to the photon replace the quantum mechanical superposed states. The probability pattern of Ψ is replaced by a simple minimization principle relating to the environmental doublet energy. The photon interacts with the environmental PDE fields entirely on the terms dictated by the environment so that the environmental energy expenditure is minimized. The supplemental fields serve to squeeze and route the photon to the particular point on the screen satisfying this criterion. And finally, the build-up of a diffraction pattern to the macroscopic equivalent is governed by the same minimization principle. In the experimental set-up the administration of all aspects of photon behavior is governed by the environmental doublets. In this sense it not only takes the place of the quantum mechanical *duality principle,* but the *collapse-* and *correspondence principles* as well. In one single sweep the neoclassical theory eliminates the need for these quantum principles. None of the principles are required; they are foreign concepts playing no role and therefore not part of the theory. The complete alleviation of the measurement problem justifies classification of the neoclassical theory as an observer-independent description of the quantum world of photons and electrons.

References

[1] Abbott L., "The Mystery of the Cosmological Constant", Scientific American, pp. 82-88, May 1988.

[2] Aharonov, Y. and L. Vaidman, "Properties of a quantum system during the time interval between two measurements." *Phys. Rev.*A.41, 11, 1990

[3] Aspect, A, P. Grangier and G. Roger, "Experimental Test of Bell's Inequalities using time-varying Analyzers", Phys. Rev. Lett. vol. 49, No.25, pp. 1804-1807, 1982.

[4] Bell, D. A., "A survey of $1/f$ noise in electrical conductors", J. Phys. C: Solid St. Phys., 13, pp. 4425-37, 1980.

[5] Bell, J. S., "Speakable and Unspeakable in Quantum Mechanics", Cambridge University Press, New York, 1987.

[6] Bell, J. S., "On the Einstein Podolsky Rosen Paradox", Physics, vol.1, No.3, pp. 195-200, 1964.

[7] Beller M., "Quantum Dialogue", The University of Chicago Press, Chicago, 1999.

[8] Bohm, D., "Quantum Theory", Prentice-Hall, Englewood Cliffs, N.J., 1951.

[9] Boyer, T. H., "A Brief Survey of Stochastic Electrodynamics", review in A. O. Barut, "Foundations of radiation Theory and Quantum Electrodynamics", Plenum, New York, 1980.

[10] Bulb, J., "Interpreting the Quantum World", Cambridge University Press, New York, 1997.

[11] Chiao, R. Y., P. G. Kwiat, and A. M. Steinberg, "Optical Tests of Quantum Mechanics", pp. 35-83, Advances in Atomic, Molecular, and Optical Physics, vol. 34, 1994.

[12] Coveney, P. and R. Highfield, "The Arrow of Time, a Voyage through Science to solve Time's greatest Mystery", Ballantine Books, 1990.

[13] Dutta, P., L. W. Eberhard and P. M. Horn, "$1/f$ noise in copper whiskers, Solid State Communications, 21, pp. 679-681, 1977.

[14] Eberhard, J. W. and P. M Horn, "Temperature dependence of $1/f$ noise in silver and copper", Phys. Letter, 39, pp. 643-646, 1977.

[15] Einstein, A., B. Podolsky, and N. Rosen, "Can a Quantum-Mechanical Description of Physical Reality be considered Complete?", Physical Rev. vol. 47, pp. 777-780, May 1935.

[16] Feshback, H., W. F. Weisskopf, "Reference Column", Physics Today, p.9, October 1988, and Letters, Physics Today, pp. 14-15, 96-104, April 1989.

[17] Feynman, R. P. , R. B. Leighton, and M. Sands, "The Feynman Lectures on Physics", Volume I, Addison-Wesley Publishing Company, Inc.

[18] Gerlach, W. and O. Stern, "Der experimentelle Nachweiss des magnetischen Moments des Silberatom", Zeitscrift Phys. 8, pp. 110-111, 1921.

[19] Gibbins, P., "Particles and Paradoxes", Cambridge University Press, 1987

[20] Goldstein, S., "Quantum Theory Without Observers—Part One", Physics Today, pp. 42-46, March 1998.

[21] Goldstein, S., "Quantum Theory Without Observers—Part Two", Physics Today, pp. 38-42, April 1998.

[22] Heisenberg, W., "The Physical principles of the Quantum Theory", Dover Publications, Inc, New York, (unabridged and unaltered republication from Chicago Press, 1930)

[23] Herbert, N., "Quantum Reality", Anchor Books, 1985.

[24] Hooge, F. N.and A. M. H. Hoppenbrouwers, "$1/f$ noise in continuous thin gold films", Physica, 45, pp. 386-392, 1969.

[25] Hooge, F. N., "1/f noise is no surface effect", Phys.Letter, 29, pp. 139-140, 1969.

[26] Johnson, J. B., "The Schottky effect in low frequency circuits", Phys.Rev. 26, pp. 71-85, 1925.

[27] Jønsson, C., "Electroneninterferenzen an mehreren künstlich hergestellten Feinspalten", Zeitschrift für Physik, 161 pp. 454-474, 1961.

[28] Krauss, M., "Cosmological Antigravity", Scientific American, pp. 53-59, January 1999.

[29] Landé, A., "New Foundatons of Quantum Mechanics", Cambridge University Press, 1965.

[30] London, R., "The quantum Theory of Light", Oxford University Press, pp. 139-141, 1994.

[31] Lévy-Leblond, J. M. and F. Balibar, "Quantics, Rudiments of Quantum Physics", North-Holland, 1990.

[32] Mackey, M. C., "Time's Arrow: The origins of Thermodynamic Behavior", Springer-Verlag, New York, 1992.

[33] Manley, J. M. and H. E. Rowe, "Some general Properties of nonlinear Elements", Proc. IRE , pp. 904-913, vol.44, 1956.

[34] Mermin, N. D., "Two lectures on the Wave-Particle Duality", Physics Today, pp. 9-11, january 1993.

[35] Peat, F. D., "Einstein's Moon, Bell's Theorem and the Curious Quest for Quantum Reality", Contemporary Books, 1990.

[36] Penrose, R., "The Emperor's New Mind, concerning Computers, Minds, and the Laws of Physics", Oxford University Press, 1989.

[37] Penrose, R., "Shadows of the Mind, a search for the Missing Science of Consciousness", Oxford University Press, 1995.

[38] Price, H., "Time's Arrow and Archimedes' Point", Oxford University Press, 1996.

[39] Ramo, S., Proc. IRE, pp. 584-585, September 1939.

[40] Robinson, A. L., "Demonstrating single Photon Interference", Science, vol 231, pp. 671-672, February 1986.

[41] Sakurai, J. J., "Modern Quantum Mechanics", Addison-Wesley Publishing Company, Inc., 1985.

[42] Støvneng, J. A. and E. Hiis Hauge, "Is quantum tunnelling faster than the speed of light?", Physics World, pp. 23-24, November 1993.

[43] Townsend, J. S., "A modern Approach to Quantum Mechanics", McGraw Hill, Inc., 1992.

[44] Van Der Ziel, A., "Noise", New Jersey, Prentice Hall, 1954.

[45] Weinberg, S., "Dreams of a Final Theory", Pantheon Books, New York, 1992.

[46] Wessel-Berg, T., "Self-consistent Theory of general periodic Circuits for Traveling Wave Tubes", International Journal of High Speed Electronics and Systems, vol.4, No.4, pp. 365-407, 1993.

[47] Wheeler J.A. and R.P. Feynman, "Interaction with the Absorber as the Mechanics of Radiation", Reviews of Modern Physics, 17, pp. 157-181, 1945.

[48] Wick, D., "The Infamous Boundary, Seven Decades of Controversy in Quantum Physics", Birkhäuser, 1995.

[49] Zukov, G., "The Dancing Wu Li Masters", Quill, William Morrow, 1979.

Index

www.ingramcontent.com/pod-product-compliance
Ingram Content Group UK Ltd.
Pitfield, Milton Keynes, MK11 3LW, UK
UKHW021506180926
475607UK00010B/37

* 9 7 8 0 7 9 2 3 7 2 5 7 8 *